The Evolution of Human Co-operation

How do people living in small groups without money, markets, police, and rigid social classes develop norms of economic and social co-operation that are sustainable over time? This book addresses this fundamental question and explains the origin, structure, and spread of complex stateless societies. Using insights from game theory, ethnography, and archaeology, Stanish shows how ritual – broadly defined – is the key. Ritual practices encode elaborate rules of behavior and are ingenious mechanisms of organizing society in the absence of coercive states. As well as asking why and how people choose to co-operate, Stanish provides the theoretical framework to understand this collective action problem. He goes on to highlight the evolution of co-operation with ethnographic and archaeological data from around the world. Merging evolutionary game theory concepts with cultural evolutionary theory, this book will appeal to those seeking a transdisciplinary approach to one of the greatest problems in human evolution.

CHARLES STANISH is Director Emeritus of the Cotsen Institute of Archaeology and Professor of Anthropology at the University of California, Los Angeles. His previous books include *Ancient Titicaca* and *Ancient Andean Political Economy*. He is a member of the National Academy of Sciences USA.

The Evolution of Human Co-operation

Ritual and Social Complexity in Stateless Societies

Charles Stanish

University of California, Los Angeles
Santa Fe Institute, Santa Fe, New Mexico

CAMBRIDGE
UNIVERSITY PRESS

CAMBRIDGE
UNIVERSITY PRESS

University Printing House, Cambridge CB2 8BS, United Kingdom

One Liberty Plaza, 20th Floor, New York, NY 10006, USA

477 Williamstown Road, Port Melbourne, VIC 3207, Australia

314-321, 3rd Floor, Plot 3, Splendor Forum, Jasola District Centre, New Delhi - 110025, India

79 Anson Road, #06-04/06, Singapore 079906

Cambridge University Press is part of the University of Cambridge.

It furthers the University's mission by disseminating knowledge in the pursuit of education, learning and research at the highest international levels of excellence.

www.cambridge.org
Information on this title: www.cambridge.org/9781107180550
DOI: 10.1017/9781316848128

First published 2017

A catalogue record for this publication is available from the British Library

Library of Congress Cataloging in Publication data
Names: Stanish, Charles, 1956– author.
Title: The evolution of human co-operation : ritual and social complexity in stateless societies / Charles Stanish, University of California, Los Angeles.
Description: Cambridge, United Kingdom ; New York, NY : Cambridge University Press, 2017. | Includes bibliographical references.
Identifiers: LCCN 2017002139 | ISBN 9781107180550
Subjects: LCSH: Cooperation. | Economic anthropology. | Social evolution. | Social groups.
Classification: LCC GN448.8 .S73 2017 | DDC 306.3 – dc23
LC record available at https://lccn.loc.gov/2017002139

ISBN 978-1-107-18055-0 Hardback
ISBN 978-1-316-63227-7 Paperback

This book is gratefully dedicated to Mr. and Mrs. Lloyd Cotsen of Los Angeles, California

Contents

List of Figures — *page* viii
Preface — x

1 The Evolution of Human Co-operation — 1
2 Economic Anthropology of Stateless Societies: The Rise and Fall of *Homo economicus* — 18
3 Conditional Co-operators: The Evolutionary Game Theory Revolution — 35
4 The Role of Coercion in Social Theory — 58
5 The Ritualized Economy: How People in Stateless Societies Co-operate — 80
6 An Anthropological Game Theory Model for the Evolution of Ritualized Economies — 166
7 The Evolution of Ritualized Economies: The Archaeological Evidence — 185
8 Epilogue: "No Beans, No Jesus" — 269

Notes — 276
References — 290
Index — 324

Figures

5.1	Locations of case studies and other places mentioned in text.	*page* 83
5.2	Ritual cycle and agriculture among the Maori.	113
5.3	Ritual cycle and agriculture among small-scale rice cultivators in Malaysia.	120
5.4	Ritual cycle in swidden agriculture in Indonesia.	122
5.5	Hypothetical reconstruction of Pecos Pueblo in the early historic period.	140
7.1	Population density proxy for prehistoric Britain.	188
7.2	Population density trend for prehistoric North America.	190
7.3	Population trend for prehistoric Santa Valley, Peru.	192
7.4	Population trend for prehistoric and historic-period Hawaii.	193
7.5	Chronological chart for the Andes.	196
7.6	Map of the Andes showing locations of sites in text.	197
7.7	Location of Caral.	199
7.8	Aerial image of Caral in the Supe Valley.	200
7.9	Photograph of Caral.	201
7.10	Solstice alignments of principal buildings at Caral.	205
7.11	Archival photograph of the Casma Valley circa 1956.	210
7.12	Site of Cerro Sechín, Casma Valley.	211
7.13	Wall depicting captives or trophies at Cerro Sechín.	212
7.14	Detail from stela panel at the site of Cerro Sechín depicting trophy heads.	212
7.15	Detail from stela panel at the site of Cerro Sechín depicting a processional figure.	213
7.16	Detail from stela panel at the site of Cerro Sechín depicting an elite personage.	214
7.17	View of the Tarapacá pampa from the east.	215
7.18	Large structure found at the site of Pircas.	216
7.19	Site of Caserones opposite the valley from Pircas.	217
7.20	Geoglyph above the Tarapacá Valley along a prehistoric trail.	218
7.21	A massive geoglyph, known as El Gigante, in the Tarapacá pampa near the site of Pircas.	218

7.22 Map of the Chincha Valley and surrounding areas. 221
7.23 An example of Paracas art. 222
7.24 Linear geoglyphs in the pampa above Chincha. 223
7.25 Linear geoglyphs and their relationship to settlements in the upper Chincha Valley. 224
7.26 Stone mound in the Chincha pampa with the platform mound of Cerro del Gentil in the background. 225
7.27 Split mound type in the Chincha pampa. 225
7.28 Circular mound type in the Chincha pampa. 226
7.29 Platform mound site of Cerro del Gentil. 226
7.30 Schematic view of Paracas platform mounds in Chincha. 227
7.31 Site area of Chococota, or El Mono. 228
7.32 Site of Huaca Soto, Chincha Valley. 230
7.33 Procession route from the highlands to settlements in the Chincha pampa. 231
7.34 Rectangular box structure in the Chincha pampa. 232
7.35 Artist's hypothetical reconstruction of the landscape of sites like Göbekli Tepe. 234
7.36 Artistic rendering of the hill of Göbekli Tepe. 235
7.37 Artistic rendering of one of the circular structures at Göbekli Tepe. 236
7.38 Location of Poverty Point. 242
7.39 Bird Mound, or Mound A, at Poverty Point. 243
7.40 View of Bird Mound at Poverty Point (background) from the first concentric ridge (looking to the mound in the background). 243
7.41 Astronomical alignments at Poverty Point. 245
7.42 Reconstruction of Watson Brake. 249
7.43 Reconstruction of Silver Glen Mound. 252
7.44 Ground plan of Dubbleby Howe. 256
7.45 Henge monument of Avebury. 259
7.46 Avebury area showing sites and barrows. 260

Preface

For well over 50,000 years, anatomically modern humans lived in small groups of hunter-gatherer-forager societies. This lifeway was the most successful adaptation in the history of our species. The biological evolution of language and the unique human capacity for symbolizing behavior allowed our species to dominate virtually all of the favorable continental habitats in the world by the early Holocene. At this critical juncture in human history, a few peoples in a few places built monuments on the landscape. These were, for lack of a better term, "special places" where nomadic and seminomadic groups congregated for periods of time. Sites such as Göbekli Tepe in Anatolia and Poverty Point in North America represent the archaeological signature of the emergence of "complex stateless societies." These societies are defined as small groups with the capacity to create successful co-operative social organizations that can, among other things, construct and maintain the monumental structures in these special places. We can reasonably infer that if these groups had the capacity to organize large numbers of people to construct such monuments, they also had a complex social structure well beyond that of the small, nomadic band typical of the Late Pleistocene.

This book develops a theoretical approach to the origin, structure, and spread of these complex stateless societies using concepts from evolutionary game theory, archaeology, and related disciplines. It asks the reader to move away from the concept of "cultural evolution" in the traditional sense to one of the "evolution of co-operation." Central to this shift is the recognition that rational actor theory, the basis of economic game theory and traditional cultural evolutionary theory, is inadequate to explain sustained co-operation among intelligent, adaptive agents, particularly people living in small groups without coercive social or political institutions. Rather, the evolution of co-operation must be understood as a type of collective action problem – getting people in your group to co-operate over time for a common set of goals even, if defecting from the group is in your immediate self-interest. This collective action problem in the historical sciences, in my view, has been most effectively dealt with using evolutionary game theory. The major problem that this book therefore seeks to resolve is this: how do people living in small groups without money, markets, policing powers, bureaucracies, social classes, and other coercive mechanisms develop norms and rules of economic and social co-operation that are sustainable over time? This is, after all, precisely the context in which complex societies evolved in the Holocene.

I argue that such co-operation is achieved by "ritualizing" the economy. These groups constructed norms, rituals, and taboos to organize their economy. These conclusions are based on a rich set of ethnographic data on stateless societies around the globe and on observations of the archaeological record. Far from being quaint and exotic customs of "primitive peoples," the elaborate rules of economic behavior, encoded in rich ritual practices, are ingenious means of organizing a society where political coercion backed by overt or subtle force is absent. In other words, in stateless societies, the collective action problem is dealt with by ritualizing certain behaviors and providing the rewards and punishments necessary to maintain co-operation. The degree to which economic relationships between members of the co-operative group were ritualized to support that co-operation is the key to success in the competitive environment of the Holocene.

This process can be understood using concepts from evolutionary game theory and allied disciplines. I propose the concept of "anthropological game theory" to differentiate it from "evolutionary" and rational actor-based or economic game theory. Anthropological game theory allows us to understand small-group behavior where social rationality – also known as "irrational, prosocial behavior" – as opposed to economic rationality, is the dominant principle of human social interaction.

Finally, the argument in this book is emphatically not an "economy of ritual." I do not in any way suggest that people were motivated by nonmaterial or spiritual goals over material necessity, comfort, and/or social prestige. We are not passive, altruistic Rousseauian angels corrupted by private property and the institutions of an oppressive state. Quite the contrary; the ritualized economy is one that uses the capacity of inherently ego-directed individuals to create co-operative social arrangements and to formalize these through decentralized but highly effective ritual practice. Rational actor theory cannot deal with ritual and taboo as organizing principles very well, but what I offer as the anthropological game theory of small groups works exceptionally well to help us understand this process in ethnographic space and over archaeological time.

This model is both informed and tested by the archaeological record. Over time, we see a cultural transmission process in which the best strategies that promote group co-operation will be selected for, or imitated by others in the broad theoretical sense as developed by a number of scholars. I refer to this process as "strategy" selection, differentiating it from group or other kinds of multilevel selection. These successful strategies are then culturally transmitted through generations in societies without formal mechanisms of enforcement until the emergence of state societies and the concomitant development of coercive social mechanisms.

I first describe the conceptual foundations for this theoretical framework. I then use examples from the ethnographic and historical record to illustrate how this ritualized economy actually works among real people. Finally, I examine the evolution of complex stateless societies in a number of archaeological case studies from four continents.

This book has been a long time in preparation. I have published many of the main ideas in several book chapters and articles (Griffin and Stanish 2007; Stanish 2004, 2009, 2013a, 2013b; Stanish and Coben 2013; Stanish and Haley 2005; Stanish et al. 2014). One reason this book has taken so long is that my arguments draw from a number of disparate academic disciplines, including anthropology, archaeology, cultural ecology, economics, economic history, evolutionary biology, evolutionary game theory, game theory, historical ecology, and human behavioral ecology. I have tried to master this vast literature as much as possible. However, it is certain that there will be errors of commission and omission. For these I take full responsibility.

I am grateful to the many anonymous reviewers of previous articles and this manuscript. I particularly thank David Carballo, Joyce Marcus, and Jerry Sabloff for their incisive and extensive comments on an earlier draft. I am also grateful to my colleagues who provided additional constructive comments and critiques over the years. In particular, I thank Elizabeth Arkush, Christopher Attarian, Hans Barnard, Jacob Bongers, Ran Boytner, Luis Jaime Castillo, Cecilia Chávez J., Larry Coben, Amanda Cohen, Randi Danforth, Edmundo de la Vega, Christopher Donnan, Alan Farahani, Helle Girey, Paul Goldstein, Mike Henderson, Kevin Hill, John Janusek, Terrah Jones, Elizabeth Klarich, Chap Kusimba, Abigail Levine, María Lozada, Kristine Martirosyan-Olshansky, Jim Mathews, Colin McEwan, Jerry Moore, Sarah Morris, Michael Moseley, Stella Nair, Ben Nigra, John Papadopoulos, Joanne Pillsbury, Don Rice, Katharina Schreiber, Carol Schultze, Matt Seddon, Jill Silton, Monica Smith, Gil Stein, Henry Tantaleán, Gary Urton, Kevin Vaughn, Lothar von Falkenhausen, Alexei Vranich, Willeke Wendrich, Ryan Williams, Karen Wise, and Colleen Zori. I acknowledge the excellent work of my research assistants Lorena Becerra and Sara Stratte and the undergraduate students in the Economic Anthropology and Cultural Evolution courses at UCLA. I thank my graduate students Jacob Bongers and Ben Nigra as well as my colleague Christopher Donnan for providing me photographs for this book. I offer a special thanks to Professor Colin Renfrew for his excellent suggestions on earlier drafts of this book.

A substantial portion of this book was written at the Santa Fe Institute. I remain extremely grateful to that gem of an institution for allowing me the opportunity for extended visits of intense and free-ranging intellectual discussions. In particular, I thank Jennifer Dunne, Mirta Galesic, Karl LaFavre, Eric Rupley, Jerry Sabloff, Paula Sabloff, David Wolpert, and Henry Wright.

I am grateful to my family members and friends, including Garine Babian; Gregg Repoff; Alexei and Amy Repoff; Jennifer Wallace; Avo Babian; Meher and Nina Babian; Palig, Anthony, and Lori Negosian; Deborah Arnold; Harris and Margaret Bass; David and Kathleen Boochever; Roger and Patty Civalleri; Tim Grosser; Dr. Bruce Hector; Mark Kielar; Mark Stanish; Jacquie Johnson; Artimisia Johnson; Alina and Tom Levy; Shannon McGarry; Dr. Barry Munitz; Charles and Ellen Steinmetz; and Walt Zipperman.

Financial supporters of my research include the Cotsen Endowments at UCLA, the National Science Foundation, the Institute for Field Research, the National Geographic Society, the Steinmetz Foundation, UCLA's Office of the Vice-Chancellor for Research, the UCLA Dean of Social Science, and the UCLA Faculty Senate as well as private donations from Debby Arnold, Harris and Margaret Bass, David and Kathleen Boochever, Dr. Bruce Hector, Charles Steinmetz, and anonymous donors.

I thank the staff of Cambridge University Press for their outstanding work, particularly Dr. Andrew Winnard, Executive Publisher of Social Sciences, Mr. Sam Shaw, Content Assistant and Mr. Abdus Salam Mazumder, Associate Project Manager and his staff.

As always, I offer a very special thanks to Mr. and Mrs. Lloyd Cotsen for their guidance and support over these many years.

Finally, I thank my dear friend, mentor, and colleague Professor Joyce Marcus for her sage advice and unwavering friendship for more than 35 years.

1 The Evolution of Human Co-operation

Sometime at the end of the Pleistocene, a few groups of mobile hunter-gatherer-foragers in the greater Mesopotamia region started to build elaborate monuments on their landscapes using wood, stone, and earth. Over the millennia, more and more of these sites were constructed over an increasingly larger region. People periodically congregated at these special places for weeks or even months at a time. This phenomenon likewise spread and independently developed throughout the entire Eastern Hemisphere over the next few millennia. The tradition of building these special places likewise died out after a millennium or so in particular areas, only to reappear in the archaeological record in different forms in neighboring regions. This phenomenon occurred independently in Asia, Africa, and the Americas. It is this process, one that I argue develops out of sustained human social interaction under certain demographic and environmental conditions, that we will explore in this book.

I refer to the small-scale societies in which people build special places as a means of organizing their economic and cultural life as *complex stateless* ones. The key difference between stateless societies and complex ones is the degree to which people in the latter co-operate on a sustained basis with others who are distantly related or even not biologically related at all. Most hunter-gatherer bands – simple stateless societies – lived in groups that averaged approximately 25 people. Complex stateless societies supported substantially larger numbers of people who interacted with each other over long periods of time. While there is some evidence that hunter-forager bands were composed of some nonrelated people, complex stateless societies were characterized by numbers an order of magnitude larger. These large numbers created complex relationships between many people. This transformation in our lifeway from small, mainly kin-based mobile groups to larger groups of distantly related and nonrelated people who came together to create monuments where they could congregate and intensify social relationships represents the origin of "civilization" in colloquial terms. Empirically, it represents one of the most radical shifts in the social and cultural organization in the history of our species. It is the core problem of this book.

This book develops a theoretical approach to the origin, structure, and spread of such complex stateless societies using concepts from evolutionary game theory, archaeology, and allied disciplines. Evolutionary game theory applies the principles of classic game theory to evolutionary questions. We understand the evolution of co-operation to be a type of collective action problem (Carballo 2013). That is, how do nonrelated people develop norms of co-operation and sustain these over time, particularly in social contexts in which political coercion is effectively absent? The key to the social co-operation in complex stateless societies is that they must effectively deal with the "free-rider" problems inherent in groups made up of ego-directed people. Overcoming these collective action problems is essential to understanding the evolution of social complexity in our species. These more successful complex stateless societies create social organizations that allow individual members of the group to benefit in ways that they cannot in smaller population sizes.

The lack of coercion in complex stateless societies is a key feature of this social phenomenon. Ethnographic data indicate that complex stateless societies certainly have social ranks and markers of prestige, but unlike leaders in state societies, those in stateless ones do not possess *coercive* power over others. This is an extremely important observation: the emergence of complex stateless societies was not a costly process in which the vast bulk of people were forced to give up resources or labor to ego-directed aggrandizers, as is argued in many more traditional cultural evolutionary models. Rather, under the appropriate conditions, an ad hoc managerial leadership will emerge to deal with the free-rider problems, on one hand, and the need to reward co-operators, on the other. This is a kind of leadership created by the group; it is not forced on the group either by aggrandizers or by environmental stresses.

These free-rider problems, if not properly and forcefully dealt with, threaten group co-operation and collective action. It is critical to note that informal social coercion exists in all stateless societies (as well as in state ones, of course), and it is manifested in taboo, black magic, and so forth. However, stateless societies are notable for their absence of institutionalized elites with power to obligate others without a substantial consensus among the community. That is, power in stateless societies by leaders is ad hoc and granted or withdrawn by the community at large. This is a critical point, and I spend the entirety of Chapter 4 addressing this issue. People who take on the organizational tasks are referred to here as managerial leaders. Leaders assume costly, prosocial roles and are rewarded with prestige and other resources. Unlike classic cultural evolutionary models in which nonstate or chiefly elites develop into coercive, hierarchical classes by taking advantage of resource and/or social stress, managerial leaders in contrast garner prestige and wealth by hosting elaborate social events such as feasts, potlatches, and other ceremonial activities. They also serve as the key mediators in structuring co-operative group activities through persuasion, not force.

The framework developed in this book is based on theoretical innovations in various kinds of evolutionary game theory in the social sciences over the past 30 years. Most significantly, we have learned that the traditional rational actor theory used in economic game theory and in the "canonical" explanations of the origins of complex society in archaeology is inadequate to understand the structure of the small groups that characterize stateless societies. Rather, we turn to newer insights into human sociality predicated on various types of prosocial behavior characteristic of small-group interaction. I refer to "irrational prosocial behavior" as "social rationality" to distinguish it from classic game theory's rational behavior. That is, a behavior that may seem irrational from an economic perspective can be quite rational from a social one insofar as that behavior is ultimately benefiting the actor. The opposite is also true. Social rationality is the cornerstone of human sociality, as we will see in the following chapters.[1] An explosion of research in the study of human co-operation in small groups over the last few decades shows that sustained co-operation between ego-directed individuals can be maintained under the appropriate conditions. The appropriate conditions include a minimal level of population density and resource availability, on one hand, and the creation of cultural norms that deal with free-rider and other collective action issues, on the other. This concept is called "homo reciprocans" or "conditional co-operation" and has been extensively developed by scholars such as Sam Bowles and Herb Gintis (Bowles et al. 1997; Bowles and Gintis 2011), to name just two of the many theorists who have worked on this theoretical issue.

We can therefore reformulate the question in a more nuanced way: how do people create complex, mutually beneficial co-operative economic organizations in small-scale societies that are not structured by markets or states and that lack money, policing power, contracts, courts, and other forms of social coercion? In collective action terms, we can ask, how did people solve the free-rider problems inherent in group behavior in these societies, and how were co-operative individuals rewarded for both co-operating with and not defecting from the group? How was non-co-operation punished? How did individuals in the group benefit more by working together than they would have by working in small, family-based units or even alone? In short, what conditions led to a few groups creating social systems of sustained co-operation at the beginning of the Holocene, and what conditions kept them co-operating for millennia up into the ethnographic present?

My analysis of ethnographic and historical data indicates that people in small groups create rules and norms to govern the production and exchange of resources, behaviors that makes sustained economic co-operation possible. These norms and rules are structured through various kinds of ritual and taboo. These rituals schedule tasks, reward co-operators, and enhance payoffs for prosocial behavior by all members of the group. They maintain fairness and punish non-co-operators. Feasting and potlatching events are an integral part of the range of reward strategies to promote co-operation. This "ritualized economy" is the means by which the emergent properties inherent in human social interaction that lead to co-operation are transformed into institutionalized norms and beliefs, sustained by costly, prosocial behaviors. In fact, these rules develop in any group of people who interact over any period of time, no matter the larger sociopolitical context.

I also contend that work in evolutionary game theory indicates that the social dynamics of small groups are significantly different than those in large ones. This is an undertheorized point in anthropology in particular and social science in general. Small-group dynamics are different for a number of reasons, not the least of which is that the role of reputation in long-term social interaction is so important to maintaining co-operation. If people in fact behave in socially rational ways based in part on their knowledge of other people's likely reaction to their actions (see Chapter 3), then small groups provide a context in which most people know each other. The ability to create social histories of most people and to use these reputations in future interactions is possible in small-group contexts in ways not possible in large groups. As Sergey Gavrilets (2015:2) nicely phrases it, "the evolutionary roots of human cognition are in our capacity to form shared goals, be committed to them, and collaborate in pursuing them and that this capacity evolved within the context of small-group cooperation." I will argue that prosocial behaviors, such as costly punishment, work in small groups but that as societies become larger and the stakes of certain behaviors increase, people's behavior trends toward being more economically rational. This process parallels the shift from stateless to state societies in deep history and archaeology – a topic for another book.

In effect, the economic anthropology of small groups and the nature of co-operation in small groups are different than in larger ones. Succinctly, we behave differently in a village than in a city due to the continual face-to-face interaction that permanently exists in the former. Small-group contexts elicit co-operation more than large-group contexts for a number of reasons, not least of which is the differential cost of enforcing and rewarding norms of co-operative behavior in each context.

Anthropology is uniquely situated to take on this theoretical challenge because it has traditionally focused on small groups that do not use modern economic mechanisms in their daily lives. Certainly, from an archaeological perspective, up to the early Bronze Age in the Old World and the later Pre-Classic periods in the Americas, all people lived and interacted in small groups in stateless societies. The principles that developed during the evolution of social complexity among our species were products of these demographic contexts. If my contention that socially rational behavior is more pronounced in small groups than in large ones is correct, then to understand the emergence and evolution of co-operation among our species, we must precisely understand the rules and logic underlying small-group behavior.

We can summarize the logic of this argument so far as to note that people living in small groups where coercion is largely absent are *conditional co-operators*, creating complex co-operative arrangements under the appropriate conditions. Conditional co-operators are *ego-directed social agents who can be induced to co-operate under the right social and material conditions*. People are intelligent and adaptive agents, but they are not always "rational" in a classic economic sense of the term. They are, however, perfectly rational in the colloquial sense, as captured in the term "social rationality." These underlying conditions center on the degree to which group co-operation benefits individuals in the group, that interaction is fair, and labor by individuals is reciprocated or rewarded in some tangible manner. The material conditions center on the degree to which the physical environment provides the means to maintain specialized labor organizations that make co-operation in a group worthwhile to individuals in that group.

Research has demonstrated beyond much doubt that people in small-group interactions engage in costly prosocial or altruistic behaviors and are willing to forgo temporary advantages to prevent cheaters from threatening group cohesion. They will co-operate to achieve goals, and they use various social tools to keep co-operation alive and well. People in successful groups recognize both the individual and collective advantage of co-operation, and some individuals in the group are willing to absorb some costs onto themselves to maintain norms of fairness in exchange for prestige.

The Economics of Small-Scale Societies as a Question of Co-operation

The work in what I call anthropological game theory provides the theoretical basis for the understanding of the formation of socially rational co-operation among ego-directed individuals in small groups and in small groups embedded in larger societies. For evolutionary biology, evolution of "irrational" or altruistic behavior is not a difficult problem to explain in groups of biologically related organisms.[2] Creatures as simple as ants and as complex as chimpanzees can imprint patterns of behavior that lead to large-scale and evolutionarily stable patterns of co-operation that are in the self-interest of each member of the co-operative group. The well-theorized concepts of reciprocal altruism and kin selection explain these phenomena. In humans as well, the tools of evolutionary biology explain in part co-operation among related individuals. However, Kim Hill et al. (2011:1286) found that even in small contemporary hunter-forager bands, most people in residential groups were not strongly genetically related.

While the degree to which modern foragers' residential patterns are similar to those of our Paleolithic ancestors is debatable, it is clear that co-operation in simple societies cannot be completely explained by evolutionary biology alone. As Hill et al. note, even in these societies, there is a large social network of individuals who are not genetically related. This empirical pattern has been noted elsewhere. This reality then explains in part the development of co-operative norms via our evolved capacities for cultural learning.

In this light, human co-operation is different from that of other social creatures in two key ways. As Sergey Gavrilets (2015:2) notes, "humans are also unique in their innate ability and willingness to cooperate at a variety of different scales." First, we co-operate with nonkin on a massive scale. Second, when one adds the unique capacity for traits like speech, symbolizing, and cultural transmission by humans, the emergence and spread of co-operative behaviors occur on an order of magnitude faster than among any nonhuman species. To restate the primary discovery in game theory of the small group – conditional co-operation – one need not invoke a Hobbesian Leviathan to explain this apparent Rousseauian co-operation among humans (Shalizi 1999). Rather, we have a theoretical framework that recognizes that emergent properties in human social interactions can create stable and complex co-operative societies in the absence of coercion but that still satisfies the evolutionary requirement that people act in their own self-interest by increasing their individual payoffs and selective advantages.

Once we understand these principles, much of what used to be considered the "exotic" and "irrational" behavior of non-Western cultures becomes eminently socially rational and intelligent, as we will see in the next chapter. Furthermore, it is a very short and easy intellectual leap to see the signal importance of this insight in modeling the emergence of complex society in the deep historical past of our species.

Making a Living in Stateless Societies: The Ritualized Economy

And here we come to the core of this book. How do people create complex, co-operative economic organizations in small-scale, stateless societies where coercion is largely absent? What rules, emerging from repeated interactions of individuals and transmitted through generations (Carballo, Roscoe, and Feinman 2014:103), allow for successful co-operation? How are these rules converted into social norms and customs that members of the group implicitly understand and support? Why do some rule-bound co-operative organizations survive and thrive while others are absorbed or die out? If you can answer these questions, you can begin to understand the evolution of the first complex societies in the archaeological record and the continued evolution of these societies up the development of coercive states.

The key to these questions, as mentioned earlier, is what David Carballo, Paul Roscoe, and Gary M. Feinman (2014:115) refer to as "coupling the logic of evolutionary cooperation and collective action approaches" (see also Feinman 2000). The collective action dilemma lies at the core of this theoretical problem. Collective action is simply a group of people working for a common goal. However, there are two basic threats to sustained collective action. First, there is a cost of co-operation that has to be offset by rewards. Essentially, groups that successfully co-operate will provide higher payoffs to individuals within that group. Second, there will always be cheaters or free-riders who will avoid the costs of co-operation while trying to obtain the benefits of the group effort. There are many instances in which individuals in a group can acquire more resources by cheating than by co-operating. However, that cheating, if done by too many people, will ultimately destroy the group and possibly the resource base on which the group is dependent. This is also known as a "tragedy of the commons" problem in collective action theory. How groups deal with the twin problems of free-riding and rewarding co-operation is the core of how complex stateless societies develop and sustain themselves in the face of collective action dilemmas.

The ritualized economy as defined here is the key to understanding how people in successful stateless societies function in the absence of coercive political institutions. They create rules of co-operative economic behavior and encode them in a variety of *ritual* practices. This economy, structured by ritual, enculturated by habit, and culturally transmitted over generations,[3] is the means by which the emergent properties of co-operation inherent in social interaction lead to "strong" forms of co-operation. These rules are transformed into norms and beliefs and are sustained by prosocial or altruistic behaviors that reward co-operation and punish free-riders. These are societies that use the capacity of inherently ego-directed individuals to create co-operative social models through effective ritual practice. These normatively organized structures of social interaction are then subject to selective pressures. One look at the ethnographic and archaeological record of small-scale societies demonstrates how much ritual constructions and behavior permeated life prior to, and outside of, state organizations over the millennia.

Once we are attuned to this insight, many comments in the rich ethnographies of small groups or "tribal societies" make sense. The celebrated E. E. Evans-Pritchard, in his book *The Nuer*, described the roles of chiefs in the political lives of these cattle-raising people in the Sudan. He notes that a particularly important activity for chiefs is the mediation of disputes, particularly lethal ones, such as blood feuds. These feuds threaten the well-being of the community, and chiefs are indispensable actors in managing this conflict. In describing the numerous events and the taboos surrounding the management of these disputes, Evans-Pritchard says, "These affairs are like a game in which everybody knows the rules and the stages of development: when one is expected to give way, when to be firm, when to yield at the last moment and so forth" (Evans-Pritchard 1940:175–176). There is no formal law code or systematic rule book in these kinds of societies. No such formalities are needed. These norms develop over time and become "second nature" to people, rules that are continually reinforced by daily social practice. In other words, these behaviors are structured by coded and ritualized norms that are instinctively understood by virtually all. These ritualized events structure people's lives, both political and economic. Once we understand them as a set of social tools that developed over time to deal with the difficulties of keeping distantly or nonrelated peoples co-operating with each other, we see evidence for such phenomena throughout the classic ethnographies of people barely touched by Western colonization.

What is the material basis of successful co-operative economic organizations in stateless society? My read of ethnography, history, and archaeology shows some very clear patterns. Co-operative groups create rudimentary economies of scale structured by ritual and taboo. Working together in specialized tasks, households can produce much more value in the aggregate than by working alone because of the efficiency in eliminating redundant tasks and institutional transaction costs, the latter term used in the sense defined by economic historian C. Douglass North. The result is that each individual in the group benefits by a much higher per capita level of resource availability. This kind of co-operation is also critical in maintaining defense against those who seek to raid villages. In a competitive environment, the profoundly non-co-operative intergroup strategy of organized violence against neighbors simultaneously creates conditions that promote intense intragroup co-operation. In such contexts, intensive co-operation to defend and to raid has a very high payoff for individuals in the group. Conversely, failure to develop strong co-operative norms places people in danger of being raided, captured, or killed.

This economic organization, however, creates a very big problem faced by all groups working in specialized labor organizations in a stateless society: the ownership of the final product in a multistage production process is not obvious, as it is in solitary production at the household level. This is true if the resources are created by specialized economies of scale or if they are obtained by raiding. This is a classic collective action problem, and it creates some very severe free-rider issues that have to be resolved if co-operation is to be maintained.

Imagine an autarkic model in which one family makes its own pottery. They can produce x amount of pottery per week. At the end of the week, they own all of these pots and can use them as they like, perhaps exchanging the pots for other commodities from other families. But if, say, five families work together in a more specialized manner, the families that make up this work group can now produce $(x)(y)$ pots per week with the same amount of labor. How does that work? The five families simply divide up the tasks and eliminate redundant transaction and other costs. The amount of labor that is reduced is enormous, because each family does not have to replicate all the tasks. As a result, in the same amount of time, the five families are able to produce substantially more pots. There is a huge advantage, of course, one that provides a large material payoff for each participant in this group activity. Not only does the surplus benefit the individuals but a portion of it is used to reward managerial leaders to maintain their earned social prestige.

However, there is a potentially huge social cost as well. Once the pots are fired and ready, there is the question of who owns the pots. In state-level societies, with coercion and/or market mechanisms, this issue is resolved through a variety of mechanisms, including customary hierarchies that predetermine the flow of resources, contracts, supply-demand pricing, enforced taxation, and even the use of slave labor. But in a stateless society without coercive social hierarchies and with no economic ones to speak of, ownership of the final product becomes a critical issue that threatens group cohesion.

These problems are resolved by the use of ritual, feasting, and taboo – effectively, a series of culturally specific rewards and punishments to keep the group together under the leadership of people the community accepts. Stateless societies are able to maintain high levels of economic co-operation by creating cultural norms that reward co-operative agents via feasting and other activities and that punish free-riders. These are the rituals that nineteenth-century Western travelers saw as irrational and a waste of time – the potlatch being the quintessential example. However, we can now see how eminently intelligent and strategic these practices are. Certain kinds of rituals are almost always associated with economic activities that require coordination and some degree of risk. Leaders step up with prosocial behaviors that simultaneously keep the group functioning. Groups of people accord ad hoc authority to leaders to enforce norms of co-operative behavior. In return, these leaders gain prestige and enhance their material and social lives. These rituals function in many ways. They schedule labor tasks. They regulate the redistribution of goods in a way that is fair and understandable by the community. The religious taboos serve to punish free-riders and allow the community to exercise some order in the way in which work and trade are conducted.

But what about the dynamic, historic component to this? Let's return to the pottery production example. If successful, the five families can now produce much more pottery in the same amount of time than five individual families working alone can produce. Over time, this group will exchange their resources for other resources not available in their village. They will acquire exotic goods that can be recycled into a political economy that puts them at even greater advantage over their peers in the rest of the village or the region. Some of the other villagers, being intelligent actors, will most likely either join the successful group or imitate the norms of co-operation developed by the first group.

Or they might create new norms that are even more effective in evoking co-operative behavior. As this dynamic works out, the most effective normative strategies will grow at the expense of less effective ones. This scenario is not a just-so story designed to fit a model. Rather, as Chapters 5–7 show, these behaviors have been documented over and over in the archaeological and ethnographic records.

In short, in the absence of state mechanisms of coercive power, ritual, feasting, and taboo serve as the means by which social norms are encoded and provide the means to coordinate activities among autonomous households. The economic organization of stateless societies is therefore best seen as a "ritualized" one, a means by which social norms of co-operation can be understood and sanctioned in a society without formal codes of behavior. Individuals acquire prestige, ritual status, and greater access to material goods by sponsoring important events for the community. In the absence of coercive mechanisms, there simply is no other way to achieve group co-operation.

The Nature of Ourselves: Are Humans Inherently Co-operative?

In developing a theoretical framework to understand the evolution of co-operation, it is necessary to begin with some fundamental principles about the nature of human psychology and sociality.[4] This is due to the simple fact that if humans are inherently co-operative, then we must explain why we did not co-operate on any significant scale until the early Holocene, given that modern *H. sapiens* goes back at least 50,000 years and probably much longer. On the other hand, if humans are inherently ego directed or selfish, then this begs the serious question of why we as a species began to co-operate at all and why we are not still living in small hunter-gatherer-forager groups. Inherently social or inherently conflictive? Implicitly or otherwise, this dichotomy has been a fundamental question of Western philosophical, economic, and economic anthropological thought since the late eighteenth century. In many ways, it has been the most significant concept looming behind discussions of the origin of complex society in archaeology and related historical disciplines.

In Western political philosophy and social science, the debate over human nature most certainly begins with Aristotle, and his views have subtly affected social science into the present century. In his classic work *Politics*, written in the middle of the fourth century BCE, he opens with the surprisingly modern observation that humans are "more of political animals than bees or any other gregarious animals" (*Politics* 1.2). Aristotle in fact believed that the state or civil society was prior to the family or individual in the sense that humans cannot live in isolation. He claims that "a social instinct is implanted in all men by nature" (*Politics* 1.2) and that individuals are not self-sufficient. In a famous aphorism, he states, "He that lives alone" is either "a beast or a god" but not a human. Humans therefore have a natural tendency to form a state or polis. As Christopher Shields (2007:353) describes Aristotle's premise: "It is in the nature of a human being to form associations capable of supporting and sustaining happiness. This is why humans have both a natural capacity and natural impulse to form a polis."[5]

Aristotle built an entire theory of the state and economy, beginning with the assumption that people are, by nature, co-operators. It is true that in his view they are co-operators out of necessity, not necessarily out of love of fellow humans, but they are inherently co-operative nonetheless. For Aristotle, the origin of co-operation, in politics and economics, is found in the nature of what it is to be human. Therefore, by necessity and natural law, people will inherently co-operate and form civil society.

Aristotle's views reigned throughout the scholastic age in the West but were systematically challenged by a number of Enlightenment figures, who saw hints of a subtle theological bias in this assumption. The theme of our species's fundamental human nature developed in earnest in the seventeenth and eighteenth centuries epitomized by the philosophers Thomas Hobbes (1561) and Jean-Jacques Rousseau (1712–1778). Their work has been caricatured over the centuries by countless scholars and popular writers. As with all great philosophers who tap into some significant insight into the human condition, subsequent writers selectively picked up passages from their work to buttress their own arguments. Both Hobbes's and Rousseau's original work is more nuanced than many later writers have suggested. However, we are not concerned with the precise thoughts of each philosopher per se but rather with the expansive tradition that each scholar represents and how that tradition was worked and reworked over the centuries. Furthermore, both Rousseau and Hobbes were working in a long intellectual tradition. Their work therefore represents the distillation of many currents of Western thought. In other words, when we speak of these two philosophers, we are in actuality referring to the traditions of thought that these two scholars systematized for later generations.

For our purposes, we can derive a basic theory of co-operation from each tradition. The Hobbesian tradition, of course, has evolved over time to be one in which humans, "in a state of nature," are assumed to be inherently selfish or ego directed. Hobbes writes:

> The right of nature, which writers commonly call jus naturale, is the liberty each man hath to use his own power as he will himself for the preservation of his own nature; that is to say, of his own life; and consequently, of doing anything which, in his own judgment and reason, he shall conceive to be the aptest means thereunto. (Hobbes 1651: Chapter 14)

From this concept of *jus naturale*, it follows that "every man has a right to every thing, even to one another's body" (Hobbes 1651: Chapter 14). In modern terminology, we would say that humans' evolved psychology is one characterized by ego-directed behavior.[6] All individuals seek to acquire those things, both material and social, that are necessary in life to survive and thrive. From these premises, Hobbes built a theory of co-operation. Since there was almost always a shortage of necessary goods, people would have to compete with each other to acquire what they wanted: "If any two men desire the same thing, which nevertheless they cannot both enjoy, they become enemies; and in the way to their end . . . endeavour to destroy or subdue one another." In such a world, "Every man is enemy to every man" (Hobbes 1651: Chapter 13), such that "there can be no security to any man, how strong or wise soever he be" (Hobbes 1651: Chapter 14).

Under such conditions, captured by the famous aphorism that life in a state of nature, where all live in continual fear, is "solitary, poor, nasty, brutish, and short," people seek to come together for their mutual protection. For Hobbes and for the centuries of followers of this tradition, the origin of co-operation comes about when people agree to transfer their natural rights of autonomy to a political authority out of fear and necessity. People do not voluntarily accept hierarchy – they are forced into new social arrangements by the exigencies of life. The notion that humans are inherently selfish and require some kind of social or political entity to maintain order and foster co-operation has a long and distinguished history in social philosophy and social science.

Rousseau developed an alternative view of human nature in modern philosophy. Rousseau saw humans as inherently good or other directed, and he viewed the arts, science, and state – i.e., civilization – as the corruptors of a benevolent nature and natural state of affairs. Whereas Hobbes believed that there was inevitable scarcity in any landscape, Rousseau argues precisely the opposite in the first part of his celebrated book *A Dissertation on the Origin and Foundation of the Inequality of Mankind* (1755): "While the earth was left to its natural fertility and covered with immense forests, whose trees were never mutilated by the axe, it would present on every side both sustenance and shelter for every species of animal." He goes on to say that the wants of the savage (humans in a state of nature) are few and easily satisfied. The notion that "the produce of the earth furnished him with all he needed" (Rousseau 1755:Part 2) created the material conditions for co-operation because, contra Hobbes, there was no scarcity to promote self-interest. Likewise, Rousseau believed that disease was rare. The world in which primitive people found themselves was a congenial and pleasant place where conflict was simply not necessary, disease limited, and where everyday needs were satisfied by nature.

In the second part of the *Dissertation*, Rousseau identifies the creation of private property as the beginning of humanity's downfall from this utopic existence. Agriculture, metallurgy, and other arts created desires for things that were inherently scarce. Property gave rise to laws, conflicts, and the inherent inequality that plagues civilized humankind. It is not insignificant that such a view goes almost directly to Friedrich Engels's Marxist theory of social evolution and represents one of the great traditions in the Western social science based upon a Rousseauian theory of human nature.

In short, in Rousseauian theory, humans are not ego directed but rather are socially motivated creatures and, as we would say today, predisposed to co-operation. The appearance of private property and civilization altered our "natural instincts," leading to resource scarcity, fostering conflict, and promoting institutions designed to protect property. This was the context in which civilization was born, complete with inequality, poverty, and the entire panorama of ills that characterize the modern human condition. The Hobbesian tradition, in contrast, is much more consistent with the concepts of *Homo economicus*, the ultimate economically rational agent. This Hobbesian tradition underwrote the political economic and cultural evolutionary theories of the nineteenth and early twentieth centuries. As we will see in subsequent chapters, this dichotomy has been present throughout the history of studying non-Western peoples. And it is a dichotomy that has been displaced by new work in evolutionary game theory.

The Evolutionary Game Theory Revolution: Conditional Co-operators

Game theory is the study of strategic decision making by rational agents. Around the end of the 1970s and into the early 1980s, scholars in very different fields began to experiment with both computer models and human subjects to understand the nature of this decision making, particularly in small groups. Somewhat surprisingly, this work challenged concepts of *Homo economicus* in the study of economic behavior and political strategies. I say "surprisingly" because at that time, most political scientists and virtually all economists worked within fairly strict rational actor assumptions of human behavior.

Theoretically, rational agents seeking to maximize their own advantages should bring about a relatively rapid breakdown of any kind of altruistic co-operation, as individuals seek advantages on each and every human interaction. Political scientist Robert Axelrod sponsored a competition to see whether there existed a set of simple rules that could produce co-operation in distantly related or unrelated and otherwise selfish agents. As reported in a now-classic article, Axelrod and W. D. Hamilton (1981) discovered that a very simple set of rules promoted sustained co-operation between ego-directed individuals. This strategy, known as Tit for Tat, was simply to co-operate on the first move and do whatever your competitor does later on. Such a strategy used by two agents in computer models worked well.

But would it work in real life? These computer games seemed to be exactly what they were called – games – and were too abstract and removed from the real world to be of value. But what would in fact surprise many, Tit for Tat likewise worked in experimental games using human subjects. Many other games – some of them decades old at the time – were also used to test these ideas. These games included the Prisoner's Dilemma, the Ultimatum Game, iterated versions of both, dictator games, and so forth. As we will see in later chapters, the hypotheses that these other games generated were also tested with observational data in the real social world. And they worked with real people as well. These games went beyond Tit for Tat and have given us a new theoretical tool kit for understanding human economic and other social behavior, particularly in small groups.

What this work revealed was that *Homo economicus*, and by extension rational actor theory, did not adequately model the actions of people, particularly in small numbers where everyone knew each other to a greater or lesser degree. In classic economic game theory there is an assumption that people will rationally maximize utility. Utility is the amount of satisfaction that an agent derives from using a good or service. In modern economics, utility is usually described in terms of price, and "rational" is understood as the maximizing of advantage in each economic interaction. This represents the basic assumption of Western economic theory for more than a century, as mentioned above, and is also known as the canonical model (Henrich et al. 2001).

Evolutionary game theory is different. It is the use of game theory principles to study evolutionary processes. Evolutionary game theory derives from classic game theory but stresses the process of natural selection of behaviors in populations of interacting individuals and groups over time. David Easley and Jon Kleinberg provide a good working definition of

> the notion of evolutionary game theory, which shows that the basic ideas of game theory can be applied even to situations in which no individual is overtly reasoning, or even making explicit decisions. Rather, game-theoretic analysis will be applied to settings in which individuals can exhibit different forms of behavior (including those that may not be the result of conscious choices), and we will consider which forms of behavior have the ability to persist in the population, and which forms of behavior have a tendency to be driven out by others. (Easley and Kleinberg 2010:209)

Experimental research in evolutionary game theory has invalidated many of the presumptions of *Homo economicus*. It demonstrates that humans can be irrationally prosocial, demand fairness, and co-operate under the appropriate conditions. This is profoundly contrary to the assumptions of the canonical model (Bowles et al. 1997; Henrich et al. 2010a), but it lies at the core of human society. As I demonstrate below, without socially rational behavior, human society as we know it could not survive.

I would go so far as to say that these insights in evolutionary game theory represent some of the most important discoveries in late-twentieth-century social and behavioral science. For more than two centuries, scientists wrestled with the theoretical assumption of self-interested or ego-directed behavior in evolutionary studies, economics, and economic anthropology. *Homo economicus* was clearly a creature who acted consistently in his or her own self-interest. Outside of economics, evolutionary biology theory was predicated upon the individual selection of organisms acting in their own self-interests, defined as staying alive, maximizing resources, and propagating their own genetic heritage. If one accepted the fundamental premise of *Homo economicus*, then individuals should behave in a selfish manner virtually all of the time. Only kin were exempted, and even here the degree of altruistic behavior was proportional to the degree of biological relatedness.

This new work was indeed revolutionary. It upended many of the principles underlying the logic of rational actor theory. Most significantly, *it showed how selfish individuals could create co-operative groups without any kind of coercive power or overarching authority*. Even in a strategy as simple as the Tit for Tat, which could be genetically imprinted on a creature as humble as an ant, seemingly "altruistic" co-operation was possible between ego-driven agents. Projected onto humans, the Tit for Tat and the results of other experiments showed a strategy that could provide bigger payoffs to all selfish parties and that did not require taking immediate advantage of each other. In short, co-operation could emerge among selfish agents under the appropriate conditions without coercive leadership.

The concept of conditional co-operators may be one of the most important social scientific concepts created in the last 40 years (see Chaudhuri 2011). It solves the contradiction of selfish human agents co-operating with nonrelated others.[7] It oftentimes references the concept of "emergent complexity," the idea that unique social structures can emerge from the interaction of numerous individual agents.[8] The new body of evolutionary game theory created over the last generation liberates us from three centuries of the philosophically -crippling dichotomy of ego directed versus altruistic, Hobbes versus Rousseau. In reality, most humans are simultaneously ego directed and prosocial under the appropriate conditions. We need much more subtle models to deal with this more complex reality, and the concept of conditional co-operation provides the basis for these models.

Even more remarkable from this perspective is the work done since the first computer tournaments in the early 1980s. Insights from experimental games in the 1990s and beyond showed yet again that people in small groups, both in the West and in the rest of the world, are generally not the *Homo economicus* of classical economics. Rather, people co-operate in ways that defy the principles of strict maximization of utility, and they will absorb significant costs to maintain and foster co-operation with their fellows. A whole new field of behavioral economics seeks to incorporate insights from evolutionary game theory into modern economic decision making. Here, nonmaximizing, suboptimal choice and downright irrational behavior are treated not as inconsequential anomalies but as existentially important components of human social behavior. These behaviors operated in the past and operate today, from the smallest non-Western society to the largest industrial nations on earth.

In short, the Rousseauian/Hobbesian dichotomy, one that dominated Western social scientific thought for 200 years, became obsolete with this research. The new understanding of our evolved human nature is summarized in the evolutionary game theory concept of *Homo reciprocans*, as coined by Herb Gintis (2000a:316). To paraphrase Cosma Shalizi (1999), this work that developed this concept demonstrated that "humans are neither the Hobbesian brute nor the Rousseauian angel." Rather, they are conditional co-operators, people able to co-operate under the appropriate circumstances. The concept of conditional co-operation is the theoretical foundation of this book: we will explore those circumstances that promote co-operation over sustained periods to create what we anthropologists and archaeologists refer to as complex society.

This game theory approach works for small groups in most cases studied in modern experiments. We also see these principles working in numerous cases in the ethnographic and historical record, again for small groups, as seen in Chapter 5. As a result, we can construct a model in Chapter 6 and apply these principles to the archaeological record for the appropriate contexts described in Chapter 7. These contexts include the many stateless societies with complex social and economic characteristics that emerged out of hunter-gatherer-forager groups around the world. Iconic sites like Stonehenge and Poverty Point are examples of massive labor expenditures by people living in stateless societies. They are special places built into the landscape and are the result of centuries or even millennia of earlier peoples refining and enhancing ritualized economies. As a new evolutionary dynamic sets in, political life is ratcheted up, as managerial leaders seeking prestige compete to build ever more elaborate monuments using the combined efforts of many groups. It is an example of hyper-co-operation at a regional scale that also feeds back upon itself: as the monuments become more spectacular, more people are drawn in and hosted, reciprocal relationships intensify, and more resources are created and distributed. More leaders obtain prestige, more social interaction occurs between people, and more resources flow throughout society. These organizations are fragile, however. Overly ambitious leaders, negative environmental changes, and competition from outside the region can cause such organizations to collapse and lead people to revert to a less complex set of co-operative behavior. This, in a nutshell, is the mechanism by which stateless economies evolve over time and that will be explored later in this book.

The intellectual history of *Homo economicus* and its relationship to the study of non-Western peoples is central to understanding the emergence of evolutionary game theory frameworks. It is to this story that we now turn in the next chapter.

2 Economic Anthropology of Stateless Societies: The Rise and Fall of *Homo economicus*

> The standard individual economic agent *Homo economicus* has been called a "hedonistic sociopath." He also has no culture at all and is far too smart. Nobody, not even exponents of "rational choice" theories, is much like *Homo economicus*, which is good for humanity but bad for those theories.
>
> – Shalizi (1999:17)

The Creation of the "Primitive" Economy

As the modern Western world encountered different cultures in their imperial adventures from the sixteenth century onward, they found people bound together by norms and behaviors that did not make sense to them. Many of the non-Western peoples accessible to Western scholars and travelers lived in small-scale societies, particularly in the Americas, Oceania, Africa, and Asia. Others lived in larger states such as China, the Ottoman Empire, and India. Sir John Lubbock, for instance, wrote, "Savages show us a melancholy spectacle of gross superstitions and ferocious forms of worship." He went on: "It must, however, be admitted that religion, as understood by the lower savage races, differs essentially from ours; nay, it is not only different, but even opposite" (Lubbock 1882:200, 202).

In all of these cases, Westerners exoticized the "other." At the core of this "orientalist" construction of other people's reality was the supposed mysticism and irrationality of their behavior. It was out of these multigenerational interactions between colonizers and the colonized that the modern concept of the "primitive" emerged. Most importantly, from the perspective of this book, it was in this globalized context where the concept of the "primitive economy" emerged. In the eyes of Western observers, it was the practical and moral opposite of the Western economy. Ethnographers and travelers alike furnished anecdotal tales of irrational behavior. Stories of people refusing money for basic goods abounded. Perfectly healthy and strong men and women refused to enter the newly created labor markets, and force – including slavery – was usually required to bring these people into the "modern" world. These "primitives" reacted in profoundly differently ways from Western peoples when offered chances to enhance their personal wealth or status, opting instead to stay with their tribes or clans and oftentimes avoiding the Western way of doing business altogether. Exotic tales of wasteful feasts, rotting yams, headhunting, cannibalism, superstitious shamanism, and other "odd" behaviors circulated through the political and intellectual classes of the West. This was a world filled with incomprehensible rituals, religious superstitions, eccentric behavior, and, depending upon your perspective, either a life of perpetual violence and poverty or a bountiful existence based upon a noble attachment to Mother Earth and her peaceful gods.

The primitives were the quintessential other. This concept of the primitive worked its way into social and political philosophy, economics, science, and Western culture in general. The myth of the primitive undergirds much of economics, economic history, archaeology, and anthropology insofar as they have a pedigree that reaches back to the nineteenth century.

Parallel to the imperial advance of European and American states, scholars like John Stuart Mill, Adam Smith, and David Ricardo studied the economies of modern nation-states. They developed several first principles to understand the political and economic world around them; these principles ultimately affected our view of the non-Western and ancient worlds. One of the most important of these principles was that of economic "rationality" – achieving one's goals in the most efficient and practical way possible. A rational person in this view was someone who maximized advantage with most every human interaction. The consumer sought the lowest price, while the producer sought the highest. The laborer sought the highest wage from the factory owner, and the factory owner sought to offer the lowest wage possible. These rational interactions in the aggregate created the market.

But rational actor theory was more than just economics. For these scholars, the entire proper Western way of acting as social beings was a network of rational individuals acting in their self-interest, which in turn created rich nations and prosperous societies. Societies in which rational behavior was manifest became wealthy. Even laborers' standards of living increased, at least in some urban areas, as did the increasingly large profits of the owners of capital. And of course, the contrary was also true: societies that "rejected" rationality were viewed as retrograde and poor – they were the "primitives."

This notion of rationality resulted in the well-known concept of Economic Man, or *Homo economicus. Homo economicus* was the epitome of this rational agent, calculating his costs and benefits of making, selling, and buying goods and services in markets spread throughout the land. Created by Mill as a mere heuristic tool, the concept became reality for a majority of scholars known as political economists, the precursors to those in the fields of economics and economic anthropology. In the world of nineteenth-century Europe and North America, the notion that humans inherently and continuously produce and the "truck and barter" out of their own self-interest seemed to be eminently logical. The libraries of this era were filled with the classic works of antiquity, and university students read of the Greek agora, the Persian bazaar, and the Roman forum. These genteel scholars could see before their eyes everyday negotiations in the local marketplace. Laborers would sell their productive skills for money; farmers would carefully calculate what goods were needed and would bargain with those same laborers for the maximum price. Great stock exchanges emerged in the capitals of Europe and America, where massive transfers of wealth took place on a daily basis. Cotton, textiles, metals, grains, coal, sugar, and every other imaginable commodity were produced and sold for cash, which was then used to purchase other goods and labor, a continual process that fueled the productive cycle. Capital and labor markets sprung up from Kenya to Kentucky, from India to Indonesia. The combined behaviors of thousands to millions of people following the rules of *Homo economicus* constituted the societies that political economists both lived in and studied. This was their world, and it made sense. And it was this commonsense understanding of the rational that formed the basis of classical economics, known as formalism in economic anthropology or as the canonical model of human behavior in the literature of the late twentieth century.[1]

This concept of "Economic Man" adversely affected our theoretical understanding of small-scale stateless societies for a century or more. In the nineteenth and early twentieth centuries, it was a form of ethnocentrism and colonialist ideology writ large over the globe. In later decades, it prevented us from understanding the degree to which people in stateless societies were not passive reactors to external factors but rather conscious agents who molded their everyday lives and physical environments in extremely sophisticated ways. Economic activity among stateless peoples was seen through the lens of the classical economist as primitive and retrograde, permeated with superstition and myth. In reality, precisely the opposite was true, and a deeper analysis of the ethnographic and historical data shows just how brilliant these social organizations were in creating the norms for sustained co-operation in economic life.

The development of the discipline of political economy, the precursor in many ways to modern economics, occurred at the height of European and American expansion across the globe. It is not surprising that scholars with this worldview saw the many non-Western economies that they read about and occasionally encountered as imperfectly evolved forms of market systems. In this perspective, the economy began with the autarkic household, where needs were met by simple, unspecialized labor applied to that which was provided by nature. Slowly, some households and villages begin to specialize and exchange in meeting places or primitive markets. Trade began as small-scale peddling; the small markets become larger fairs. Towns then developed on the landscape, and long-distance trade evolved into some kind of mercantilism, which in turn became full-fledged capital and labor markets in the European sense.[2]

Countless encounters between Westerners and the "other," however, revealed that the expectations of *Homo economicus* were often not met in faraway lands. For the nineteenth-century scholar adopting the canonical model of economic behavior based on rational actor assumptions, the failure of non-Western peoples to adopt Western rational behaviors became the basis of the anthropological concepts of "primitive" tribal society. In the mid- to late nineteenth century, buttressed by the dominant philosophical concept of "progress," Western scholars developed typologies of societies based at their core on economic and political criteria. In this view, the poorest of these tribal societies were also those most bound by irrational custom and religious superstition. The wealthiest societies, in turn, tucked away their religious worship in cathedrals one day a week, while the rest of their time was spent maximizing profits from their industries, trade, and labor. Classic positivist thinking saw a progression in human society: society was initially permeated by myth, then by religion, and finally by science. This process paralleled an intellectual shift in society from pure superstition to pure rationality in the so-called evolutionary philosophy of the time. Such a scheme, of course, paralleled the evolution of society from savages through barbarians to civilizations in the frameworks developed by Henry Lewis Morgan, Edward Burnett Tylor, Herbert Spencer, and others.

Political Economy

It is not an exaggeration to say that the Hobbes/Rousseau dichotomy described above was a major feature of the intellectual context in which the discipline of political economy was created. The intellectual history of economic anthropology is likewise intimately tied to political economy in the nineteenth century and the subsequent emergence of economics at the turn of the twentieth. In fact, the field of economic anthropology was not formally recognized until the first decade of the twentieth century, but speculation on the nature of non-Western societies can be found throughout the earlier discussions of political economists and early anthropologists.

Political economy was the precursor to modern economics. This discipline sought to understand, among other things, why some nations were wealthy and others were poor. As John Stuart Mill famously said, "Political economy was the science that professed to teach how a nation is to get rich." More precisely for Mill, political economy was the discipline that "informs us of the laws which regulate the production, distribution, and consumption of wealth. . . . Political Economy, it is said, is to the state, what domestic economy is to the family" (Mill 1874 [1844]:essay V10).

One of the first systematic works in the political economy tradition is *An Inquiry into the Nature and Causes of the Wealth of Nations* by Adam Smith, written some 70 years before Mill's celebrated works on the same subject. Smith's classic has been extensively analyzed over the last two centuries and need not be repeated in any great detail here. What is important for us is that he began with several premises that drew off earlier and contemporary scholarship in social and natural philosophy. His assumptions and conclusions had a strong influence on later political economists and economists and on our notions of stateless societies.

Milton Myers (1983), among others, sets out the strong case that Smith accepted the Hobbesian premise that all people act out of their own self-interest. For Smith, the division of labor in society leads to an increase in the "productive powers" of each individual, which in turn benefits the entire nation. As Salim Rashid (1985:332) points out, the division of labor for Smith in essence incorporates self-interest that, if coordinated properly with the appropriate incentives, will lead to a greater good. The following quote by Smith captures this link between the division of labor and self-interest and develops a crude theory of redistribution and the barter economy:

> Whoever offers to another a bargain of any kind, proposes to do this. Give me that which I want, and you shall have this which you want, is the meaning of every such offer; and it is in this manner that we obtain from one another the far greater part of those good offices which we stand in need of. It is not from the benevolence of the butcher, the brewer, or the baker that we expect our dinner, but from their regard to their own interest. We address ourselves, not to their humanity, but to their self-love, and never talk to them of our own necessities, but of their advantages. (Smith 1937 [1776]:Chapter 2)

From Malthus, Smith assumed that there is an inherent scarcity of goods in nature due to the tendency of populations to grow up to carrying capacity of the land and other productive forces. And, like Aristotle, Smith recognized that no human is self-sufficient but by necessity must rely on others: "Man has almost constant occasion for the help of his brethren, and it is in vain for him to expect it from their benevolence only" (Smith 1937 [1776]:Chapter 2). This inherent necessity to be economically gregarious is coupled with what Smith famously calls a "propensity in human nature" to "truck, barter, exchange one thing for another" (Smith 1937 [1776]:Chapter 2).[3]

> This division of labour, from which so many advantages are derived, is not originally the effect of any human wisdom, which foresees and intends that general opulence to which it gives occasion. It is the necessary, though very slow and gradual consequence of a certain propensity in human nature which has in view no such extensive utility; the propensity to truck, barter, and exchange one thing for another. Whether this propensity be one of those original principles in human nature of which no further account can be given; or whether, as seems more probable, it be the necessary consequence of the faculties of reason and speech, it belongs not to our present subject to inquire. It is

common to all men, and to be found in no other race of animals, which seem to know neither this nor any other species of contracts. (Smith 1937 [1776]:Book 1, Chapter 2)

This early great work in political economy was built upon the edifice of Hobbesian views of human nature. For Smith, co-operation through the division of labor was essential to the well-being of nation-states. His entire book is permeated with the notion that division of labor increases the productive powers of society. Co-operation derives from the inherent desire of people to engage in exchange, both a product and a cause of the division of labor.

Economic Man and Non-Western Peoples

It was in this environment that the concept of *Homo economicus* developed and flourished. To restate, the primary principle of Economic Man (EM) was that of the inherently self-motivated or ego-directed individual. People are inherently prone to antisocial behavior, while civil society keeps them co-operating, or at least not harming each other. This is the famous "social contract theory" that developed out of Enlightenment thought, as we just saw with Hobbes, in which one gives up autonomy and freedom of action to the state for security.[4]

Based on this assumption, political economists deduced that humans inherently maximized utility. As mentioned above, utility is the capacity of a tangible good or intangible service to satisfy the needs and wants of individuals. In simplistic terms, needs were those goods and services necessary for the biological and social reproduction of life, while wants were those goods and services that enhanced a person's life. The concept of utility was not explicitly developed until the later nineteenth century, but the idea was clearly emerging in political economic theory. In this view, people naturally sought to enhance their material and social well-being with virtually every human interaction. This was considered rational behavior. "Rational" in this sense means that to acquire or produce those goods and services, people make choices that provide maximum utility given a suite of alternatives.

EM (contra Rousseau for humans in a state of nature) also assumed that there was a universal scarcity of needs and wants. This assumption traces its intellectual history directly to Malthus, who argued that population growth would always outstrip the capacity of the earth to support that growth. This was a very important assumption, though it appears to the contemporary mind to be somewhat self-evident. In a world full of uncertainty, where food, clothing, and other needs were limited, people would have to work on a daily basis to survive and thrive.

Smith, in fact, was not the creator of Economic Man (that would come a generation later), although he certainly accepted most of the principles that underlie this concept. Historians generally point to Mill as the political economist who created the concept (Persky 1995:222).[5] What Mill did was to create an abstraction of a person who "is a being who desires to possess wealth, and who is capable of judging the comparative efficacy of means for obtaining that end" (Mill 1874 [1844]:essay V38). This abstraction relied heavily on Smith's work, as well as that of other political economists of the day, including the extremely influential David Ricardo.

Yet what is less known is that while Mill developed this concept as a means to create a science of political economy, he himself was quick to point out (in the same paragraph, in fact) that this was not an actual description of human behavior. It was rather simply a necessary concept in the construction of a science of human economic decision making, no different than that for physics or geometry: "Not that any political economist was ever so absurd as to suppose that mankind are really thus constituted, but because this is the mode in which science must necessarily proceed" (cited in Redman 1997:341). In spite of this, EM became a mainstay for economists and most anthropologists and would greatly influence the way in which stateless societies were perceived in the late nineteenth and twentieth centuries.

Given the universal scarcity of needs and wants, it was also clear that Economic Man had to inherently "economize" – that is, to make strategic decisions to derive the greatest self-interest from any human interaction. These decisions are also economically rational. Furthermore, the proponents of EM presumed that people had close to complete knowledge of supply, cost, and demand regarding goods and services. If they did not have this knowledge, then the economizing behavior would deviate from maximization principles.

The concept of EM also assumed that exchange of goods is inevitable, a fact that derives from the natural scarcity of needs and the logical outcome of people specializing to provide one another, at mutual benefit, the desired goods. One of Smith's contributions was to argue that humans naturally buy and sell as a matter of course (truck and barter). In short, the principles of classical political economy were bundled into a universal theory of human economic behavior: producing, exchanging, and consuming in society. Formally synthesized by Mill as a heuristic device, EM – the rational, maximizing, ego-directed, and economizing agent – became the model of all human society for generations of anthropologists and economists; most importantly for our discussion, it became the idealized agent of civilized society.

Along with EM, two other intertwined intellectual currents shaped the West's view of "primitive" peoples in the nineteenth century. The first was the concept of "progress," an Enlightenment ideal of continual movement toward perfection. The second was the development of the evolutionary theories of Herbert Spencer and his intellectual compatriots (see Freeman 1974). In this "evolutionary philosophy" of the time, all of nature naturally changes from homogeneity to heterogeneity. Culture, proceeds from a primitive to a civilized state. As Herbert Spencer (1900:367) originally put it: The world is guided by a process characterized by "matter" transforming from an indefinite, "incoherent homogeneity to a definite, coherent" heterogeneity. It is this process by "which all progress, organic or other, essentially consists."[6]

These intellectual currents – evolutionary philosophy, Economic Man, and progress – constituted the basis of the first coherent and widely accepted theory of human cultural evolution in modern scholarship. In the nineteenth century, figures such as Henry Lewis Morgan, H. J. Sumner Maine, and E. B. Tylor developed evolutionary theories that derived from the philosophy of Herbert Spencer.[7] The well-known evolutionary sequence of savagery, barbarism, and civilization forms the backbone of this model. These theories likewise saw a progression from societies with "primitives" in a state of nature, who did not understand or appreciate property rights, to civilized people in societies in which EM emerges.[8]

This view is perfectly represented by an article called "Economic Man – A Definition" in the flagship journal *American Anthropologist* in 1902:

> This wonderful change from the primitive man with his fire, shelter, clothing, weapons, and tools – all of which he alone could move from twenty-five to one hundred miles in a single day – to the man in our own city with property-interests of millions scattered through a dozen states, is due to the rise, development, and predominance of what may be called the economic sense. Economic sense expresses itself in a unique mental attitude toward consumable goods. It is the invariable possession of the economic man – it is the mark of the economic man. (Jenks 1902:201–202)

Echoing the dominant view of the time, Jenks went on: "Let economic man be defined as one who, for future gain, produces or traffics in consumable goods; and let natural man be defined as one who produces to satisfy only immediate wants" (Jenks 1902:202). Here is a vivid expression of the concept of the Rousseauian primitive tied to the rhythms of nature contrasted with the civilized person with an "evolved" capacity for economic rational logic. Jenks not surprisingly goes on to describe different stages of culture in a classic Spencerian evolutionary process, with the degree to which one stores and produces wealth being the measure of progress.

The view was buttressed by various travelers in reports that circulated on the edges of scholarship but that had a strong effect on academic thinking. Nineteenth- and early-twentieth-century Western travelers encountered tribal peoples around the world who simply confounded them. The linking of excessive gift giving and waste with ritual and magic among non-Western peoples was one of the strongest metaphors for the primitive in this period. To take one example, seventeenth-century traveler Nicolas Perrot (1911:88) describes a funerary feast, which was form of potlatching among the Hurons of the Great

Lakes of North America. He bemoans the fact that the people give away so much wealth that "they reduce themselves to . . . an extreme state of poverty." He goes on in the same page to note that various French visitors have helped them realize that "these useless extravagances . . . were ruining their families."

There are so many examples of gross ethnocentrism and outright racism by Western travelers, who, beginning with their first encounters, failed to grasp even the most basic social principles of the people that they described. Henry Ling Roth (1968 [1896]:221) summarizing the accounts of nineteenth-century travelers to Borneo: "The Kayans have a curious, if somewhat childish, custom, of foretelling whether an absent friend is proceeding further from home or likely soon to return." Another writes, "The Dyaks are troubled with many superstitions" (Roth 1968 [1896]:221). This is quickly followed by the writer calling them "deluded" for using insects as omens.

Regarding other aspects of their rituals and belief system, Roth recounts another traveler who disparages the Dyak culture: "However absurdly these omens lead the human race, they steadily continue to follow and believe in such practices. Faith predominates and hugs huge wonders, and tenaciously lives in the minds of the ignorant" (Roth 1968 [1896]:230).

The concept of the "primitive," therefore, developed in the West, with myth and ritual almost always at the core of this "refusal" to conform to the principles of EM. It was a profoundly ethnocentric and racist worldview that was empirically wrong. These so-called primitive societies were supposedly riddled with myth and superstition because they were not like us. The savages, it was said, were ruled by irrational customs. Their economies were undifferentiated, production was for subsistence, and exchange was limited to simple reciprocity. In reality, it was their own rational actor assumptions that made these travelers and scholars incapable of understanding what they saw in these "exotic" societies.

The Malinowski Revolution and the Challenge to EM

The idea of "economic anthropology," as a field distinct from economics or political economy, did not develop fully until the beginning of the twentieth century. One of the earliest uses of the term can be found in the proceedings of the American Association for the Advancement of Science meeting in 1902. As reported in the journal *Science*, the heading "economic anthropology" was used in a report on Section H activities by Lindley Keasbey. He explicitly adopted a formalist mandate for anthropologists: "By applying the economic principles of utility and utilization, the anthropologist should be able to establish the first stages of industrial development and determine the essential characteristics of primitive culture" (Keasbey 1903:286). Such a quote betrays a virtual wholesale adoption of the then newly forming science of economics, which concentrated on the quantitative analysis of resource allocation in capitalist society.

EM dominated this academic environment. It soon became known as classic economics and later would be referred to as the canonical model. It was in this intellectual context that the celebrated Bronisław Malinowski started his meticulous fieldwork with some of the world's most "exotic" peoples. Starting in the early part of the twentieth century, Malinowski studied the Trobriand Islanders of Melanesia, a culture about as different from Europe as one could find on the planet at that time. Adopting a functionalist theoretical framework, Malinowski argued that the Trobriand Islanders (and by theoretical extension all non-Western peoples) were far more rational than given credit for by his anthropological and economist colleagues.[9] The superstitions and rituals of these tribal groups – including, most importantly, those of making a living – served them well in their daily lives. Malinowski attacked the core principles of the dominant social Darwinist theories and demonstrated with precise ethnographic data that the tribal peoples of the world were acting in perfectly good and rational ways, even if they were not the ways of *Homo economicus*. A legion of fieldwork by anthropologists followed. These scholars demonstrated that behind the behaviors viewed as odd from a European or American perspective were eminently intelligent means by which people could structure their economic lives. And they organized their life in ways far different from, but equally effective in their own setting, as ours in the Western world.

Malinowski, along with the equally prominent Franz Boas, actually spent long periods in the field living with non-Western peoples and set the standards for the great ethnographers of the twentieth century. Both of these anthropologists developed deep sympathy and respect for the people they studied. Most professionals took very brief (and comfortable) trips to visit "natives" or used naturalists,' imperial bureaucrats,' and military accounts of non-Western cultures. They then analyzed them from home, divorced from the actual realities of human behavior in a culture unlike their own. These armchair anthropologists, as they are disparagingly called today, were responsible for much of the misconceptions about stateless economies.

Malinowski in contrast developed a huge corpus of data from which he drew conclusions, perhaps modified his a priori assumptions, and actually experienced and recorded human behavior in a profoundly non-Western culture. From this work, Malinowski articulated a new anthropological theory that substantially deviated from classical economics based upon his analysis of the sophisticated exchange system of the Trobriand Islanders. As he poetically put it: "All this shows how entirely the real native of flesh and bone differs from the shadowy Primitive Economic Man, on whose imaginary behaviour many of the scholastic deductions of abstract economics are based" (Malinowski 1966[1922]:61–62).

It is important to note, however, that Malinowski worked in an environment that was ripe for his kind of ethnographic and theoretical tour de force, which changed the view of the stateless economy. What is not greatly appreciated is that there was already considerable dissension against EM even at the turn of the twentieth century. Malinowski did not create this anticlassical economic anthropology out of whole cloth. Rather, scholars in fields as wide as anthropology, psychology, social psychology, economics, and history objected to this pillar of classical economic theory.

For instance, in "The Rationality of Economic Activity," an article published in 1910 and still cited today, Wesley Mitchell attacked the very foundations of *Homo economicus*:

> The chief point of this criticism is that the classical economists assumed man to be "a reasonable being who always intelligently seeks his own good or is guided in all his activities by enlightened self-interest." The truth is, Mr. McDougall asserts that "mankind is only a little bit reasonable and to a great extent very unintelligently moved in quite unreasonable ways." That is, the economists have committed "the intellectualist fallacy." (Mitchell 1910:197)[10]

Even in economics, there was widespread caution against the canonical model. Walton Hamilton formally launched an attack on classical theory in 1918 at the American Economic Association meetings. This movement was called institutionalism, and it survived into the 1940s, to be resurrected again in the 1970s by economic historians such as C. Douglass North. As Geoffrey Hodgson (2000:318) summarizes the arguments of the time, Hamilton in reality built a kind of economic anthropology developed by Malinowski and others:

> Institutionalism makes extensive use of ideas and data from other disciplines such as psychology, sociology and anthropology in order to develop a richer analysis of institutions and of human behavior. . . . The notion of individual agents as utility-maximising is regarded as inadequate or erroneous. Institutionalism does not take the individual as given. Individuals are affected by their institutional and cultural situations. (Hodgson 2000:318)

Working in this environment, Malinowski used his data to directly challenge prevailing assumptions about the nature of the non-Western economies. For instance, he noted that "[political economist Karl] Buecher comes to the conclusion that the savages . . . have no economic organisation, and that they are in a pre-economic stage – the lowest is that of the individual search for food, the higher ones in the stage of self-sufficient household economy" (Malinowski 1921:1–2).[11] As we saw above, these views were consistent with the evolutionary theory of the day.

Malinowski rejected these views wholeheartedly. Rather, he found that the Trobriand Islanders were quite sophisticated in the way they managed their economy. He describes Trobriand exchange as "far from being casual or surreptitious" but "carried on according to very definite and very complex rules" (Malinowski 1920a:98–99):

> The Kula involves a complicated system of gifts and countergifts, in which the social side (partnership), as well as the rules of give and take, are definitely established and regulated by custom. It must also be emphasized that all these natives . . . have both a word for, and a clear idea of, barter (gimwali), and that they are fully aware of the difference between transactions at the Kula and common barter. The Kula involves the elements of trust and of a sort of commercial honour, as the equivalence between gift and countergift cannot be strictly enforced. (Malinowski 1920a:98–99)

> All the facts . . . lead us to the conclusion that primitive economics are not by any means the simple matter we are generally led to suppose. . . . Instead, we find a state of affairs where production, exchange and consumption are socially organised and regulated by custom, and where a special system of traditional economic values governs their activities and spurs them on to efforts. (Malinowski 1921:15)

He then argued that his data from the Trobriand Islanders represented "the majority of the primitive peoples known at that time" (Malinowski 1921:1–2) and described them as having "several features remarkable in their bearing upon questions of primitive economies" (Malinowski 1920a:97). Given this, Malinowski was able to articulate a new anthropological theory that deviated from the emerging field of economics. The principles of his new theory were the antithesis of EM.

For Malinowski, needs are generally satisfied in primitive economies. This is a very contentious point right up to today in modern economic anthropology. It is not, however, as Malinowski (1966[1922]:58) says, the case of the romanticized Rousseauian Noble Savage "waiting till the ripe fruits, so bountifully supplied by generous tropical Nature, fall into his mouth." Rather, the natives work hard to grow their food in their gardens, and they have to fish and collect wild foods. On the other hand, they are not in a Hobbesian condition of constant want and conflict. His fieldwork showed that the Trobriand Islanders produced substantial surplus. He writes: "In gardening, for instance, the natives produce much more than they actually require, and in any average year they harvest perhaps twice as much as they can eat. Nowadays, this surplus is exported by Europeans to feed plantation hands in other parts of New Guinea; in olden days it was simply allowed to rot" (Malinowski 1966[1922]:58). As Stuart Plattner (1989:13) aptly notes about the tradition that Malinowski started, they simply did not accept the idea that scarcity was part of the human condition in these "primitive societies."

Malinowski makes these points many times, highlighting each one with more carefully collected data. Regarding fishing, for instance he says (Malinowski 1918:90): "As a rule, the catch is plentiful, often so plentiful that part [of the catch] must be thrown away, as there are no means of preserving the fish." In this case, virtually all subsistence was assured, and surpluses were common.

Malinowski broke with EM in that people in non-Western societies are socially directed, not ego directed. One classic quote of Malinowski's is that "[the Trobriand Islander] is not primarily guided by his desire to satisfy his wants, but by a very complex set of traditional forces, duties and obligations, beliefs in magic, social ambitions and vanities" (Malinowski 1966[1922]:62). This is a critical observation that directly attacks the canonical model. Malinowski is quite clear that it is, again, not some romantic notion of the "native in a state of nature." He points out many instances of selfish behavior, where people constantly try to avoid social obligations, gift exchanges, and the like. However, the general lack of scarcity of subsistence needs creates an environment where people are motivated by social needs. Therefore, in non-Western societies, people do not maximize material wants but rather seek to satisfy social wants. People truck and barter, but they do so not for profit but for social prestige or for the community.

Malinowski concluded that primitive people, or savages as he called them (the term did not carry the pejorative meaning that it does today), were indeed driven by custom, but such custom was rational in its own world, even if it was not economically rational in ours. Malinowski argued forcefully that one had to understand the native economy in its own terms, and not in terms of classic economic theory of the West. He says that someone "accustomed to think in terms of current economic theory will naturally apply the conceptions of supply and demand to labour" and then use these to understand the native economy. He then goes on to make the central anthropological point: "If then you measure him by moral, legal or economic standards, also essentially foreign to him, you cannot but obtain a caricature in your estimate" (Malinowski 1966[1922]:157).

In short, for Malinowski, non-Western peoples like the Trobriand Islanders were the antithesis of Economic Man. They indeed sought to survive and thrive, but not by maximizing advantage on each interaction and seeking to accumulate wealth. These so-called primitive peoples were entirely rational people, in the commonly understood and colloquial sense of the term, using intelligent means to achieve their ends. The Trobriander Islanders sought prestige and community solidarity through ritual and other social practices, much of which was mediated through elaborate exchange ceremonies such as the Kula. He says in numerous instances in his famous book *Argonauts of the Western Pacific* that prestige and status are primary motivations: "Work and effort, instead of being merely a means to an end, are, in a way an end in themselves. A good garden worker in the Trobriands derives a direct prestige from the amount of labour he can do, and the size of garden he can till" (Malinowski 1966[1922]:60).

For Malinowski, it is patently self-evident that people in stateless societies like the Trobriand Islands are not rational in the EM sense but also are eminently intelligent, dealing with their world on their own terms: "Thus, in several aspects, the Kula presents to us a new type of phenomenon, lying on the borderland between the commercial and the ceremonial and expressing a complex and interesting attitude of mind" (Malinowski 1966[1922]:513).

There are a few weaknesses in Malinowski's encyclopedic work. He was perhaps too adamant that the people were not motivated by selfish gain. He tended to ignore (though he commendably reported) the large domestic trade that occurred alongside the Kula. He overestimated the ease with which people made a living and made some racist comments in his private diaries. However, Malinowski was brilliant in his empirical contributions and theoretical insights. He demolished the racist notions that the primitives were lazy and unintelligent. He demonstrated the sophisticated means by which people in Melanesia structured their lives to ensure their economic and social well-being. And of particular interest to us, he illustrated the connection between religion, myth, magic, and the economy. There are literally scores of observations as to how the economic and the ritual are intertwined. In a summary of his greatest work, he states, "Economic enterprise and magical ritual form one inseparable whole, the forces of the magical belief and the efforts of man moulding and influencing one another" (Malinowski 1966[1922]:515).

It was Malinowski who effectively created the concept of the "embeddedness" of the economy in non-Western peoples. People eschewed EM, including the psychological motivations behind such "primitive" economies, and instead produced and exchanged goods not through negotiation and price but for social reasons. This essential insight was taken up by economic historian Karl Polanyi, and became the theoretical basis of the substantivist school of economic anthropology.

Malinowski's work did not make the science of economics change its theoretical predilections in any significant way. Far from it. By the turn of the twentieth century, in fact, economists had effectively killed the labor theory of value preferred by anthropologists at that time in favor of supply and demand, favored by economists, as the principal means by which value was established in all economies. Supply-demand models actually work quite well in developed market economies, where money and constant trading allow prices to be set, all enforced by state-sanctioned contracts and judicial systems. However, in stateless societies without money, markets, and judiciaries, the inability of supply-demand mechanisms to explain much of human economic behavior became increasingly evident. A new generation of economic anthropologists, led by the great Raymond Firth, sought to merge Western economics with the empirical reality of stateless or peasant societies. These "neoformalists" tried to modify economic principles created to understand capitalist societies into concepts useful for nonmarket, peasant, or tribal societies.

The work of Malinowski and the neoformalists from the 1930s to the 1950s set the stage for the intellectual divide in economic anthropology known as the formalist versus substantivist debate.[12] This debate lingered on for a half-century or more. As it continued, the positions on each side consolidated around a number of core principles. In the broadest terms, substantivists drew off

Rousseauian and Marxist traditions (principally Friedrich Engels for precapitalist societies), seeing people as fundamentally "other directed" or inherently social in their social, economic, and political lives.[13] The development of the family, private property, and the state was responsible for pushing people away from the supposedly natural qualities of compassion and sociality, turning them into competitive, antisocial beings focused on their own self-interests. Formalists, in contrast, drew off the Hobbesian, Malthusian, Darwinian, and classical political economic traditions in which people were seen as fundamentally ego directed. Substantivists insisted that each economy had to be evaluated in its own terms, while formalists felt that the many principles in economics were applicable to both Western and non-Western groups. Throughout this period, theory in economic anthropology effectively stagnated in my view, due in large part to these deep epistemological differences, which ultimately derived from the classic Rousseauian–Hobbsean dichotomy that characterized more than 300 years of Western philosophy and social science.

The Formalist Reaction to Substantivist Economic Anthropology

As LeClair (1962) says, Malinowski in reality started the fashion in anthropology to bash any use of Western economic theory to study non-Western peoples. This position culminated in the substantivist school and produced a reaction on the part of scholars like Raymond Firth, Douglas Oliver, and others who also went out and spent long periods with other cultures. However, unlike Malinowski and Polanyi, who avoided economic concepts, they were quite familiar with and used Western economic theory. Firth in particular understood modern economics well. He was not ready to abandon all economic principles but wanted to adapt them to non-Western cultures if possible.

Raymond Firth describes the theoretical position that some general concepts in economics can be used by anthropology and can apply cross-culturally. In his classic 1972 review article in *Man*, he responded to the substantivist position: "It has now become clear that the original issue, as it took shape in dispute between 'substantivists' and 'formalists' . . . as to whether economic theory could be applied to primitive economies, was largely sterile. The issue was rather where, how far, and with what modifications and additions economic theory could be found appropriate to interpret 'primitive' systems" (Firth 1972:470).

Firth went on:

> The central theoretical position of Karl Polanyi . . . was that economic theory, invented to explain market phenomena, could be of little use in the interpretation of other systems. This came as no great surprise to many anthropologists. The firmness, if not dogmatism of the argument may have been resented by some whose own field experience had led them to more modified conclusions, but the basic problem was common ground.

Holding, as many of us would still, that scarcity, maximisation and choice are meaningful economic concepts over a great range of non-market phenomena, a series of basic theoretical issues about allocation of resources and provision of goods and services in communities lacking a pricing mechanism were – and remain – widely open. (Firth 1972:468, emphasis added)

A Summary of the Formalist Position

In short, the formalists, while not as ideologically coherent an academic school as the substantivists, developed some basic principles that are recognizable as an alternative approach. They agreed, for instance, that the ethnocentrism of the classical economists was indeed wrong. It followed that in fact different people had different needs and wants, which were created by culture. Of course, the needs that were molded by culture were fewer than the wants, but even needs were affected by different cultural norms. These different needs and wants were not irrational in an economic sense, in the same way that the French eat horses and the English find that odd simply means that culture and history create different wants in different peoples. The utility of goods and services is not universal, but the use of the concept of utility can be universally applied.

The formalists felt that the principles of economics were universally applicable and could shed great light on how other economies work. The definition of economics as simply the allocation of scarce resources to alternative ends was a universally applicable frame of reference to begin any study. They agreed with the classical economists that all societies faced scarcities, and they essentially adopted a Malthusian view on this point. They rightly pointed out that virtually all ethnographies available to them at the time, including Malinowski's and especially Firth's, indicated this fact, even if Malinowski himself downplayed it. All people had to economize to meet their needs – that is, to make choices between alternatives in a given social context. They also agreed that all societies did, in fact, maximize their wants, even if these wants were not all material and were culturally contingent. So, if people seek social prestige and are willing to trade something else of value for it, including their own labor, then in economic terms *prestige* has a utility.

A fundamental point here is that while the theory of maximization is central to classical economics, the neoformalists accepted that what people maximize is culturally determined. The classical economists were wrong to think, ethnocentrically, that all people around the world would want to maximize what an English or Scottish person wanted to maximize. However, once you accept that perfectly rational people can have different goals and maximize very different things, then the principle of maximization is universally applicable. The study of the economizing behavior of Western capitalism is Western economics. The study of the economizing behavior of Papua New Guineans is PNG economics.

The study of the economizing behavior of all peoples, not just in the West, in this sense is a universal framework that can be applied to all economies in space and time. From this perspective, Western economics was in fact merely a subdiscipline of a larger study of economics with universal applicability. We can take this one step further and argue that economic anthropology is the universal discipline, with Western economics simply a subset.

The formalist–substantivist debate profoundly affected the way in which we understood and modeled the emergence of complex stateless societies. The most significant effect from our perspective here centers on why people developed more complex economic institutions. Both traditions rely on a key assumption that I reject: that force or social coercion is responsible for the emergence of rank and social hierarchy in the shift from egalitarian hunter-gatherer societies to ranked ones. This assumption is a cornerstone of classic evolutionary theory in archaeology as well as being foundational for postmodern theory. Coercion, both environmental (population pressure, ecological shifts) and social (aggrandizing behavior by elites) are clearly present in the rise of state societies. However, as I will now show, this assumption is incorrect in the formation of complex stateless societies.

3 Conditional Co-operators: The Evolutionary Game Theory Revolution

> People have a propensity to cooperate, they respond to the cooperation of others by maintaining or increasing their level of cooperation, and they respond to defection on the part of others by retaliating against the offenders, even at a cost to themselves, and even when they cannot reasonably expect future personal gains from such retaliation. In particular, when other forms of punishment are not available, individuals respond to defection with defection. . . . *Homo reciprocans is thus neither the selfless altruist of utopian theory, nor the selfish hedonist of neoclassical economics. Rather, he is a conditional cooperator whose penchant for reciprocity can be elicited under the proper circumstances.*
>
> – Bowles et al. (1997:4–5, italics added)

The game theory revolution effectively reinvigorates the economic anthropology of the small group and allows us to move beyond the tired substantivist-formalist debate (and see Carballo, Roscoe, and Feinman 2014). It teaches us that in the aggregate, people are neither Hobbesian hyperrational and selfish actors nor other-directed Rousseauian utopians. People are overwhelmingly ego directed but are simultaneously *conditional co-operators*. Or as Herb Gintis et al. (2005:20) put it, there is a "predisposition to contribute to a cooperative endeavor, contingent upon the cooperation of others."

This sounds like a contradiction but it is not. People will co-operate and act altruistically when it is in their longer-term self-interest, as opposed to an immediate one, and this co-operation takes the form of other-directed behavior.[1] These new concepts make the substantivist-formalist debate effectively irrelevant in that it shows how the ego-directed individual will behave in socially co-operative ways under the appropriate circumstances. These appropriate circumstances must include a longer-term payoff for co-operative behavior for individuals participating in group strategies.

In this sense, the definition of a rational agent as simply a decision maker who acts in pursuit of his or her objectives, both short and long term, can be universally applied, even if those acts are altruistic and prosocial. This view recognizes that "objectives" are culturally and historically conditioned and that these objectives are at times not framed as simple, immediate material self-interest.[2] People universally seek a better way of life for themselves, but they do so in a myriad of ways that vary well beyond the artificial confines of Economic Man. In short, culture matters, and it shapes economic behavior and the forms and nature of co-operation in small societies.

Firth's observation about the Tikopia nicely defines this theoretical position:

> The Tikopia are not without the desire to increase their wealth or to see that reciprocity is obtained for a gift. In some circumstances, as in the acquisition of European goods, they display a keenness which amounts even to greed. But this desire for acquisition operates within the sphere of their social values. Put another way it may be said that the particular code regulating their use of material goods inhibits to a considerable extent the profit motive per se and limits the extent of production. (Firth 1965:60)

Co-operation is therefore the central focus of our analysis. Why do people co-operate? Game theory and its allied disciplines of over 30 years of research have given us a set of concepts to explain how humans co-operate in small groups. This research involves simulations, experimental games, and theoretical deductions. In the aggregate, the work has provided us a new way to view human interactions in small groups and provides a very powerful means of modeling the emergence and functioning of co-operation in stateless societies. As Sam Bowles and Herb Gintis (2011:110) rhetorically ask, "For at least 95% of the time that biologically modern humans existed, our ancestors somehow fashioned a system of governance that without the assistance of governments avoided the chaos of the Hobbesian state of nature." Game theory and its related fields of evolutionary and anthropological game theory provide the theoretical framework for understanding this process in stateless societies.

What Is Game Theory?

In simplest terms, game theory seeks to study conflict and co-operation between rational decision makers. Classical or economic game theory studies human interactions as two or more people compete for resources or some other social advantage. Human social interaction in this sense is a "game" played between two or more people. Research can take the form of a controlled experiment in which each player has control over his or her actions. While the term "game" superficially sounds trivial, it succinctly refers to human interactions and strategies, with individuals acting in their own self-interest. The analysis of these interactions provides great insight into human social behavior. It provides models for understanding how people actually define and control their social worlds.

A key element of game theory is to define the strategies people use to negotiate their behaviors with others. Players adopt strategies to achieve a positive payoff. Payoff is the result that accrues to the individuals involved in the social interactions or "game." Strategies are courses of action, and reactions to the parameters of the game and to other players' actions. Strategies ideally are designed to maximize a player's payoff. Game theorists make observational studies, create computer simulations, and construct actual experimental games to test people's behavior in social interactions.

Most historians trace the beginnings of modern game theory a classic paper in 1928 by John von Neumann. However, it was his collaboration with Oskar Morgenstern that led to the book *Theory of Games and Economic Behaviour*, published by Princeton University Press in 1944 (Morgenstern 1976). This book analyzed co-operative games with multiple actors and was considered a major advance in our understanding of strategic decision making by individuals. The book has been described as one of the great intellectual achievements of the twentieth century, and it spawned a revolution in economic theory.

The great mathematician John Nash was responsible for one of the key concepts in game theory, aptly named the Nash equilibrium, which he published in the early 1950s. A Nash equilibrium exists when all players in a competitive game know all others' strategies and do not alter their own insofar as other players do not alter their strategies. This situation is considered optimal, in that there is no additional strategy that could enhance any player's payoff. While seemingly abstract, it is a concept with enormous real-world applications and is one of many tools useful for understanding the behavior of people in stateless societies.

Classical game theory derives from classic economics and assumes that people are generally rational and knowledgeable about their environment. The history and nature of game theory has been described in substantial detail for more than 50 years. For our purposes, it is sufficient to note that game theory provided social scientists a means to mathematically analyze the behavior of multiple rational agents as they strategized to maximize something of value. Thus game theory explicitly derives from classical economics, relying on the concept of *Homo economicus* as its central assumption.

"Rational" is a loaded term. In classical economic theory, it betrays the assumption that all people in all places and times will always evaluate benefits and costs of any action in similar ways and adopt the optimal strategy. People are considered to be ego directed, acting in their self-interest. Actors were assumed to have "unlimited information processing capacity" (Gintis 2000b:xxiii). Classical game theory is useful in a number of circumstances, such as the buying and selling of commodities. But as we saw in Chapter 2, while *Homo economicus* is a useful tool in a limited set of social interactions, it is a poor assumption for much of the behaviors that we observe in small groups. Even here, however, a whole new field of behavioral economics has challenged the utility of classical game theory assumptions of hyperrationality.

Real social life is substantially more complex than allowed for in the assumptions that underlie classical game theory. As a result, new means of modeling human interaction have been refined and modified to better explain a wider variety of human behavior in space and time. As described above, the most significant result of this new work is that the canonical assumption of rationality is not supported by experimental data. A new body of theory demonstrating that some people are irrationally prosocial, particularly in small groups, augmented game theory experiments based upon rational actor theory. Irrational prosocial behavior refers to altruistic acts that do not immediately benefit the actors but that help other individuals or groups. As we will see, the concept of irrational prosocial behavior is central to understanding co-operation, particularly in small groups. The issue then is how to keep people from cheating on the one hand and co-operating on the other. Co-operating in a group that is contributing to a public good is understood as a problem of altruism in theoretical biology and game theory.

The Theoretical Problem of Altruistic Co-operation by Individuals

What is fascinating is that at least some people in effectively all cultures so far studied will engage in costly prosocial or altruistic behavior at no gain and even at some cost to themselves. This may seem obvious to sentient ethnographic observers, but it is a profoundly important fact of human social life for social theory. The existence of altruistic behavior in human society, however, is a highly contentious issue in the evolutionary biological literature. It is, and will continue to be, a major divide in arguments between and among theoretical biologists and theoretical social scientists. As Michael Zaggl notes, "Cooperation is one of the basic elements of social life. It is essential for emergent social phenomena, such as the formation of families, groups, and societies. *However, evolutionary forces counter cooperation*" (Zaggl 2014:197, emphasis added). From a classic evolutionary biological perspective, therefore, altruistic behavior, particularly among nonrelated people (nonkin), simply does not provide a payoff, and it should be selected out.

At the risk of oversimplification, the debate about altruistic behavior can be broken down into two partially overlapping dichotomies: (1) individual (or gene) selection versus group selection; and (2) whether co-operative behavior is culturally or genetically based. On top of these two dichotomies are some tangential but important issues that follow from the arguments of each position.

Individual or gene-level selection is the classic evolutionary formulation in which those with an advantageous characteristic will reproduce more than those without it. Over time, the gene pool will shift toward those advantageous traits, and evolution will proceed as a selection process in which gene frequencies change. Certain kinds of seemingly altruistic behavior occur, and these are accounted for by kin selection or reciprocal altruism.[3] In these cases, deferred benefits outweigh costs to the individuals involved in these behaviors and so-called altruistic behavior can survive in a population.

Group or multilevel selection, in contrast, argues that there are selective pressures that work on the group. Altruistic or prosocial behaviors that benefit the group will increase the fitness of that group. Group selection theory effectively says that selection works on variation in the group as well as the individual. As a general rule, most evolutionary biologists dismiss group selection as theoretically not viable because they cannot produce models in which altruistic individuals will survive over time unless the altruism is directed toward relatives.[4]

The idea that we have a genetic predisposition to co-operate has been developed by evolutionary theorists such as Herb Gintis. He writes:

> The past few decades have seen the massive accumulation of evidence in favor of the view that human beings are inherently moral creatures, and that morality is not a simple cultural veneer. Humans are born with a moral sense as well with a predisposition to accept and internalize moral norms [of] their society, and often to act on these moral precepts at personal cost. (Gintis 2012:2)

Others argue that, in contrast to innate predispositions, over time and through trial and error, people develop "social norms and informal institutions that are capable of domesticating our innate psychology for life in ever-expanding populations" (Henrich et al. 2010c:1480). This is effectively the anthropological or social scientific response to this question. We focus on the ability of the evolved human brain to create strategies among members of our species that other animals cannot create. These strategies of co-operative behavior override or at least "trick" the evolutionary strategies (known as heuristics) that evolved in humans in the Paleolithic. Once these strategies evolve, they create their own cultural evolutionary dynamic requiring a set of principles that does not contradict but certainly goes beyond evolutionary biology. Why do men volunteer to fight against others? The risk is small in total numbers, but the consequences of failure are enormous – removal from the gene pool in a most irreversible manner. Why do people invest in children they have adopted? There is no genetic payoff to this, but it is a cornerstone of human sociality. Such prosocial behaviors between nonrelated people are magnified in small groups and require explanation.

As one reads the evolutionary literature, it becomes evident that one key difference between the various disciplines is the definition of "payoff" for success. Evolutionary biologists identify reproductive success as the key indicator – the number of offspring that an individual has and therefore the amount of genes that are passed on. In game theory with iterated components – that is, a process whereby agents interact and compete over time – economists focus on material success. Those who are most successful acquire more resources. Finally, anthropologists who focus on the small group, particularly in nonmarket societies, focus on prestige or status.

It is an unfortunate fact that reproductive success and resources are easy to measure, whereas prestige or status is less quantifiable. What is prestige? Is the poor shaman who is central to any society more or less prestigious than a rich chief? In potlatching, for instance, we know that one obtains status by *giving* resources, not by acquiring, although that sets up debt obligations for a future date. But, is an ill-defined debt obligation more or less valuable than a beautiful copper object? What about the modern demographic transition in which the richer people in industrialized societies become, the fewer children they create? This is hardly conducive to reproductive success.

We see throughout the ethnographic record of stateless societies that chiefs have no real coercive authority on their own. Rather, they derive this authority from the group, and this authority must generally be used for the good of the collectivity as perceived by that collectivity. Aggrandizing behaviors by chiefs occur, but there are substantial sanctions against such behaviors, as seen in the ethnographic record. Chiefs receive social prestige, have larger households, and have possible access to a broader range of nonsubsistence goods, but in many cases they (or their households) also till their own fields and build their own houses. The materialist assumption that wealth is the prime motivator is indeed part of the story, but it is not the whole story. Prestige and status are intangibles that likewise motivate people as much as wealth. This is particularly true in societies without developed markets, currencies, means to store wealth, and class structure designed to protect privilege.

The key to this theoretical dilemma, I believe, is the existence in stateless societies of what I have referred to above as a managerial elite or leader. A managerial leader is created by a group and is accorded prestige. But managerial leaders do not have coercive power beyond persuasion; nor do their households control significantly more wealth than other households. They are key to solving co-operation problems. The acceptance of an "elite" is considered a viable solution for all co-operating members of a small group, including those who do not have offices of leadership. It is not a cost borne by the rest of the population as in state societies, in which elite classes are able to coerce labor and resources from others. *Rather, it is recognition by the group that the existence of a "coordinator," accorded certain privileges for the effort to maintain fairness and norms of reciprocity, can be beneficial for the vast majority of the group's membership.* This is elegantly expressed in an article about rank in nonstate societies:

> An alternative body of thought in political philosophy and the social sciences, however, has suggested that greater differentiation by social rank may in some cases be preferred over more egalitarian arrangements by both higher- and lower-ranking individuals as a solution to the key challenges of life in social groups, such as conflict over resources,

> coordination failures, and free-riding in cooperative relationships.... The existence of facultative, cooperation-facilitating leadership roles in otherwise egalitarian small-scale human societies... supports this conjecture. (Hooper, Kaplan, and Boone 2010:634, emphasis added)

In this perspective, payoffs for co-operation and prosocial behaviors involve both material and nonmaterial utilities. Leaders emerge not as simple aggrandizers but as facilitators of co-operative behavior among a group, who in turn earn nonmaterial prestige as their payoff for this extra "altruistic" effort.

In this sense, co-operation in humans organized by managers is central to resolving the collective action problem. As Paul Roscoe aptly notes:

> The promise of collective action is that people can realize goals that they could not – or could not as efficiently – realize alone; the problem lies in getting people to contribute to the public good when to do so would exact a personal cost. Individuals then have an incentive to capitalize on the benefits of collective action without contributing to it, a state of affairs that, left unchecked, would lead to the collapse of cooperation and the prospects of collective benefit. (Roscoe 2009:70)

The collective action problem is key to understanding the emergence of leadership in small, stateless societies that lack coercive mechanisms. Christopher von Rueden et al. (2014) discuss the challenges faced by leaders in egalitarian societies. They observe that the resolution of the collective action problem requires such leadership, particularly in heterogeneous societies. These challenges include what they call coordination, or the need to resolve different needs and wants of group members as well as dealing with the free-rider issue briefly discussed above and dealt with at length in the following chapters.

Has Our Evolved Psychology Changed in the Last 10,000 Years?

Most sociocultural anthropologists and archaeologists assume that the basic evolved psychology of humans has not changed much in the last 10,000 years, and certainly not in the last 5,000. Colin Renfrew aptly sums up the archaeological perspective:

> For the archaeologist, the challenge is to understand how human societies, through their interactions with the material world, came to bring about the transformations in life and culture over the millennia drawing upon a hardware (the human genome) that may not have changed significantly in the space of 10,000 years. (Renfrew 2008:2041)

There is abundant indirect evidence for this. First, 10,000 years is not very long time in evolutionary history, and for a species that interbreeds widely, it would be difficult to imagine a substantial change in fundamental evolved psychology in this short time. The most conservative (youngest) estimate for the evolution of fully developed language is 40,000 years ago, with most physical anthropologists arguing for a much earlier appearance (Aiello and Dunbar 1993).

Likewise, there is no difference in the capacity for complex social interactions among existing populations on the globe today. The earliest recorded texts in both hemispheres are immediately recognizable in form and substance. The elaborate art produced in the Upper Paleolithic and certainly in the Neolithic suggests that the people producing it were quite similar to us in intellectual capacity. The behaviors that we detect in the archaeological record – war, peace, religion, and politics, among others – look similar to modern behaviors. The population-specific differences that developed recently – sickle cell anemia, lactose deficiencies, and so forth – are directly tied to reproduction and disease resistance, two features subject to the severest kinds of selection and rapid evolutionary change.

On the other hand, some scholars of evolutionary game theory suggest that indeed some subtle genetic differences affect the degree to which we co-operate. This theory derives from evolutionary biology, and as one text nicely puts it, "Evolutionary biology is based on the idea that an organism's genes largely determine its observable characteristics, and hence its *fitness* in a given environment. Organisms that are more fit will tend to produce more offspring, causing genes that provide greater fitness to increase their representation in the population" (Easley and Kleinberg 2010:209). In this reasoning, social behavior is a genetically based "observable characteristic" like any other trait and is therefore subject to natural selection.

The debate remains somewhat open, but I fall into the camp that our basic evolved psychology has not changed much vis-à-vis our capacity to create complex social arrangements. New data could certainly alter this interpretation, but at present I favor interpretations that see our species as effectively genetically similar in this capacity over the last 300 generations or 10,000 years. While I advocate a selectionist mechanism, it is a selection on the culturally constructed strategies of interaction, and not on biologically driven ones. Mine is at base an economic logic, not a biological one, even though the selective mechanism uses a biological analogy.[5] As we will see in the following chapters, "anthropological" game theory provides us a third alternative, but that alternative is based more upon human recognition of strategic utility of seemingly altruistic behavior and less on a biological determinism.

I fall into the theoretical camp that sees humans as having some unique characteristics from other social animals, including higher primates, that require concepts beyond evolutionary biology to fully understand human co-operation. These unique traits derive from our highly evolved mental capacity for language, cognition, and ability to maintain sustained interaction with others based upon our ability to differentiate co-operators from non-co-operators. Michael Tomasello et al. (2005:675) articulate this position well: "that the crucial difference between human cognition and that of other species is the ability to participate with others in collaborative activities with shared goals and intentions: shared intentionality." They point out that such an evolved capacity is "species-unique forms of cultural cognition and evolution, enabling everything from the creation and use of linguistic symbols to the construction of social norms and individual beliefs to the establishment of social institutions," a profoundly important insight that is central to the argument presented in this book.

Christopher von Rueden et al. (2014) likewise note that "A defining feature of humans relative to other primates is our tendency to act upon opportunities for mutual benefit." They point out that "intentionality" is key here, a unique capacity of human groups. Eric Smith (2003:401) notes that "language clearly increases the likelihood of solving coordination games and appears to lower the cost of monitoring and enforcement in other payoff environments." In other words, we humans use our evolved psychological capacity to create "social tools" that serve, among other things, as the basis for sustained conditional co-operation.

Bowles and Gintis also argue that such altruistic co-operation is unique to humans, and as such, requires a set of theory that goes beyond simple assumptions of *Homo economicus*:

> While the high frequency of altruistic cooperation in humans relative to other species could be an evolutionary accident, a more plausible explanation is that altruistic cooperation among humans is the result of capacities that are unique to our species and that strongly promote our relative reproductive fitness. Thus we seek an explanation of cooperation that works for humans, but which, because it involves capacities that are unique to humans, does not work for other species, or works substantially less well. (Bowles and Gintis 2008:953)

Other research in cultural transmission theory, as described below, suggests strongly the great diversity in human social organization that has developed of the last 10,000 years is due more to cultural innovations and selection than it is due to our evolved psychology. This is an important point adopted in the anthropological game theory framework developed in this book and described in detail below.

The Empirical Basis of Conditional Co-operation

A quick search of the academic literature provides a huge bounty of work on the nature of co-operation (or non-co-operation) in small groups. The experiments and observations have been made largely in Western societies (WEIRD, which stands for Western, educated, industrial, rich democracies; see below), and this is a problem. However, additional research has been conducted in non-Western contexts as well. The combined data indicate certain properties of small-group behavior are evident in all groups. There is considerable variation between groups, with different cultures using different strategies. We will now briefly explore these new developments. The most powerful of these experimental games are the Prisoner's Dilemma (PD) and the Ultimatum Game (UG).

Prisoner's Dilemma

The PD is one of the classic games in the history of the discipline. It was developed in the 1950s and is still used.[6] In this game, people are placed in a situation where co-operation will be the optimal outcome for all players. For example, two prisoners placed in separate cells who co-operate by not talking will both go free for lack of evidence. However, each individual could do better or "less worse" than the other by being the first to not co-operate, or defect: If each prisoner keeps quiet, they both can go free, but if one talks, he or she partially gains at the expense of the other. The experimental data clearly indicate that defection is the norm in the regular PD, as people cannot trust their fellow players not to defect and therefore settle for a suboptimal outcome. In this case, it would be a plea agreement for a lesser sentence in return for defecting on your partner. This game is a type of "tragedy of the commons" situation, in which individuals acting in their own self-interest benefit themselves in the short term but hurt the group and eventually themselves in the long term. In other words, if each individual chooses a maximization strategy, everyone eventually loses more than they would have gained by co-operation. The PD demonstrates a classic canonical outcome: Individuals acting in their own self-interest will trump co-operation, even if the latter is in everyone's interest. It is a "rational" outcome with socially (from the perspective of the group) "irrational" results.

The one-time PD offers great insight into a particular kind of ideal social situation. However, it poorly reflects the dynamics of social life. When time, learning, and history are added to the mix – that is, when an evolutionary component is included – the results of the PD dramatically change.

Iterated Prisoner's Dilemma

The simple, one-shot PD game shows that the best strategy is to defect, even though mutual co-operation would provide a better payoff. In a more complex experiment, the Iterated Prisoner's Dilemma (IPD), the game is played over an extended period by the same players (Axelrod 1980). These players know the other players, and they can remember their previous behaviors as to defection or co-operation. The IPD is a powerful game. As one of its most creative practitioners has noted, "What the Prisoner's Dilemma captures so well is the tension between the advantages of selfishness in the short run versus the *need to elicit cooperation from the other player to be successful in the longer run*" (Axelrod 1997:6, emphasis added).

As just mentioned, the outcome of one-shot PD games is rationality – everyone acts in their own self-interest. The iterated version of the game produces different results. In running simulations of the IPD, Axelrod (1980) found that the highest payoff for players was a simple strategy called Tit for Tat. The strategy was to simply co-operate on the first move and to do whatever the other player did on each subsequent move. Co-operation emerges between competing individuals because co-operation is in each player's self-interest. This Tit for Tat is in reality a form of simple reciprocity. Each agent effectively reciprocates the behavior of the other agents. A basic "social history" of the players develops very quickly, and this is the basis for future reciprocal behavior over time. As Daniel Nettle and Robin Dunbar (1997:94) point out in referring to Axelrod's work, "Cooperation can emerge in a population of unrelated individuals where there is a possibility of reciprocity. However, it is viable in fitness terms for individual A to help individual B only if there is a high likelihood of A's meeting B again in the future, giving B the opportunity to reciprocate. This continuity is essential in securing cooperative social relations." As we will see in later chapters, this observation is extremely important and underlies arguments for the role of population densities in the emergence of initial complexity in stateless societies.

The Ultimatum Game

Along with the IPD, the UG is one of the most useful means of understanding the nature of human interaction and studying deviations from the expectations of classical economics. The UG has been utilized since the early 1980s and is one of the most popular experimental tools we have to analyze human behavior in small groups (Güth, Schmittberger, and Schwarze 1982). The UG consists of two kinds of players: proposers and responders. Each proposer is given a quantity of money. He or she can offer part or all of that amount to a responder. If the responder accepts the offer, both players can keep the divided amount. If the responder declines, neither player receives any money.

There is a vast literature on the outcomes of UG experiments around the world. Virtually all of these experiments indicate that expectations of the canonical model (*Homo economicus*) are not met. As Nowak, Page, and Sigmund (2000:1773) describe it, "The rational solution, suggested by game theory, is for the proposer to offer the smallest possible share and for the responder to accept it. If humans play the game, however, the most frequent outcome is a fair share." Rational actor theory is quite clear in what the obvious choice is for responders in the UG: they should accept any amount of money, since there is virtually no cost to them and there is at least a marginal benefit. However, in practice, responders are willing to give up free money if the offer is not deemed fair.

The UG is one of the most intensively used tools in the experimental kit of co-operation theorists. In all cultures studied, people do not conform to the rational actor or EM expectation. The desire for "fair" outcomes, whether to maintain reputations for future interactions, to punish potential free-riders, or simply as a result of some kind of innate sense of justice, is a cornerstone of human co-operation. Where face-to-face interaction happens on a daily basis, the demand for fairness is intense, as demonstrated in the vast UG literature.

The Elements of Co-operation: Fairness, Reputation, Reward, and Punishment

David Carballo (2013:11) notes that contemporary work in evolutionary game theory and collective action theory emphasize four mechanisms that most encourage co-operation: reciprocity (fairness), reputation (social history), reward, and retribution (punishment).

Fairness

Norms that enhance fairness among strangers are likely causally interconnected with the diffusion of several kinds of [co-operative] institutions. – Henrich et al. (2010c:1480)

The perception of fairness is a cornerstone of co-operation in a group in which coercion, as defined in the previous chapter, is largely absent. To paraphrase an article that Kevin Haley and I published in 2005 and to reiterate some points made in the last chapter, people have strong propensities for behaviors that can threaten co-operation, particularly in a small group. These behaviors include free-riding, envy, and spiteful punishment. In such contexts, it is critical to create conditions in which the norms of fairness are rigidly and visibly followed. Adherence to these norms, by both leaders and followers, is an essential component to the success of any strategy in such a social context. In other words, consistent adherence to norms of fairness and free-rider punishment stand as fundamental and necessary conditions for the emergence and maintenance of co-operative organizations in the absence of coercion. Aspiring leaders who are perceived to be unfair will be left without followers. Leaders must strictly adhere to norms of reciprocity to keep their status positions. In other words, they must persuade people to work co-operatively, providing material incentives and maintaining appropriate ideologies for reciprocity.

Stefan Kohler (2013) produced some fascinating data from a UG experiment in a number of villages in Zimbabwe. He conducted the experiment in 20 different villages, 14 of which were "traditional" in that they had been in existence for a considerable period of time. The six other villages were newly settled ones, created by government land reform. The results indicate that the respondents from the resettled areas were significantly more fair and rational than their counterparts in the established villages. These results indicate a couple of important things. First, the increased fairness in the settlers suggests that norms of co-operation are strong in stressful situations, like resettlement. That is, these newly founded villages created these norms as a strategy to deal with potential competition from existing populations. Second, the data indicate quite clearly that culturally similar peoples can exhibit varying degrees of prosocial behavior. These different strategies are quite conscious and derive from real-world actors who comprehend their utility. Fairness and hence co-operation were more valued in the newly founded villages because of their more precarious position. As David Rand et al. (2013:2581) so aptly put it, "In an uncertain world, fairness finishes first."

In this light, it is quite fascinating that researchers using MRI technology have in fact demonstrated increased neural responses during UG experiments. Sanfey et al. (2003:1755) show how the rejection of an unfair offer "significantly heightened activity in anterior insula for rejected unfair offers suggests an important role for emotions in decision-making." While such a result seems patently self-evident to any anthropologist or sociologist who has studied small-group interactions, it does confirm experiential observations that negotiations between individuals in small groups can be highly emotional. Since these emotions derive from biological processes controlled by a complex genetic substrate, they should be subject to selective pressures.

Martin Nowak, Karen Page, and Karl Sigmund (2000) ran a series of simulations using an iterated UG. They discovered an "irrational" human emphasis on fairness in games that most approximate social reality (and see Rabin 1993). "Irrational" in this case refers to what would be expected in an economic exchange in which each actor seeks to maximize resources. The irrational behavior included the punishment of cheaters; people were ready to hurt cheaters even at a cost to themselves. Nowak, Page, and Sigmund (2000) found that without information on players' reputations, evolution always leads to the predominance of economically rational actors – that is, offering little and rejecting nothing or offering a lot and accepting little. However, with the addition of players' social histories, fair strategies dominated. The ability to remember what a fellow actor will do and to pass this information on to others is critical, and it represents the adaptive capacity of human beings. As the authors summarize, "When reputation is included in the Ultimatum Game, adaptation favors *fairness over reason*. In this most elementary game, information on the co-player fosters the emergence of strategies that are *nonrational, but promote economic exchange*" (Nowak, Page, and Sigmund 2000:1774, emphasis added). They convincingly argue that fairness, like co-operation, will evolve only if players have information on how others have behaved in the past. Reputation is key. Only humans have this capacity to remember the reputation of many others over long periods. Thus the UG is another example of the unique capacity of ego-directed humans to create co-operative social institutions in the absence of coercion.

The literature on the role of fairness in human interactions is vast (e.g., Bowles and Gintis 2011; Chaudhuri 2011; Henrich et al. 2010c; Rabin 1993; Rand et al. 2013, to name but just a very few). The key take-away from this work for this discussion is that fairness plays a central role in eliciting co-operation among self-regarding or ego-directed agents. Any model of the origins of complex society out of small, nomadic bands must incorporate this fundamental empirical insight from evolutionary game theory research.

Reputation (Social History)

> It has been shown that fairness can be favored by natural selection if agents can recognize their partners' strategies . . . or have reputations that carry from game to game. Here it pays to reject a stingy offer today so that others will make you higher offers in the future. Without a sufficiently high expectation of future interactions and a sufficiently strong reputation system, however, fairness collapses. – Rand et al. (2013:2582)

The work by Nowak, Page, and Sigmund (2000) discussed in the last section emphasizes that fairness and reputation go together as factors that promote co-operation between unrelated individuals. A key element of the IPD and other iterated games is that players must have some assurance that they will meet the people they are interacting with on a consistent basis. In the experimental games, this means there is a reasonable probability that the players will meet again. In actual social situations, this means that people who interact understand that they will also meet again, not just in the immediate future but days, weeks, or even years later. The ability of humans to remember the history of interaction with dozens of others is a unique capacity of our species. It alters the cost-benefit calculation from the simple PD, where defection is the norm, to a situation more reflective of human social interaction. It also demonstrates how co-operation can emerge from ego-directed individuals, again under the right circumstances. In this instance, the circumstances include the ability to remember social histories and a social environment in which there is reasonable expectation of meeting people in the future.

David Levine and Wolfgang Pesendorfer (2002) likewise discovered that when players are fully aware of how another player will behave based upon earlier interaction, much greater co-operation results in games like the PD. Again, the key here is social histories, or how people have reacted in the past, serving as a guide to probable future behavior. James Andreoni and John Miller report a similar role for information on fellow actors. Using a finite repeated PD game, they found that cheating or defection was the dominant strategy for both players with no information on each other. However, if players had some information, "it could be in each player's best interest to pretend, at least for some time, to be an altruistic player in order to build a reputation for cooperation" (Andreoni and Miller 1993:571).

Why is a small-group context different from large groups in regards to reputation? A major reason is that to develop a social history of individuals, actors need to be able to remember the faces and behaviors of people they have interacted with over time. There is obviously a limit as to how many people can interact before it becomes impossible for one person to retain accurate social histories in his or her mind. After this limit is met, a new set of institutional mechanisms is required to keep the social histories accurate.[7]

Like fairness, reputation and social history are necessary components of any model of the origins of complex society.

Reward

Rewarding co-operation is obviously an essential way to keep a co-operative group co-operating. This sounds self-evident on the surface, but it is important to emphasize from a collective action theory perspective, in contrast to a canonical model of resource stress. In the latter, such stress forces people to co-operate, giving up autonomy and accepting a costly level of hierarchy – a pure Hobbesian/Malthusian view. Rather, recognizing that reward is a key to co-operation focuses our attention on factors that promote more social complexity by lowering the costs and increasing the benefits of more intensive social interaction.

The role of reward versus punishment, as described in the next section, is debated in studies of co-operation and collective action (e.g., see Carballo, Roscoe, and Feinman 2014:107). Rewards come in a variety of ways, and they are very culturally contingent. As Kenneth Clark and Martin Sefton (2001:52) note, in experimental studies, "subjects are willing to forego material gain in order to reward those who have treated them favourably." In other words, people engage in costly rewards to encourage group co-operation. In the next chapter, I will demonstrate that highly co-operative groups that specialize labor tasks among their members achieve economies of scale that serve to reward all members of that group. But rewards can be both material and nonmaterial, with prestige and the enhancement of social status as important as economic rewards in many cultures.

My read of the ethnographic literature is that in stateless societies the role of reward is more important than sanction, though punishments are certainly part of the normative set of social tools that people use to keep the group co-operating. One could profitably argue that the shift from stateless societies to those with state mechanisms parallels a shift from more reward-based strategies to punishment-based ones. Of course, by definition, state societies are those that incorporate coercion between and among different social groups.

Punishment

A number of mathematical models suggest that retribution against defectors and free-riders – along with retribution of individuals who do not punish such transgressions (i.e., second-order free-rider problems or third-party sanctions) – can effectively stabilize norms of cooperation within a population. – Carballo (2013:12)

A substantial body of evidence supports the proposition that the punishment of free-riders is a key to maintaining a co-operative social group, particularly among nonrelated people (e.g., Boyd and Richerson 1992; Egas and Riedl 2008; Fehr and Gächter 2002; Henrich et al. 2006). Experimental and theoretical work demonstrates clear variability in the way in which norms of fairness and punishment work, but that "all populations demonstrate some willingness to administer costly punishment as unequal behaviour increases, [and] . . . the magnitude of this punishment varies substantially across populations" (Henrich et al. 2006:1767). Ernst Fehr and Simon Gächter (2002:137) note that "altruistic punishment of defectors is a key motive for the explanation of cooperation." The role of altruistic punishment in the maintenance of co-operation is one of the most theorized concepts in the field.

Other work demonstrates the importance of punishment as an adaptive strategy.[8] Bowles et al. (1997:2–3) note that "important forms of cooperative behavior are commonly observed" both in controlled observations and in the laboratory. The traditional rational actor model is insufficient. They note that people are traditionally "irrationally pro-social," a behavior that does not conform to economic maximization models as found in classical economics. They also note (Bowles et al. 1997:5) that defection (also referred to as free-riding) "is significantly alleviated if there is an opportunity for costly retaliation." Retaliation benefits outweigh other social costs in long-term runs. They note that the experimental evidence supports the proposition that "if costly retaliation opportunities are combined with communication opportunities almost no defection occurs and, therefore, no resources are wasted for retaliation" (Bowles et al. 1997:5–6). They argue, "A predisposition to cooperate and to undertake costly punishment are probably related phenomena," and they refer to them as "reciprocal fairness."

Gintis (2000b:2, 2000c) refers to this phenomenon as strong reciprocity, defining it as a predisposition "to cooperate with others and punish non-cooperators, even when this behaviour cannot be justified in terms of self-interest, extended kinship, or reciprocal altruism." He contrasts "strong" reciprocity with its "weak" counterpart. The latter is known in the literature as reciprocal altruism. He argues that reciprocal altruism is too weak to explain the evolution of co-operative behavior, because when a group is threatened, the optimal behavior for self-interested agents is defection: "Precisely when a group is most in need of prosocial behaviour, cooperation based on reciprocal altruism will collapse" (Gintis 2000b:6). Again, in this model, the role of punishment is central to the evolution of co-operation in a group of self-interested individuals.

An article by Sarah Mathew and Robert Boyd (2011) reiterates that punishment is very important is small-scale societies. They write that their work on the Turkana of East Africa "challenge[s] current views that punishment is unimportant in small-scale societies" (2011:11375). Their results indicate that co-operation at the larger scale of ethnolinguistic units, enforced by third-party sanctions, could have a deep evolutionary history in the human species:

> Turkana sustain costly cooperation in warfare at a remarkably large scale without political centralization, formal coercive institutions, or even a political system based on descent. . . . The punishment of free riders plays a role in sustaining this large-scale cooperation. Punishment is based on informal third party sanctions. Although cultural norms specify how free riders should be treated, the enforcement of these norms requires the participation of the community at large, and no particular individuals are invested with the authority to implement sanctions. (Mathew and Boyd 2011:11379)

This behavior is opposite what one would expect from classical economic assumptions about rational and ego-directed human behavior, but it is an empirically verified and theoretically sound observation in social science that simply cannot be ignored (see Boyd et al. 2003). Following Fehr and Fischbacher (2003:790), "When punishment of non-cooperators and non-punishers is possible, punishment evolves and cooperation in much larger groups can be maintained." As with the other mechanisms described in this section, punishment in the form of taboos, ostracism, exile and even execution are necessary components of any model of the origins of complex society among self-regarding agents in a noncoercive social context.

The Problem with WEIRD People

One of the criticisms of experimental game theory is that subjects tend to be from the Western world and therefore not representative of the many global cultures, past and present, that have evolved in the last 10,000 years. Many of the WEIRD people are undergraduates in research universities. While the cultures differ – Japan, Israel, the United States, Brazil, and so forth – there is always a suspicion that undergraduates in universities are not representative of the bulk of humanity, an observation that is difficult to refute by anyone who has taught some university classes. This is a problem, and this issue of so-called WEIRD people biasing our understanding of humans' evolved psychology has become a prominent source of debate. Henrich and colleagues have published a series of papers arguing that WEIRD populations are not good representatives to characterize human behavioral choices under many circumstances (Henrich, Heine, and Norenzayan 2010a, 2010b).

In 2001 Henrich et al. (2001, 2004) reported on UG experiments among non-Western peoples. They found that even in non-Western societies, the expectations of the canonical models of rational actor theory or *Homo economicus* were not met. However, they did in fact find much greater variation between groups in the range of rejections and offers in the various instances of the experiments. They argue that degree of market integration is one factor. Somewhat ironically, they found that offers were fairer with the higher the amount of market participation of the studied group. Effectively, they suggest that Western-style marketing behaviors require people to develop the norms of sharing necessary for markets to function: "The rationale for market integration as an explanatory variable is that the more frequently people experience market transactions, the more they will also experience abstract sharing principles concerning behaviors toward strangers of which the UG is an example" (Henrich et al. 2001:76).

Henrich et al.'s research also suggests that "payoffs to cooperation" vary between groups more than previously suggested. People who are organized by family groups with high degrees of self-sufficiency, for instance, will experience lower payoffs from co-operation than groups with higher levels of economic interdependence. This is consistent with the discussion of economies of scale found in the next chapter.

The implications of Henrich et al.'s research have sparked vigorous debate. In their cross-cultural work, two conclusions of this study stand out for our purposes. First, *all* groups studied do not follow the canonical model of *Homo economicus*. Second, in spite of this consistent deviation from canonical models, there is considerably more variation in these cultures than that found in WEIRD societies. As Dan Jones (2010:1627) says, "[non-Western peoples] tend to place less importance on choice, be more holistic in their reasoning, and be less concerned with seeing themselves as above average."

This work does not reject the proposition that there are fundamental evolved similarities in our species: "We expect humans from all societies to share, and probably share substantially, basic aspects of cognition, motivation, and behavior" (Henrich, Heine, and Norenzayan 2010b:62). They emphasize the fact that people in large, industrialized societies are different than those in small-scale ones and that the use of WEIRD subjects to define our evolved psychology skews our understanding of that psychology. It is hard to disagree with such a conclusion. The question is one of degree. How far do WEIRD people deviate from the norm?

Of particular relevance here is that Henrich, Heine, and Norenzayan (2010b:65) suggest that non-Western peoples actually stress fairness *less* than Western peoples. They argue that recent empirical work challenges the conclusions of Nowak, Page, and Sigmund, and others. This is a debate that is just beginning. Alexandra Maryanski (2010:104), in a commentary on the Henrich,

Heine, and Norenzayan articles, echoes critiques claiming just the opposite, that "WEIRD populations may be more compatible with humans' evolved nature than are most traditional societies."[9]

In a manuscript describing the results of experiments contrasting US undergrads (the epitome of WEIRD) with Fijians, Henrich and Henrich (2004:6) conclude that "Fijian villagers and Emory [University] subjects make very similar offers and that the effects of punishment seem to have similar effects in both groups." Here we begin to see that it is a matter of degree. The big difference that they detect is that Fijians are much less likely to punish than are U.S. undergrads. They experimenters found that Fijians share without the need for punishment, most likely reflecting particular norms that developed in their particular cultural setting (Henrich and Henrich 2004).

Even though Fijians and other cultures do not respond in manners precisely similar to WEIRD cultures, we know that *all* cultures do not follow the canonical model for small groups. In fact, an abstract summarizing experiments from 15 societies notes, "The findings of these experiments demonstrated [that there is] . . . no society in which experimental behaviour is consistent with the canonical model of self-interest. Indeed, results showed that the variation in behaviour is far greater than previously thought" (Henrich et al. 2004:abstract).[10]

In fact, we can state that accumulated research indicates that punishment, reward, and fairness are norms expressed in various ways by all societies. A whole series of factors, ranging from cultural histories to exogenous circumstances such as market influence, will alter the relative expression of these norms. However, norms that contradict the canons of *Homo economicus* are the mainstay for understanding small-group co-operative behavior.

Experimental and theoretical research has therefore demonstrated that the assumptions of *Homo economicus* and rational actor theory, the canonical model, are inadequate to explain small-group behavior in modern society and, by implication, similar behaviors throughout our human history. We have established beyond virtually any doubt that humans can be "irrationally" prosocial, demand fairness, and co-operate under the appropriate conditions. This is particularly true in small groups, precisely the social context in which complex stateless societies developed.

Raising the Stakes: The Principle of Proportional Rationality

> The ultimatum game has generated considerable interest because experimental evidence strongly rejects the standard game-theoretic predictions. A limitation to this general result is the possibility that experimental results are an artifact of small stakes. – Cameron (1999:47)

The results from 20 years of UG experiments are almost uniform in rejecting the canonical model of rational actor theory. This has had a profound effect on our understanding of economic behavior and makes the formerly "exotic" behavior of non-Western peoples substantially more intelligible by Western observers.[11] There are, however, critics of these experiments, particularly orthodox economists. While conceding that prosocial behavior exists, they maintain that (1) the proposers are basically asked in these experiments to give away something of value that they did not work for, effectively an unanticipated gift, which is rare in actual social interactions; (2) responders are offered a sudden windfall of this gift, which also is rare in normal human social interactions; and (3) all of these experiments are low stakes, and therefore there is little actual risk and reward in playing the game.

In response, experiments have been conducted to assess the effects of windfalls and stakes in the UG. Experiments such as those done by Roger Berger and colleagues were designed to deal with the windfall issue: "The experiment was created to avoid effects of windfall gains," and the experiment "confirms other-regarding behaviour in the ultimatum game" (Berger et al. 2010:1). This work indicates that the nature of the payoff did not affect the outcome and therefore supports the nonrational actor conclusions of previous work with the UG.

Some experiments focused on stakes (Cameron 1999; Carpenter et al. 2005; Munier and Zaharia 2002). This work supported the earlier findings as well: as the stakes went up, the effect of fairness remained constant. Proposers continued to offer fair amounts, and responders continued to reject unfair offers. These results are profoundly counterintuitive – one can reject 10 percent of $20 easily enough, but would one reject 10 percent of $100,000 so readily?

The main criticism, of course, is that while the stakes may be raised by an order of magnitude, the difference between $5 and $50 is, for pampered undergraduates in rich societies, not very much of a difference at all. Likewise, studies conducted in developing countries, where the purchasing power of the money is considerably higher, still rely on relatively small sums.

Using more realistic conditions in my view, Steffen Andersen and his colleagues (2011) conducted a UG experiment in a poor part of India with a range of 20 to 20,000 rupees, a factor of 1,000. In this study, they indeed found that stakes do significantly matter.[12] At the low stakes game, rejection rates were equal to observed studies in other UG experiments around the world. At the highest stakes, however, they found only one rejection out of 24 responders. What is curious here is that the offers were proportionally lower in the high stakes than in the lower stakes, but acceptances reached almost 100 percent. In this case, we observe that there is an absolute amount of money – a significant percentage of an average wage – that shifts behaviors toward the canonical model.[13]

Similar results obtained in other studies. Julie Novakova and Jaroslav Flegr conducted a questionnaire study among science undergraduates in the Czech Republic. They found "a significant relationship between the amount at stake and the minimum acceptable offer" in both the UG and Dictator Games (DG) (Novakova and Flegr 2013:1). In other words, even though the subjects in

Prague were university undergraduates and in a different economic situation than Indian farmers, the results were the same. What is even more fascinating is that this study was not done with real money. Rather, the students were asked to give an honest and confidential answer to how much they would offer and accept in the UG and how much they would offer in the DG. We can presume that were the scientists able to provide actual money, the results would be even more significant in that the stakes would be tangible and a great factor.

I run an informal UG game in my own classes on economic anthropology. At stakes of $20, the results always conform to other experimental results with responders demanding at least $8 or so. However, when asked hypothetically if the stakes were $100,000, almost all respondents were willing to take as low as 5 percent, many even much less. Upon post study discussion, one constant that emerges is the idea that receiving even 5 percent of $100,000 is still a respectable action. The perception is that getting a free $5,000 would not damage a responder's reputation, but accepting anything less that half of a trivial amount would indeed damage reputation. This proposition requires much more study. Unfortunately, this is a kind of systematic study that most funding agencies would be unlikely to support given its prohibitive cost. But as we see in the India study above, such experiments are sometimes possible and support the power of stakes in decision making in small-group contexts.

What these studies show is that the *stakes matter*. On a small scale and for small stakes, people around the world engage in economically irrational but socially essential ways, with norms of fairness dominating the social interaction. As the stakes go up, however, and we can presume that as the size of the group or social network increases over the threshold of 300 or so actors, people shift to more canonical and other-directed behaviors.

There is, to repeat again, a difference between large group and small-group dynamics, between high stakes and low. For anthropologists, this is the difference between stateless societies and larger, more complex ones. The kinds of societies that economists study are different than the kinds that most anthropologists traditionally study, and the major differences are the size of the group and the stakes involved. We cannot take lessons learned from the study of large firms and project them onto small societies. The nature of human interaction shifts in these different contexts. For small societies, Western and non-Western alike, we need a different set of theoretical tools to fully understand economic and social behavior. These tools, I suggest, are called anthropological game theory.

Co-operation without Coercion in Small Groups: Anthropological Game Theory

I propose a term for the field that studies noncanonical behavior of people in small groups. I call this anthropological game theory or AGT. AGT is a set of principles that explains the behavior of people in small groups as distinct from those in larger ones. It is based upon a multilevel selection mechanism, non-rational actor assumptions (the predominance of social rationality over economic rationality), and stresses that variation in the ethnographic record since at least the end of the Pleistocene is due in large part to cultural factors, and not only our evolved psychology. As Henrich et al. (2010b:1480) note that "modern prosociality is not solely the product of an innate psychology, but also reflects norms and institutions that have emerged over the course of human history." In this sense, AGT stresses an economic logic – a primary source of variation between groups emerges not just from biological differences in our evolved psychology (as in some evolutionary game theory frameworks) but also from those norms of behavior that emerge from sustained human interaction. Within the AGT tradition, cultural transmission models based on group or "strategy" selection mechanisms are therefore necessary to explain the evolution of human co-operation over a millennial time scale. These strategies are created by people and altered over time through subtle trial-and-error mechanisms and are subject to selection like any other technology.[14]

AGT research studies the strategies that people use to maintain co-operation beyond household and kin group in groups of people numbering under about 500, usually far fewer. AGT relies heavily on studies of non-Western peoples, using WEIRD populations as just one of many kinds of societies studied. Our work shows that people are still ego directed in the broad sense of the term,[15] but the nature of interactions in small groups differs from that in large institutions, where the stakes and size create the conditions for the predominance of canonical behaviors.

Norms of fairness and reciprocity permeate these small-group dynamics, more so than they do in large-group contexts. Furthermore, as anthropologists, we know that these norms are expressed differently in different cultures and may not be readily understandable by an untrained WEIRD person. As with language, we all have an evolved psychology that allows us to learn whatever languages we are immersed in at an early age. And, also like language, there is tremendous diversity in the ways in which we express norms of fairness and reciprocity. There are no genes for Russian or French, and there is no set of genes for any particular norm of fairness and reciprocity. The evolution of norms of fairness comes from a constant interplay between our evolved psychology and cultural effects, all mediated at some level by selective pressures on individuals and on group strategies.

Given the vast amount of work in evolutionary game theory, we can assume with a high degree of probability that people in small groups throughout history and prehistory did not behave rationally in the classic economic sense. Rather, people today and in the past were *adaptive* in the evolutionary sense used by theorists such as Robert Axelrod. As he describes it, "Individuals are not fully able to analyze the situation and calculate their optimal strategy. Instead, they can be expected to *adapt their strategy over time based upon what has been effective and what has not*" (Axelrod 1997:14, emphasis added). Our highly complex evolved psychology allows us to strategically form alliances with people beyond our kin group and to create norms and customs that perpetuate these in the group.

Most likely more specific evolved innate psychological dispositions (such as the capacity for language) underlie these strategies as well. But more important are the cultural forces that develop over time and that shape the way we create and follow norms of co-operation. The specific ways (like particular languages) are culturally constructed and are effective strategies for negotiating the myriad social realities that confront humans in both small and large groups. These strategies can be conceived as "social technologies" that allow people to survive and thrive in their world. And most importantly, these social technologies are differentially effective in promoting co-operation. Because they vary in their capacity to sustain co-operation, these social technologies are subject to selective pressures in competitive environments.

As Robert Boyd et al. (2003:3531) point out, "Unlike any other species, humans cooperate with non-kin in large groups. This behaviour is puzzling from an evolutionary perspective because cooperating individuals incur individual costs to confer benefits on unrelated group members." Selection occurs at the level of the strategy, but this is not a group selection in the classic sense. It is better understood as a kind of strategy selection

Anthropological game theory is predicated on the empirical observation and theoretical expectation that co-operation in small groups emerges out of sustained interaction between ego-directed individuals. People co-operate (and decide not to co-operate) because it is in their self-interest. As such, the question of coercion in social theory is central. People are not forced to co-operate but rather an economic logic allows them to assess the costs and benefits of co-operation. We now look to the ethnographic and historical record to illustrate that coercion does not emerge in any significant way until we have larger societies with institutionalized forms of leadership, what we traditionally refer to as archaic states.

4 The Role of Coercion in Social Theory

A cornerstone of canonical thinking in archaeology and evolutionary anthropology is that people will adopt more complex forms of "sociopolitical integration" only if they are forced to by social or environmental pressures. It is assumed that all forms of rank are negative for the bulk of the population. Such theory was intimately linked with economic anthropology of the twentieth century. In both the substantivist and formalist traditions as applied to long-term cultural change, "force" is a central causal agent in the rise of complex stateless societies. As we will see, this is an incorrect assumption, a view understandable by the evolutionary game theory revolution of the late twentieth and early twenty-first centuries.

The formalists' positions derived from classical political economy with its assumptions of the validity of *Homo economicus*. This was a world populated by agents who are by nature inherently ego directed. Life in stateless societies was viewed as short, competitive, and difficult. The state emerged out of an egalitarian society as a necessary but negative means of social control. In this world, humans inherently acted for their own advantage, and when left to their own devices, the forces of supply and demand would regulate the economy. Classical political economists essentially assumed that our ancestors, the Hobbesian savages, were effectively domesticated by the state, with market exchange emerging as one of the great pacifiers of our inherently conflictive nature. This naturally lead to hierarchies forming in which the bulk of people were forced to accept social hierarchies in which one class of elites coopted social power.

In contrast, the utopian substantivists assumed a world where people are by nature other directed and relatively peaceful, a world where social values trump material ones. The rise of patriarchy, private property, and the state destroyed the principles of reciprocity and the generalized sense of equality that had characterized "primitive" or stateless society. It therefore follows from this view that people would not want to change this utopic state of affairs. As in the formalist position, elites arose and coopted power as well.

In both of these traditions, therefore, it is difficult to explain why people would voluntarily give up social autonomy to permit an elite group to develop in their society. Why would people allow others in their society to have any kind of control over their lives if it were not necessary? As Allen Johnson and Timothy Earle (2000:203) rhetorically ask, "Why [would] food producers forgo leisure in order to generate a surplus in the first place . . . why are people willing to accept the burden of supporting Big Men, their expensive feasting, and public displays of wealth and status? Why do people abandon a condition of political equality for inequality?" These reflections echo Marshal Sahlins who reviewed the classic economic anthropological position regarding the creation of surplus. From a material and economic perspective, surplus is a precondition for hierarchy, a position that few would disagree with. Sahlins famously goes on to say that there are only two ways to create the increased production necessary to have surplus: induce more people to work or get people to work more (Sahlins 1972:82). Both choices are coercive and assume that people would not voluntarily give up autonomy or resources unless they were forced to do it. In these views, coercion is central to the development of rank and complex economies. As we will see below, this view is incorrect, but it was a mainstay of economic anthropological theory in both the substantivist and formalist tradition.

In cruder versions of coercive theories, people were overtly forced by others in a kind of Hobbesian brute-strength competition in which the strongest emerge as the chiefs. These chiefs then go on to consolidate their power and force others to their will, creating a class-based society in which a few thrive and most are exploited. In the more sophisticated models, force was indirect and subtle, taking the form of population pressure, climatic shifts, the rise of a generalized interethnic conflict that required hierarchies for protection, and the like.

Social Contract Theory

The issue of coercion in social theory is fundamental to fully appreciating the sharp theoretical contrast argued in this book. To fully understand the issue, we must briefly travel to the nineteenth century. The era provided some of the first comprehensive theoretical frameworks for understanding the development of society throughout prehistory and early history. Central to notions of cultural evolution was an inevitable shift from egalitarian early societies to hierarchical state ones, of course. This general process was viewed as a largely negative one in one respect – people went from being "free" (in a Enlightenment sense) to being bound by rules and norms created and used by an elite to their own advantage. Many theorists in the early to mid-nineteenth century were deeply influenced by this kind of social contract theory.

This theory, as mentioned above, argued that the origins of civilization rest upon individuals voluntarily surrendering liberty to some kind of political authority to protect their physical safety (Hobbes) as well as to protect other rights (Rousseau). Other notable social contract theorists included Immanuel Kant and John Locke, who argued for various kinds of social compacts between people and sovereigns. A common theme in all these theories is that the adoption of "civilization" is therefore a compromise between the loss of freedom and the gain and/or retention of other benefits. At best, rulers were seen as a necessary burden to prevent even worse things in life; at worst, hierarchies were viewed as inimical to humanity's welfare.

The social contract theories lost ground in the late nineteenth century to utilitarian (John Stuart Mill, Jeremy Bentham), Marxist (Karl Marx, Frederick Engels), and social Darwinist ones (Herbert Spencer, Walter Bagehot, William G. Sumner). Virtually all these traditions continued to be influenced by the idea from social contract theory that the shift from egalitarian "savages" to "civilization" represented a loss of freedom and liberty. The scholars looked around the world at the great ancient and non-Western states and correctly saw exploitation by the ruling classes of the vast bulk of working peoples. Virtually their entire empirical experience reinforced the view that coercion was necessary to explain the origin of rank, chieftaincy, and, ultimately, the state.

The relatively poor empirical record of stateless societies in the nineteenth century reinforced this view. The non-European world was effectively divided by evolutionary anthropologists into egalitarian peoples and "barbarian" ones – the latter, like the Aztecs, Incas, ancient Egyptians, Babylonians, and others, having despotic regimes, underdeveloped economies, poor writing systems, inadequate systems of money and trade, and so on. These latter characteristics were of course considered to be the hallmarks of "civilization." If one society in the past had big bureaucracies, armies, engineering works, and the like but lacked these other attributes, then it became the middle stage between small bands or village societies and fully developed civilizations. Therefore the "intermediate" societies in this context were those that were fully "despotic" and ruled by exploitative classes.

As legions of ethnographers like Malinowski traveled the globe and lived with and studied stateless peoples, the evidence consistently contradicted this view of coercion permeating smaller societies living in villages with chiefs. Ethnographers studied tribal societies that had a rich tapestry of rank, chieftaincy, and the like. Yet, as we will see below, indigenous peoples themselves over and over informed Western observers that their chiefs did not have coercive powers over them. The ethnographers, in turn, observed the very same, with some chiefs extolling their supposed "vast" powers to anthropologists and then, after the interview was over, returning to digging their own yam or maize gardens. In other words, what the ethnographers witnessed were men of rank using the ideology and rhetoric of power but consistently failing to exercise that power in any arbitrary way. These chiefs did not fit the preconception of despots, yet they clearly enjoyed material privileges derived from the leadership position. They did not act like kings (and occasionally queens) are supposed to act, yet at the same time they spoke like petty kings. Throughout these rhetorical exercises, the "commoners" did not contradict these statements.

With hindsight, we can see what happened. First, the Westerners were too ethnocentric to accord to non-Western ancient empires and states the rank of "civilization," and they therefore conceptualized these groups as representative of the intermediary stage between "primitive" and "civilized." The ancient states of course *did* in fact have coercive hierarchies and entrenched despotic classes that exploited other people. But these were not, as it turns out, appropriate analogies for smaller chiefdoms and other stateless societies. It was too broad a brush based upon too poor a database and some false assumptions. The conceptual typology used was a dichotomy between egalitarian savages and coercive states. The middle societies, "barbarians" or their equivalent, were simply immature coercive states in this view. They were conceived to be despotic as well, albeit in a reduced form. It was only the egalitarian societies, what is referred to here as simple stateless ones, in which "men were truly free."

Second, as we will see below, the people in such intermediate societies (our complex stateless ones) accorded ad hoc authority to leaders in order to draw on that organizing power when necessary, particularly when coordination of labor and application of taboo was necessary. Western ethnographers often did not make the distinction between a "first among equals," whose authority was invoked at the behest of the community, and a bona fide elite with institutionalized power that exploited people. A chief was a petty king who simply had reduced powers. The nature of stateless societies, as we would say, was severely undertheorized. This has affected our view to the present day.

Most anthropological archaeologists working within evolutionary paradigms – selectionist, adaptationist, cultural ecological, and the like – view the development of rank as a cost to the bulk of society. This is also a fundamental assumption in any Marxist or Marxist-inspired theory, such as political economy and most postprocessual theory. In this view, the shift to more complex political organization must therefore be forced, either by exogenous circumstances, such as population pressure or environmental resource stress, or by other people coopting norms of egalitarian social structure.

There is also a strong Western ethnocentrism here. The idea that rank is inherently immoral is also an essential part of one of our Western metanarratives since at least the late eighteenth century. The Western world is composed largely of parliamentary democracies that emerged out of a centuries-long struggle between monarchies and the bourgeoisie, workers and capital, aristocracies and the gentry. The republican democracies that we have constructed in the last two centuries absolutely depend upon an ideology of egalitarianism and meritocracy (even if the reality is quite far from the ideal).

Marxist and postmodern theory, also firmly constructed and constituted within this Western ideology, fundamentally presupposes an original egalitarian, peaceful society that was usurped by elites and their systems of property and patriarchy. Hierarchy in any sense is bad, pure and simple, and people can only be forced to accept it. To repeat, in state societies this is true but in trying to understand the origin and operation of complex stateless ones, this assumption is incorrect.

Coercion in stateless societies is rare. We are confronted with ethnographic data that consistently indicate that people in complex stateless societies (formerly referred to as chiefdoms) do not give up any significant autonomy with the adoption of modest rank within their midst. In fact, I argue to the contrary that under the appropriate conditions, people in stateless societies create these complex social organizations for their own individual and collective benefit. The concept of societies possessing "rank in the service of the collectivity" successfully answers the contradiction in the ethnographic record of stateless societies. Such a concept directly challenges our ethnocentric bias, but it is a concept based upon a virtually unassailable body of data collected for more than 200 years.

To emphatically repeat myself, I do not want to give the impression that hierarchies in premodern *states* were noncoercive. Rank beyond the stateless society does indeed involve coercion. In some very complex stateless societies (formerly referred to as complex chiefdoms) as well, we see a constant tension between elite groups trying to establish some coercive social mechanism and kin groups resisting that process. The transition to state societies represents the emergence of clear instances of coercive authority in society, a struggle between emergent elites and commoners, between a new political class and the traditional kin-based corporate group, over the nature of authority in society. It is in these rare sociopolitical formations that we see the efforts of an ersatz managerial leadership work to create a class that serves its own interest, and not one bound by the traditional norms of the collectivity. Archaeology teaches us that complex stateless societies rise and fall with some regularity. It also teaches us that the nonelite in most instances successfully resist attempts by an emerging class to dominate. States rarely emerged, but when they did they had the classic hallmarks of coercive hierarchy and rank. However, complex stateless societies did not have institutionalized mechanisms of coercion available to chiefs or leaders to use at will in their own self-interest.

The Ethnographic Data on Coercion in Stateless Societies

A common rhetorical device in archaeology is to decry the use of ethnographic comparison when it does not give you the "right" answers. Yet, virtually all anthropological and archaeological theorists of any viewpoint use it when such comparisons support their ideas. I will modify that fine tradition with the difference being that I also note contrary data when I find it.

Lamont Lindstrom (1981) provides a short but very incisive discussion on the terminology of nineteenth- and early-twentieth-century anthropologists in their description of Melanesian leaders. In his account, these observers misapplied the term "chief," a point that generally applies to many ethnographies of the time. He writes:

> People soon realized the men thus described did not measure up to the idea encapsulated in the word. The romantic notion, entertained by Europeans, of strong, singular, tyrannical, omnipotent, polygynous, decorated, gracious, wealthy, and hospitable native chiefs was stretched to fit the realities of Melanesian leadership systems at threat of conceptual self-destruction. (Lindstrom 1981:902)

The corpus of ethnographic data that supports the proposition that leaders in stateless societies – Big Men, chiefs, and so on – do not have coercive power is vast. These data include observations from early ethnographers such as Malinowski and Firth, observations from some of the first Western travelers to encounter indigenous peoples, and later work on societies at the margins of colonial states. It is often rightfully noted that the interaction of non-Western peoples with colonial powers exacerbates internal hierarchies as it alters indigenous socioeconomic structures (and see Pels 1997). It is instructive in this regard that the people in indigenous communities of effectively stateless societies studied by ethnographers still maintained strong control over their leaders, even in these colonial contexts.

In examining the ethnographic record, we return to one of anthropology's most celebrated figures. Malinowski's observations of Trobriand chiefs constantly note that they do not have the ability to force people to act. All is, for Malinowski, the force of custom:

> Every man knows what is expected from him, in virtue of his position, and he does it, whether it means the obtaining of a privilege, the performance of a task, or the acquiescence in a *status quo*. . . . The chief's authority, his privileges, the customary give and take which exist between him and the community, all that is merely, so to speak, the mechanism through which the force of tradition acts. *For there is no organised physical means by which those in authority could enforce their will* in a case like this [canoe building]. Order is kept by direct force of everybody's adhesion to custom, rules and laws, by the same psychological influences which in our society prevent a man of the world doing something which is not "the right thing." (Malinowski 1966[1922]:158–159, emphasis added)

E. E. Evans-Pritchard's 1940 classic ethnography on the Nuer likewise emphasizes the fact that chiefs have little coercive power outside of the brief raiding parties. Regarding the leopard-chief, Evans-Pritchard (1940:172) tells us, "It might be supposed that this functionary has a position of great authority, but this is not so." He goes on to note that there is no real "government" in Nuer society, meaning that there are no formal positions of power. The leopard-chiefs had no judicial authority. He notes that the early British administrators of this colonial holding complained that there were no positions of authority or people of sufficient influence to construct a viable administrative structure. His summary quote (1940:173) highlights the role that chiefs play – one of persuasive mediator lacking any coercive authority: "On the whole we may say that Nuer chiefs are sacred persons, but that their sacredness gives them no general authority outside specific social situations."

A critical function of the Nuer leopard-chief is to mediate blood feuds, one of the "specific social situations" that Evans-Pritchard mentions. The chiefs' homes are sanctuaries, where avenging family members cannot enter. He also negotiates with each side to see if the dead man's family will accept payment. If successful, the chief delivers cattle to the offended families and acts as a go-between for what can be a very lengthy blood payment process. A major observation by Evans-Pritchard is that the chief's intervention is only as good as the parties' desires to reach an honorable agreement. The chief is "the machinery which enables groups to bring about a normal state of affairs when they want this end" (Evans-Pritchard 1940:175). The chief in this sense is acting in a prosocial manner to maintain group co-operation. His reward is status, of course, but he invests substantial effort to prevent the collapse of the norms of co-operative behavior that otherwise prove so effective.

Another African ethnography, by Edward Winter, makes the point that formal coercion is virtually absent in the Amba people of central Africa.

> The Amba system depends to a very small degree upon the use of force and other coercive measures and throws very great responsibility upon subjective feelings of loyalty, and motivations tending towards a willingness to subordinate personal interests to the well-being of the collectivity.... Ancestor cults are often given a prominent place in the discussion of solidarity within such units. The Amba practise an ancestor cult and it is undoubtedly true that ceremonies carried out in connection with it exert a strong cohesive influence. (Winter 1958:144)

Winter, like many of his contemporaries in African social anthropology at the time, was a structural-functionalist in the tradition of A. R. Radcliffe-Brown. In this light, the observations of David Tait among the Konkomba of West Africa are equally interesting: "There is no supra-Elder authority to impose a solution on recalcitrants of a district. Within the district there is no organized force which can be employed to ensure acquiescence; there is no coercion other than the ritual and moral power of the Elder and the diffuse sanctions of kinsmen's disapproval" (Tait 1958:193).

Thomas Håkansson writes about a precolonial stateless society in Tanzania, noting that coercive institutions were absent and that local descent groups were independent of chiefly authority:

The power of the chiefs was circumscribed (Rother 1906). The Lutheran missionary Oldewage reported that "the chief's influence depends on his powers of persuasion and speech. Direct orders are usually ignored" (Ankermann 1929:15). A chief could not make decisions without discussing matters with his council (chila ya njama), which was composed of male elders from each commoner clan in the chiefdom. They selected the chief's successor from the male members of his lineage and had the power to depose him (Ankermann 1929:36). The chief's authority in judicial procedures, leadership in warfare, matters of cultivations and in the allocation of land derived from his ritual position. It was as ritual experts that the chiefs obtained tribute in the form of a portion of crops, livestock, and sugarcane beer from each family. (Håkansson 1998:275)

Zoë Strother makes a good case for chiefs with little power in the Eastern Pende of central Africa. She notes that a chief's influence is not based on any kind of coercion but is instead based on the power of oratory, persuasion, and reputation. She draws a sound comparison with observations about Amazonian chiefs:

Clastres argues that chieftaincy can be divorced from coercion when "power" roots itself in prestige, persuasiveness, and a reputation for fairness. According to his model for decentralized societies in Amazonia, "despite his lack of authority, the chief enjoy[ed] an enviable social status." He was first and foremost a peacemaker and orator, charged with the impossible task of keeping the community together, held to exacting standards of generosity. He was remunerated in wives and deference and was closely watched for signs of authoritarianism. (Strother 2004:273)

She then compares this model of chieftaincy with those among the Eastern Pende:

All of this was equally true for the Eastern Pende chief in the 1980s. . . . Because he was dressed and housed differently from everyone else, and because the Eastern Pende treated the office with respect, it was easy to overlook the fact that *the chief was seldom able to issue a direct order unless first securing consensus among his advisers or, for weighty issues, among the population at large in public meetings*. (Strother 2004:273–274, emphasis added)

Aylward Shorter (1972:120) writes that in the Kimbu culture of western Tanzania, in traditional life "the chief was not a despot. He ruled together with a council of elders." In the same book he later notes that chiefs "owned" the whole country and controlled the distribution of valued items (Shorter 1972:137). This apparent contradiction makes sense if one understands the difference between ad hoc power granted to a leader and coercive power appropriated by a ruler. This concept is illustrated by Shorter's description of the work of a previous ethnographer, who published at the turn of the twentieth century that "the chief's actions are often determined by the council of elders, however, more through fear of them. . . . [He, the ethnographer] believed that the chief was theoretically not responsible to anyone for his actions, *but the practice was obviously very different*" (Shorter 1972:120, emphasis added). Shorter goes on to describe how chiefs were regularly deposed in the mid-nineteenth century when the council of elders became dissatisfied with the existing chief.[1]

Denis Werner describes the nature of chieftaincy in the Mekranoti-kayapó of central Brazil. Among the traditional characteristics, Werner notes, "Like leaders in other simpler societies, Mekranoti leaders do not carry much authority. They cannot force people to obey orders, but must limit themselves to giving suggestions or advice" (Werner 1982:137). People in this society can transfer loyalty to other chiefs if they please. The basis of leadership revolves around men's societies and the aspiring chief's ability to keep a faction. Both men and women can be chiefs, with the latter joining and running women's societies. Chiefs do not get material rewards, and work allocation studies by Werner indicate that there is little difference in expenditures of labor between chiefs and followers.

Joanna Overing (1993:31–32), working in the Orinoco River valley in Venezuela, writes, "While the leader of a Piaroa community had no powers of coercion over work, and little weight in the daily organization of economic activities, it was his duty . . . to maintain the high morale of his community." The traditional area of the Piaroa is in the middle Orinoco, in the highlands above the savannas. They were quite isolated until the 1960s, when substantial resettlement programs moved them into the savanna flats (Zent 1999) barely a few years after Overing conducted her research.

John Burton (1984:243) tells us about the Tungei of Papua New Guinea: "Tungei big-men, however, were and are not tyrants or dictators. They had the power of sway that comes with accurate appraisal of a situation and were forceful orators, but no more. There was no overall tribal leader." As we will see in Chapter 4, Tungei people organized very complex stone quarrying operations requiring substantial co-operation, all without coercive chiefly authority.

Jesse Byock (1986) details the many constraints on early medieval Icelandic chiefs, a social structure that represents a shift from a state society (Norway) to a stateless one for a few generations after immigration. He writes that they were "an elite with limited power. They were neither a commanding, territorial nobility, nor tribal leaders, rather they functioned as leaders of interest groups composed of thingmen or followers drawn from among the farmers" (Byock 2003:235). Byock goes on to report that the chiefs were not "war leaders so much as that of political middlemen adept at molding often unrelated backers into relatively short-lived political groupings" (Byock 2003:235).

Glenn Petersen (1982:117) writes about the leaders in a Micronesian culture: "Ponapean chiefs today have no basis in law for any sort of coercion whatsoever, and have had none since the early twentieth century. Their rule is one of moral force, derived from the good will of the community and the chief's own personal ability to manipulate and make use of it." Petersen does say that people report that in earlier times, chiefs owned the land, implying that they had more power. This is possible, or it is possible that it simply reflects ideological consistency with a concept of a chief. Regardless, the ethnographic record is clear that their chiefs did not have coercive powers.

John Caughey III describes a similar situation on the island of Truk in the Caroline Islands:

> The status of the chief is the "highest" position in a district. Everyone should employ a humble manner and be diligent in their use of "respect etiquette"... when dealing with a chief. The position of chief, however, does not carry much in the way of real power.... The chief cannot, for example, order individuals to contribute labor or valuables for some project. As the lineage leader he has such power over persons of his own lineage but he does not have such power over persons of other lineages in the district.... The main function of the chief is to call meetings when matters of significance to the entire district arise. At such meetings the chief and the other elder men of the district together decide what should be done.... Each [leader] represents his own interests and the interests of his group. (Caughey 1970:113)

This description likewise reinforces the fact that chiefs derive real prestige from their positions. They have social obligations, and in return they are treated with deference and respect, have larger households, and control more resources. But they cannot tell people what to do. As the first among equals, they articulate community wishes in a way that is wrapped with moral persuasion.

An early ethnography of the Andaman Islanders was conducted from 1869 to 1880 by Edward Horace Man. He reports that the chief has authority over subchiefs, but "his power, like theirs, is very limited" (1883[1975]:41). He notes that the chief is responsible for mobilizing people in hunting, fishing, and migrations and is called upon to adjudicate disputes. However, "he possesses no power to punish or enforce obedience to his wishes" (Man 1883[1975]:41).

The list goes on and on. Bruce Trigger (1963:153) notes that among the Huron of the seventeenth century, the "leaders were able to exert few overt sanctions." Roy Rappaport's (1967) work on the Tsembaba of Papua New Guinea illustrates yet again chiefs without coercive power. Meyer Fortes tells us that among the Tallensi of Ghana, "indeed the Chiefs and Earth-priests have no powers comparable to what we would call political authority even in their own groups" (Fortes 1962:65).

The ethnographic record is replete with examples of chiefs or leaders in stateless societies who are reported to lack coercive powers. One of the confounding factors in our perception of chiefly power in an intermediate society, as alluded to above, centers on the effects of colonialism, both Western and non-Western. Expanding empires as a general rule do not like stateless societies. They prefer some indigenous power structure that they can coopt for their own needs. We see in the historical and ethnographic record a process over a couple generations in which chiefs who traditionally lack coercive power are transformed into agents of an expanding polity. This transformation involves shifts from traditional social relationships between chiefs and communities to ones where chiefs and their assistants acquire coercive power, backed by the authority of the larger political entity. This used to be known, in the theoretical literature, as one form of "secondary state formation." There is no need to go into great detail here. Suffice to note that all of our examples above come from Oceania, Africa, and early historical America, where ethnographers were able to analyze chiefly political forms prior to any significant incorporation of these polities into formal agents of colonial powers.[2]

There is virtually no doubt that stateless societies described as "chiefdoms" or "tribal" that have people with clear rank existed and functioned without significant coercive political institutions in most instances. This is not some historically contingent phenomenon. It appears to be a structural feature of settled village societies around the world. In other words, there is something in the emergent properties of people interacting in their own self-interest over time that promotes the creation of noncoercive positions of rank, usually ad hoc and transitory, in groups of people organized beyond the migratory forager band. This emergent property arises as a way to keep the co-operative group co-operating in the face of the centrifugal forces.

Most contemporary theory as used in archaeology, what I refer to as the canonical model of cultural evolution and chiefdom studies, therefore assumes that people were forced into more complex levels of sociopolitical integration, either by factors largely beyond their control – environmental and demographic stress – or by other people – ego-directed and aggrandizing elites in the development of any kind of complexity beyond the hunter-gatherer-forager band. Exogenous conditions of life forced people to accept higher levels of sociopolitical organization to survive. In short, hierarchy was considered inherently negative because it was rightly assumed that people would not assume the costs of an exploitative elite without coercion

We have tested these theories over the last two generations in scientific archaeology. Many of the factors originally hypothesized to explain this evolutionary process work well for the development of state societies but have proven insufficient for explaining the initial expressions of social complexity in both the Americas and the Eastern Hemisphere. The key reason is that both ethnographic and archaeological data strongly suggest that coercive social power by one group over another did not develop historically until state societies took hold in the early Bronze Age in the Old World and in the Preclassic, Classic, or Middle Horizon periods in the Americas. The ethnographic data of stateless societies presented above likewise reinforces this conclusion – social rank is indeed evident, but these are hierarchies at the service of group co-operation, hierarchies that can be quickly reconstituted with other people if the community desires to change leadership. After many years of analysis, I have come to the conclusion that coercive theories for the development of early complexity in the archaeological record simply do not work. I find that the analysis of leadership in "tribal" or nonstate societies by Paul Roscoe, a sociocultural anthropologist, is closest to the ethnographic and archaeological evidence:

> My argument is that the Big Man was not, as the stereotype would have him, an economic entrepreneur, nor were the seeds of elite control inherent in his most visible relationships with his followers. *First and foremost, he was a manager, the initiator and principal organizer of large-scale community displays*. In some areas, these displays took the form of conspicuous material giveaways with which the Big Man is stereotypically associated. They could also, however, take the form of large-scale exhibitions of singing and dancing. And under yet other circumstances they took the form of monument building, a phenomenon almost entirely overlooked in anthropology but with obvious relevance to archaeology. (Roscoe 2012:41, emphasis added)

Universal Coercive Aggrandizers as Explanation?

One very prominent theory on the cause of sociopolitical complexity directly challenges my argument about the lack of social coercion in intermediate or complex stateless societies. This theory assumes that all groups have aggrandizers in their midst and that these individuals, under the right circumstances, will create inequality using feasts and other strategies to obligate others. In this view, all societies have highly ego-directed and aggressive individuals, almost always males, who relentlessly pursue their material and biological self-interest. It therefore follows that complex stateless societies were coercive.

The celebrated anthropologist and archaeologist Brian Hayden has advanced this theory most vigorously over the past decades. In his latest book (2014), he summarizes his fundamental assumption:

> Thus, we can expect that at least a few people who were highly motivated and determined to pursue their own self-interests (and change social norms to suit their own goals) occurred in all large populations, even in the most egalitarian cultures, and that they took risks to achieve benefits for themselves. These are the individuals that I refer to as "aggrandizers" or "triple A" personality types: ambitious, abrasive, aggressive, accumulative, aggrandizing people. (Hayden 2014:17)

Hayden's argument is quite elegant, providing a theoretical link between small band societies (even nonhuman primate societies) and the kings of states and titans of modern corporations. To repeat because it is so important, fundamental to this theory is the assumption that there is a normal distribution of aggressive/altruistic men in any society (Hayden 2014:15). He argues that aggressive men in hunter-gatherer society are constrained by resource limits, technology, and ideologies of egalitarianism. He argues that under the appropriate conditions of resource surplus in complex hunter-gatherer and horticultural societies, aggrandizers can emerge and create sociopolitical hierarchies. They achieve this by manipulating surpluses to indebt people through feasting. As

Polly Wiessner aptly describes it, in this model, "Aggrandizers strive to gain control of strategic resources or manipulate the production of others through debt and contract, coercion, marginalization, or exploitation made possible by restricted mobility. Though power comes from several sources, primacy rests in material processes" (Wiessner 2002:234). This control allows them to accumulate wealth and to obtain "many wives and to have many children" (Hayden 2014:20). In short, under the right conditions of resource surplus, a raw Darwinian competition erupts; a few males compete for prestige, women, and wealth. Hayden's work is a classic form of rational actor theory in economics wedded to a raw ecological and materialist epistemology.

Hayden's theory is influential, and his recent theoretical tour de force – *The Power of Feasts* – articulates his position extremely well. He sets up a dichotomy between traditional kinds of theories about the structure of society. On one side are those based upon principles of economics, cultural ecology, and evolutionary biology that assume a Hobbesian world of ego-directed and self-regarding actors playing out a rational competition in most human social interactions. The other end of the dichotomy includes what he and others call communitarian theories. Here, people who organize feasts and other "altruistic" acts, for instance, are motivated by psychological satisfaction. Feasts can be likewise be used as leveling devices to redistribute resources. He disapprovingly notes: "Communitarian or functionalist advocates view feasting as serving communal goals and promoting social solidarity" (Hayden 2014:7).

Hayden refers to complex stateless societies as transegalitarian. These are complex hunter-gatherer and/or small horticultural and village-based agricultural societies that have evidence of Big Men or chiefs. I would also agree to a certain point that in *state* societies, feasts substantially serve the interest of elites and their retainers. There, a model of naked self-interested behavior by dominant groups is an effective way to explain much in the historical and archaeological record (think of the Moche or Maya, among many others).

In short, this is an empirical question: Do the bulk of people in stateless societies allow alpha males to coerce most other people in the group? Do chiefs or lesser chiefs (Big Men) in complex stateless societies acquire coercive power over others, or is that power accorded to leaders by the community? Can the people of a stateless society withdraw that authority at will, or is it institutionalized and effectively nonnegotiable? Is the authority claimed by chiefs in stateless societies used to benefit the aggrandizer, the members of the group, or both?

I believe that the review above substantially answers that empirical question, and it is an emphatic *no* that leaders in stateless societies have coercive power. Behind the ideological facades of rhetorical power, we have just seen above countless ethnographic accounts of the rejection of coercive elites in stateless societies. There are indeed aggrandizers in stateless societies, but the social and material conditions of life preclude them from exercising systematic control over others. Aggrandizers indeed pursue their self-interests, but they must adjust to the realities on the ground. That reality is, in short, that aggrandizers must earn prestige and some advantage by hosting feasts, conducting rituals to regulate complex economic activities, and engaging in costly prosocial behavior that benefits not only themselves but the group as well.

I believe that the ethnographic and historical examples described in the section above indicate that chiefs and Big Men do not have coercive authority. However, Hayden likewise produces his own list of societies with leaders who are aggrandizing and coercive. There is no theoretical middle ground here between our two theoretical frameworks for nonstate societies. Chiefs are either aggrandizers who act in their own self-interest against the good of their faction or community, or they act in their own self-interest by organizing co-operative ventures that benefit both themselves and their faction. This is a fundamental epistemological difference. It is therefore worth parsing Hayden's sample of ethnographic data a bit more.

In the first instance, we see that Hayden uses two different kinds of societies to illustrate his point about aggrandizing behavior. The first are indeed complex stateless societies as described in this book. The second, however, are small-scale subgroups in modern states that use the mechanisms of those states to manipulate social norms. These latter examples are not appropriate as analogies to stateless societies in my framework because once an external entity with coercive state power exists and which can influence individual decisions, the cost-benefit matrix shifts to the advantage of leaders. This is, in fact, how states grow, a process at one time known as a kind of secondary state formation. These are fascinating case studies in the anthropology of colonialism but make poor analogies to premodern or early colonial societies.

In the first set of examples used by Hayden, I will discuss the ethnographic data using his same sources, when available. We will see that these data in fact challenge the argument that chiefs have any coercive power beyond that which the group accords. Hayden breaks down his discussion into simple hunter-gatherer, complex hunter-gatherer, agricultural, and state societies. I will look just at the complex stateless societies that he discusses that are relevant to the theoretical arguments in this book.

Hayden lists the Abelam of Papua New Guinea as an example of a transegalitarian society with coercive Big Men. He (2014:186) notes, "The big men were the entrepreneurs of Abelam society." He states that "elders occupied important roles in public affairs . . . whereas the big men of the clan provided the organization and financing." These "entrepreneurs," however, had to have permission of the elders to conduct their business. This latter fact alone suggests that the Big Men did not have independent coercive power but relied on the elders to approve and constrain their behaviors. The elders in turn, were responsible to their factions and had to act accordingly.

According to linguist Richard Scaglion: "Abelam (or Ambulas), the Papuan language with which I am most familiar, is a language of the Ndu family. *It has no term for 'chief.'* The closest equivalent would more precisely be translated as 'big-man,' both literally and figuratively. . . . A person of influence in the village might also be referred to as nemaan mban (important one). *The term suggests influence but not authority or power*" (Scaglion 1996:7, emphasis added). This discussion parallels the discussion of Lindstrom in his linguistic parsing of the word chief in Oceania. If this is accurate, then at least linguistically, the Abelam speakers did not recognize coercive power as traditionally understood in the word "chief"; rather, they recognized the concept of the socially influential man as implied in the term "Big Man," the term that Hayden appropriately uses.

This linguistic observation is backed by primary ethnographic data. Ethnographer Phyllis Kaberry (1941a, 1941b) describes "entrepreneurs" in Abelam society based upon observations prior to World War II:

> There is no system of hereditary chieftainship, but each clan has its important men (nzmandu), who as elders exercise much influence over the younger members, and who apportion the land for cultivation. Some of these men, through ambition and initiative, have acquired outstanding prestige in the yam and *tamberan* [men's society] cults, and they are able to mobilize a larger group of men as labour for certain economic and ceremonial activities. They are known in other villages; they are treated as honoured guests at the ceremonies, and they are mourned by the surrounding region when they die. Their names are remembered, and cited by their descendants, and sometimes assumed by them when they reach a similar age. Some have a reputation as carvers and painters and their services are called upon by other villages.
>
> . . . The important men preside over public affairs – the construction of new houses – *tamberans*, initiation ceremonies, the making of large gardens for the large yams; the display of their harvest; the maintenance of peace; and the handing on of social and ritual traditions. (Kaberry 1941a:240)

This is a classic description of a manager, not a power-wielding king. Kaberry (1965:351) further notes that a Big Man in Abelam society "cannot command force to secure compliance with his orders." She goes on to note that he has to get the support of elders before he can carry out a project. This is likewise a classic description of Melanesian Big Man behavior. Young men, and sometimes young women through proxies, recruit followers and mobilize family members to host feasts and other ceremonies. These ceremonies can be competitive in the sense that successful hosts increase their social and material well-being. There is, in fact, no evidence whatsoever that Big Men in Abelam society had coercive power.

This description of the Abelam is quite similar to Mervyn Meggitt's description of the Mae Enga of the western highlands of Papua New Guinea. Regarding traditional leadership in the society, he says, "There are no hereditary chiefs or formally elected political leaders" (Meggitt 1974:168). In another article he notes, "Followers demand and create leaders rather than *vice versa*" (Meggitt 1967:20). It is difficult to be clearer here – it is the community that creates leaders, almost certainly because leaders serve a coordination function. Among these groups, it is clear that these young men, the "entrepreneurs," have some influence due to their ambition and success in feasting and exchange, but there is no detectable and enforceable coercion involved in these social relationships. All debt obligations are part of a reciprocal cycle, not a one-way ticket to bondage. Leaders have followers only insofar as they can provide ceremonies, social events, and/or other benefits.[3] As Michael Allen (1984:33) says about leaders in the eastern Ambae in Melanesia: "Youths often attach themselves to leaders; and in return for various services they hope to advance their own careers." Clearly, they are not forced to attach themselves, and this describes a common process of leaders attempting to secure followers in various activities.

Polly Wiessner describes the nonmaterial benefits that Big Men in Enga society gained for organizing feasts. One was the "prestige that accrued from their ability to coordinate the production, cooperation, and participation necessary to stage cult performances and feasts" (Wiessner 2001:127). She notes that such prestige was more than just emotionally satisfying. It was fame. And this fame was "only accorded to those who offered strong benefits to the group as a whole" (Wiessner 2001:127). Again, this observation reinforces the concept that leaders in stateless societies must function to help maintain group co-operation if they also want to pursue their self-interests. Again, these are managers, not tyrants.

Working in the highlands of New Guinea as well, Paul Wohlt notes, "Before [the colonial period], there were fewer pigs and they were concentrated around a few individuals. Now there are more pigs and everyone can make exchanges and be bigmen" (Wohlt 1978:98). We are not entirely clear what the precolonial period was like, but certainly in the ethnographic present, effectively "everyone" can be a Big Man. Successful leadership here also depends upon a man recruiting followers, but this is without question a voluntary process. Because it is voluntary, other people decide to ally with any particular leader out of their own self-interests. As Wohlt (1978:276) says, when the relationship between leader and potential follower, in this case a migrant, is working, "both host and migrant benefit." In another case, Wohlt (1978:279) says, "It may thus be that it is more advantageous for the would-be migrant to pick a different host to maximize his assistance in exchange endeavors." This likewise reinforces the notion that the relationship between potential leader and potential follower is a negotiated one. This is much more consistent with the model of achieved, temporary, and noncoercive leadership that we see in the ethnographic literature.

Regarding the Enga, Flannery and Marcus (2012:104) describe them perfectly as "a society whose leaders had achieved renown but no real authority."

Hayden himself notes on several occasions the voluntary nature of followers. For example, "When a leader wanted to hold . . . [a feast] he invited all the people he wanted to involve as supporters to a meal" (Hayden 2014:188). He goes on to say, "If an individual accepted this invitation . . . he could not refuse to support the organizer." Hayden notes that this is similar to the way in which leaders work in Southeast Asia. He also says that activities other than feasts, such as raids, were structured in the same manner (Hayden 2014:188). Of course, the key phrase here is if that person "accepted this invitation." It is hard to describe this statement in anything but noncoecive terms. These observations again illustrate the voluntary nature of the leader–supporter relationship.

The Kuma people of the New Guinea highlands also are one of the case studies in Hayden's book. He notes correctly that the leaders act in their own self-interest. Yet earlier ethnographers make it clear that these leaders do not have coercive power: "On the one hand, an important duty of the authorized leader is to maintain law and order. On the other, he is only first among other leaders. His 'followers' may either co-operate with him and consolidate his influence, or compete with him to supplant his power with their own" (Reay 1959:117). The "authorized" leaders are the strongest in this society, according to principal ethnographer Marie Reay. They command labor, but this is through clear norms of mutual benefit, reciprocity, payments, and so forth.

The same is true for the Melanesian group on Goodenough Island studied by Michael Young and detailed in the book *Fighting with Food* (1971). This island is just southwest of the Trobriand Islands in eastern Papua New Guinea. This fascinating custom of competing with feasts is led by leaders but is clearly a group activity that requires assent by the followers of the Big Men: "The person wishing to initiate *abutu* [a challenge feast] consults his *fofofo* [unconditional followers] to determine their willingness to help" (Young 1971:197). Young notes that these people are unlikely to say no. But what is critical for the success of the challenge is the assent of two other groups, kin-related supporters of the challenger and *tabotabo*, described as "residual helpers who stand in a variety of relationships to the big man doing the challenge" (Young 1971:190). Likewise, Young notes that if the challenged Big Man cannot rally his supporters, there is no competitive feast, and it is a loss by default. In each of these instances, Big Men require group participation. The groups can be hostile factions within a village or even, on rarer occasions, between entire villages. The feast is as much a contest to see which group can accumulate more resources, indicating the relative strengths of the political factions, as it is a competition between two aggrandizing males.

The Core of the Issue: A False Dichotomy

Hayden creates a stark Hobbesian/Rousseauian dichotomy by contrasting what he implicitly sees as a fuzzy logic of "cognitive, cultural normative, social and psychological gratification explanations" with the more *realistic* "ecological, political economy, or cultural materialist" positions (Hayden 2014:7). I agree with Hayden on at least the point that there is a normal distribution of saints and angels in all societies. And I agree that it is exogenous conditions that permit certain kinds of more complex social organization to develop. However, Hayden sets up the false dichotomy that was discussed in an earlier chapter. It is either a Hobbesian world of pure ego-directed struggle or a Rousseauian noble life other-directed sharing and peace. This logic forces him to argue that because there are aggressive people in any society who act in their own self-interest (true), there must be substantial coercion in all societies, including stateless ones (false). Trapped in this false dichotomy, there is no other way to characterize the behavior of aggrandizers as they manipulate social norms to control people. In this view, social control develops in tandem with even the most modest increase in complexity as defined by indicators of rank.

He breaks down theories of feasting into four categories: functionalist, costly signaling, communitarian, and aggrandizer. As mentioned, he and others who accept his basic assumptions use a classic Hobbesian/Rousseauian dichotomy between ego-directed/coercive assumptions of human behavior and romanticized, other-directed/altruistic views of human nature. Of these four categories, the first three come from different traditions, but all have in common the characteristic that the actions of chiefs or leaders benefit the group. In these models, the actions may or may not benefit the leaders, but certainly there is a benefit to the community as a whole. Only the latter theory, his theory of aggrandizers, is one in which only the leaders benefit and the one that resists the temptation to adopt utopic assumptions. In his view, it is a sad but hard truth that represents the human condition – anything else is simply naive.[4]

Hayden defines any behavior that benefits the group, even if there is a corresponding benefit to the leaders, as "altruistic" and therefore epistemologically suspect. The communitarian models are the most extreme in the degree to which the community is helped, and they represent a Rousseauian ideal. With a logic forced by this dichotomy, any deviation from altruistic behavior must necessarily imply coercion. This is evident in some of Hayden's quotes: "Wohlt (1978) similarly observes that big men had to convince their supporters that they acted in the interests of supporters, despite the fact that big men covertly acted in their own interests" (Hayden 2014:183). He refers to "intense individual maneuvering for self-benefits, especially to acquire food and wealth (in particular, women and pigs)," "flagrant departures from altruism," "ruthless calculation of ultimate advantages," and so forth (Hayden 2014:181–182). There are many instances in which he characterizes leadership in a stateless society as almost pure ego-directed behavior.

This is, in my view, a straw man argument, because neither the pure Rousseaean nor Hobbesean world has ever existed. It too is a product of post-Enlightenment Romantic ideologies of the West. We can begin with the assumption that all people act in their immediate or long-term self-interest and that ego-directed behavior underlies almost all social action. However, as introduced in the previous chapter and as discussed in great detail in the following one, research in collective action theory and evolutionary game theory makes the Hobbes/Rousseau debate moot. People are conditional co-operators. Ego-directed individuals will engage in prosocial (altruistic) behavior if there is a payoff to them or their kin, particularly in the short or intermediate term.

And there is indeed a payoff for both the followers as well as the followed in stateless societies. But in the absence of coercive social mechanisms, any leader in a stateless society who wants to garner prestige and resources must by necessity deal with the collective action problem. As ethnography and history show, the only effective way to do this is with intensely prosocial rewards combined with behaviors designed to minimize free-riding and defection. A good leader in a stateless society who keeps the group together will reap the rewards. But he or she cannot force people to work. Otherwise, we know from theory and observation that the group will fragment or fission off and establish another group.

It is also important not to make the simple mistake of assuming that the existence of conflict or "war" is evidence of coercion in stateless societies. As I will describe in the following chapters, organizing for a raiding party by a small group is one of the most intense forms of *co-operation* between unrelated or distantly related people that we see in the historical or ethnographic record. Conflict between groups in fact intensifies co-operation *within* the group. As we will see, this is a well-established feature of collective action theory.

It is true that raiding occasionally leads to the capture of individuals who become enslaved, even in stateless societies. The circumstances of different cultures are highly varied regarding this social phenomenon. In some cases, "enslaved peoples" are more like indentured servants and become members of the community after a period of time. In other cases, they are even killed. Captured women are often incorporated into households, and their position is ambiguous. But what is critical to note is that the kinds of slavery noted in non-Western stateless societies differ fundamentally from the chattel slavery of many ancient and almost all modern states.

What Is Coercion?

This discussion obviously begs the question of what is coercion? The *Oxford English Dictionary* defines it as "the application of force to control the action of a voluntary agent." This is a good starting point. We can refine the definition a bit for our purposes here to this: the state in which someone is forced to do something that is contrary to his or her self-interest.[5] At one level, the existence of magic and taboo can be considered as a form of coercion, and in fact we will see that groups use such practices to prevent free-riding and to enforce other social norms. However, there is an essential difference between state-backed coercion for the benefit of an elite group versus informal coercion used by the community against free-riders or people who violate common norms of acceptable behavior. When most social scientists speak of coercion, they imply some kind of politically enforced physical sanction against people controlled by a subgroup of society. This kind of coercion does not exist in stateless societies. There is, to be sure, good evidence of slaves obtained as war captives in non-state societies. This is undoubtedly a form of coercion. However, the slaves are not part of the co-operative group and represent a tiny fraction of the total household labor.

In my perspective as developed in this book, the shift from complex stateless societies to state societies is effectively a shift from noncoercive to coercive means of leadership.[6] Read any historical or ethnographic account of premodern *state* societies, Western or non-Western. We see a rich and historically deep set of institutional norms that force large numbers of people into behaving in ways that are contrary to their self-interest. Where the texts exist, we see that people generally understand the repressive relationships they find themselves in but are incapable of resisting, except in brief "revolutionary" moments where the social order is disrupted. States are characterized by forms of serfdom, slavery, ideological delusions about the proper "order" of the world and so forth. They have large population densities in a conflictive social environment that prohibit people from leaving a political system or group. They have repressive policing and military organizations; state societies are effectively structured by hierarchically organized classes that are bound together in largely one-way exploitative relationships. Populations are large, people do not generally control their own total subsistence or security, and mechanisms of force permeate society. People must involuntary give up their labor and/or goods or face serious, life-threatening sanction in the form of overt and subtle coercion.

In contrast, read the ethnographic examples in virtually any stateless society. Selfish leaders exist in stateless societies, but authority is simply not coercive if that authority can be withdrawn by the group, or if people can simply choose not be part of the faction. The ethnographies are permeated with instances where people decide not to follow a leader. In virtually no instances in stateless societies can we find examples of households being forced into sustained involuntary behavior against their self-interest by other people, as we see to the contrary in states. It is true, as Hayden and others point out, that social norms can be manipulated and people can find themselves indebted to others. But the key difference is that this debt obligation is part of a larger web of reciprocities mediated by noncoercive negotiations between factions and subgroups within the society. This simply is not the case in states.

This chapter leads us to question some basic tenets of canonical theory. The observation that coercion is largely absent in complex, stateless societies poses a serious and fascinating theoretical challenge. How does complex society emerge in such a noncoercive environment? We have established that people are not forced to accept social complexity prior to state emergence. What if it were possible to demonstrate that rank and hierarchy could emerge from egalitarian groups without coercion? That is, what if you could demonstrate that ego-directed individuals could develop large-scale and stable co-operation among unrelated people over time without overt coercion? What if, in fact, these hierarchies for the benefit of the group were headed by chiefs with little de jure power but who rather were accorded ad hoc authority at the behest of the community? What if the emergence of a managerial hierarchy, accepted by small groups as a means to organize production and exchange for their own individual and collective benefit, provided such a selective advantage that groups would embrace these new organizations? In other words, what if we could demonstrate either that people did not have to force others to accept hierarchy or that exogenous factors (population pressure, climate change, and so forth) were not theoretically necessary to force people to accept hierarchy? What if our own ethnocentric bias against rank, born of our culture's 200-year history of political philosophy and two millennia long steel cold reality of exploitation of the many by the few, is not a good basis for understanding people who are not like us and who live in social contexts that we can only read about and barely imagine from our own cultural experience?

Game theorists have very successfully provided models for the emergence of complexity among ego-directed actors in small groups in noncoercive contexts, precisely the kind of cultural context found in the earliest expressions of complex society in the archaeological record. Based upon a simple set of principles developed in experiments and computer modeling, we can show how such co-operation can develop among otherwise selfish individuals. This is the game theory revolution, which finally allows us to understand, with mathematical precision, how selfish actors can create ad hoc figures of authority in order for those same selfish actors to survive and thrive in a socially competitive environment. We are not faced with the stultifying Hobbesian or Rousseauian choice. It is this theoretical revolution that lets us understand how complex stateless societies function. These concepts in turn teach us how such societies emerged in the history of our species, beginning at the end of the Pleistocene and continuing in fits and starts over ten millennia.

5 The Ritualized Economy: How People in Stateless Societies Co-operate

Evolutionary game theorists have provided us with a rich corpus of empirical observations and theoretical concepts to explain how people in stateless societies survive and thrive in their environments. The previous chapters in this book have been theoretically consistent but by necessity they were very abstract. We now have to translate these abstractions into empirically verified behaviors from known stateless societies in the ethnographic, archaeological, and historical records. We have a wealth of information on many stateless societies from around the world, from nineteenth-century Oceania to eleventh-century Iceland and many lands in between. This information includes details on the social organization of many kinds of co-operative organizations found in these groups. We know how people co-operate to plant and harvest their food; how they fish, hunt, and manufacture objects for domestic use and trade; how they exchange their surplus; and how they share this extra wealth with others in their group; how some use this wealth to indebt others; and how they interact with people far from their villages. While there is a huge diversity in cultural practices in these behaviors, people around the world have developed a limited number of common co-operative strategies.

This chapter distills this empirical knowledge and then reformulates these concepts into anthropological game theory terms. We will look at the nature of production and exchange in stateless societies, most of which require various levels of co-operation. We note that reciprocity is the fundamental socioeconomic mechanism of co-operation in virtually all societies, and numerous dyadic and multiple reciprocal relationships in complex stateless societies form the backbone of sustained co-operation among people. There is abundant empirical confirmation that the lowest and most basic unit of production and reciprocal exchange is the household, composed of related individuals and occasionally incorporating nonrelated ones, sometimes called fictive kin. Societies become more "complex," from this perspective, as the number and intensity of social relationships dedicated to surviving and thriving increase and form more complex levels of co-operative behavior between larger and larger numbers of actors.

The Regulation of Specialized Labor through Ritual and Taboo

And thus religion and practical aim are fused together. – Raymond Firth, *The Work of the Yam*

Success in most economic pursuits depends in the eyes of the natives upon the performance of effective magic. – Malinowski, *Fishing in the Trobriand Islands*

In reviewing the vast ethnographic and historical record of stateless societies, I conclude that intense ritualizing behavior constitutes the mechanism by which rules and norms successfully structure co-operative economic relationships. I agree wholeheartedly with Katherine Spielmann (2002:203), who notes that "in small-scale societies ritual and belief define the rules, practices, and rationale for much of the production, allocation, and consumption in an individual's life." This is an empirically derived conclusion from ethnography, history, and archaeology that is consistent with the broad game-theoretic literature outlined in this book. Ritual is not necessarily religious. Rather, "ritualized" refers to the degree to which abstract norms of proper behavior are encoded and signaled in societies. Echoing Malinowski's concept of customary behavior, people "instinctively" know how to behave, and they are rewarded and punished according to how well they follow these norms.

To put it another way, economic co-operation in stateless societies depends upon the creation of *ritualized norms of behavior, sanctioned by taboo* and sustained by the intensified production of resources not feasible in household-level production. These norms must reward, punish, create social histories, and follow the behaviors defined by theoretical work in anthropological game theory. I can find no other social mechanism that can maintain the norms of fairness required for sustained co-operation among nonrelated individuals and groups where institutionalized coercion is absent.

High levels of co-operation are therefore sustained by ritual and taboo. Expensive dance and performance rituals, for instance, by people in stateless societies used to strike outsiders as odd in the nineteenth and early twentieth centuries, but they can easily be understood using costly signaling theory as a means of reinforcing group cohesion and co-operation. Ritualizing the work calendar schedules the way in which specialized laborers from autonomous households fairly coordinate their tasks; rituals schedule barter markets and periodic fairs as well as lavish feasts and Kula-like exchanges; taboos and other forms of intangible punishments deal effectively with the public goods and free-rider problems that inevitably arise. People have experimented with and created a vast array of social relationships – what we can call social technologies – to take advantage of the benefits and to mitigate the costs of sustained co-operative behavior in societies without the coercive power of states.

As we saw above, the development of the rational actor theory of classical economics created the intellectual climate in which the concept of the "primitive economy" was able to develop. A cornerstone of that concept was the fact that religion and belief permeated the many non-Western cultures that colonialists and visitors experienced in the nineteenth and twentieth centuries. A century of serious anthropological research reveals that so-called primitives were hardly the irrational Noble Savage of the eighteenth-and nineteenth-century imagination. Rather, these societies were the product of many generations of selection that produced eminently successful and sophisticated ways of surviving and thriving. The core of these adaptations was the capacity to produce and exchange resources to provide for biological and social needs and wants without the use of money, markets, or forms of coercion.

Once we go beyond rational actor models to embrace new theoretical insights from anthropological game theory, we can much more fully understand these societies and the underlying logic that guided their behavior. Anthropological game theory helps us understand how the seemingly irrational behavior of stateless societies is actually eminently adaptive, successful, and quite rational. This theory shows us that ego-directed individuals will indeed co-operate under the "appropriate circumstances" in the absence of coercion. Throughout the millennia, people created and tested different ways of organizing themselves to reproduce their culture and to produce material resources.

What we will see in this review is that stateless societies worked quite well to structure co-operation between individuals in the absence of coercion. And they worked very well to provision their people and to maintain high levels of co-operation between kin and nonkin alike. Ritualizing economic behavior, backed by taboo, fits extremely well the theoretical need to reward co-operators and punish defectors in game theory. Unfortunately, only a small number of ethnographers have studied societies largely unaffected by the spread of Western society; once market mechanisms kick into an economy, much of the organizing function of ritual becomes replaced with currencies, price-making market mechanisms, and so forth. However, a common theme runs through the observations of virtually all primary ethnographies: ritual or "religion" was a core element of daily life (Figure 5.1).

Refining Some Concepts: Ritual, Magic, and Religion

That religious beliefs and ritual practices can structure our daily life is hardly a new idea in social science theory. Max Weber famously proposed his theory linking shifts in religious belief and the emergence of modern capitalism in the West in his book *The Protestant Ethic and the Spirit of Capitalism*. The basic premise of this iconic book is that the traditional view of work prior to modern capitalism was that it was a drudgery to be avoided, merely a means to

Figure 5.1. Locations of case studies and other places mentioned in text.

other ends. Those other ends included the pursuit of leisure and social activities. People "traditionally" worked only as much as necessary to maintain life. As Stephen Kalberg (2002:lxxvii) writes: "Work is viewed as a necessary evil and only one arena of life, no more important than the arenas of leisure, family, and friends. 'Traditional needs' are implied; when fulfilled, then work ceases."

The revolution that took place in some Protestant circles was to redefine work as a virtuous and godly activity in and of itself. Adherents to ascetic Protestantism developed a strict set of religious norms that became a "methodical-rational organization of life" that blended psychological rewards with a strong work ethic focused on material success (Kalberg 2002:xxviii). These norms were ingeniously guided by new religious principles developed by radical theologians that tied concepts such as predestination and divine will to social conformity and economic success in life.[1]

Weber was of course not a functionalist in the same theoretical camp as Malinowski and others who followed, but his theories clearly show how the ideological and the material can affect each other.[2] Of particular relevance here, his work demonstrates how a shift in religious beliefs and practices can alter economic behavior in profound ways. The core of Weber's work, from our perspective, is that ideology is causative and not epiphenomenal, at least vis-à-vis economic behavior.[3]

Weber's ideas profoundly affected anthropological research and theory on non-Western societies. Scholars in this broadly conceived Weberian tradition focus on the gifts, feasts, organizations, and reciprocities that "ritually" bind people together in local and regional networks of mutual aid in complex societies. Mayfair Yang (1989, 1994) stands out here, describing elaborate ritual exchange (what she refers to as gift exchange) in the Chinese countryside that helps people with the practicalities of paying for weddings, funerals, and major household activities. This is of course an example of ritual exchange within a strong state society, replete with multiple instances of power and coercion that permeates social life. In that regard, Yang also sees these relationships as having a political function. They are a means to insulate and protect people from the many instances of power that emanate from central state authorities. In spite of this larger political context, we see how people in small subgroups within a large society use ritualized exchange to deal with risk, costly events that require accumulation and ritualized means of debt discharge that benefit both the individuals and the group as a whole.

Another perspective on the role of ritual and religion in the economy is more utilitarian and functionalist, and sees such ideologies as a means to organize production and facilitate exchange. Again, religion is not epiphenomenal, to use the old pejorative term, but it serves clear and obvious material ends. However, religion in and of itself in this tradition is not a central motivator either. Religion and economic behavior are correlative (empirically) and codeterminative (theoretically) and must be understood as such. I follow the lead of Firth (1970:19), who saw religion as "a field of intricate strategies operated by people . . . for ends which range from very mundane and ego-centred to other worldly and vicarious."

The underlying theoretical framework that structures the argument in this book is therefore located firmly within the functionalist tradition, but with the caveat that culture and history are also determinative factors in the precise configurations that people create to organize their lives. From an epistemological perspective, we can derive commonalities from the study of all economies around the world and over time using a comparative approach. However, the scale of the analysis will determine the degree to which one can make empirically verifiable universalizing statements. At the broadest scale of comparative analysis, there simply are a limited number of viable ways in which people can successfully organize their economic lives. Within those constraints, much variation will be determined by historical trends and cultural choices, and therefore at the smallest scale of analysis, say the community, we see substantial variation within those constraints.

The edited volume by E. Christian Wells and Karla L. Davis-Salazar titled *Mesoamerican Ritual Economy* provides a Weberian discussion of the role of ritual and its effect on the economy and therefore deserves some mention. The concept of the "ritual economy" in this work can be viewed as a case of ideological causation in economic practices. They define the ritual economy as "the materialization of values and beliefs through acquisition and consumption for managing meaning and shaping interpretation" (Wells and Davis-Salazar 2007:3). In this view, belief and people's spiritual views stimulate economic behavior. This concept falls within the broad utopic and substantivist economic anthropological assumptions about non-Western peoples – that is, non-Western peoples are motivated by nonmaterial goals and cannot be understood by Western economics. Many of the applications in *Mesoamerican Ritual Economy* refer to state societies in which religious ceremony indeed is a motivation for economic production and exchange and are therefore not relevant to the complex, stateless societies that are focus of this work.

I reject the concept of the "ritual economy" in the sense used by Wells and Davis-Salazar for stateless societies insofar as it implies that people, in the aggregate and over any serious length of time, can ignore the material and social constraints of making a living in a competitive environment. This is not what Wells and Davis-Salazar say, but it is implicit in other "false consciousness" theory. Economies based on nonmaterial, ideological principles (that is, utopian principles) have a very short half-life in any competitive environment. This is not to say that people are not *capable* of creating such organizations, and in a few cases in history people have indeed tried. However, I will demonstrate that such experiments in ideologically driven or ritual economies cannot survive in the competitive environment that has characterized the human condition from at the very least the beginning of the Holocene, and most likely throughout our existence as fully modern humans.

A fascinating article by Katherine Spielmann (2002, and see Spielmann 2008) argues for a "ritual mode of production." The fundamental point of the argument is that people in stateless societies, what she calls small-scale societies, effectively produce for feasts and other communal events. This is similar in theoretical content to the ritual economy of Wells and Davis-Salazar for Mesoamerica. Here there is a subtle but very important distinction between production for subsistence and production for social reproduction: "I suggest that a primary motivation for subsistence intensification and craft specialization in small-scale societies lies not in subsistence provisioning but in the demand for items critical for social re-production" (Spielmann 2002:195).

The argument that Spielmann makes is quite viable in that it can be reformulated in economic terms as: People produce resources for a variety of reasons, including ritual ones. Feasts and intense social interaction have utility, as the economist would say. I part slightly with Spielmann's arguments on the role of ritual in production. As with the concept of the "ritual economy," there is a sense that people produce and consume for ritual purposes as a means in and of itself. I would fall in with most evolutionary theorists and economists, who properly see this ritual practice as part of a larger system of group coordination, production, exchange, and competition.

We therefore have to differentiate between ritual, magic, and religion in this perspective. I use definitions that draw from traditions established by Malinowski, Emile Durkheim, Raymond Firth, Max Weber, Roy Rappaport, Catherine Bell, and others and that are consistent with the theoretical framework of anthropological game theory. There is a vast literature on these concepts in a variety of social scientific disciplines. Some scholars focus on symbols; some on behavior; others on performance, beliefs, norms, and so forth. Again, I will focus on only that which is relevant to economic co-operation as defined in this book. Naturally, these are not rigid categories, and there is always some overlap in how these are manifest in any particular culture in space and time.

Ritual

We can now turn to some definitions. At the outset, it is important to note that "ritual is not synonymous with religion." As Yorke Rowan (2011:2) succinctly puts it distilling a consensus among several scholars of religion: "All ritual is not sacred, and ritual does not represent the totality of religious belief... ritual is not solely the performance of religion."

What is ritual? I define ritual as any kind of *patterned and predictable behavior* that expresses and instantiates a larger cultural norm. Stanley Tambiah (1985:185) defines ritual as "constituted of patterned and ordered sequences of words and acts, often expressed in multiple media." These definitions are consistent with the OED entry of "an action or series of actions regularly or habitually repeated" (OED, def. 2a). Ritual may be sacred – behavior in a temple to appeal to or appease the divine – or profane – behavior at a scheduled dinner, battle, agricultural activity, or other quotidian pursuit. Lars Fogelin (2007:58) notes, "Sports, theater, and other activities – what Bell (1997:Chapter 5) labels ritual-like activities – are difficult to distinguish from religious ritual." This quote highlights the important fact that much ritual is not religious.

The key to the concept of "ritual" as used in this work is that it refers to habitual behaviors. Rituals are patterned behaviors that people instinctively understand and can follow as is necessary. Rituals vary from culture to culture and change in the same group over time. Catherine Bell (1997:138) nicely describes a number of ritual properties. These are formalism, traditionalism, disciplined invariance, rule-governance, sacred symbolism, and performance. These properties can be seen in a wide range of ritual behavior, from small stateless groups to large, transnational states.[4] She furthermore proposes a typology of rituals that I find useful because it outlines a range of behaviors that we can properly see as true "ritual": (1) rites of passage; (2) calendrical rites; (3) rites of exchange and communion; (4) feasting, fasting, and festivals; (5) rites of affliction; and (6) political rites. Of these six types, the secular ones are most relevant to this discussion, although rituals that are performed within a magical or religious sense as described below could also have facilitated group co-operation.

It is critical to note that virtually any commonly occurring social behavior can be ritualized, though of course not all of our social experience is ritualized. In effect, humans are quite creative and if necessary will ritualize virtually any kind of social interaction through the construction of norms and rules. As my colleague Paula Sabloff has suggested, perhaps a more accurate term would be "habitualized behavior" instead of ritual. Religious concepts can be a tool to habitualize or ritualize people's behavior, but religion itself is not necessary. Quotidian, nonreligious ritual likewise structures much of our lives.

One example of a sacred set of rituals from our own society is the Christian Mass and its many manifestations over the centuries and around the world (and see Stanish 2012). It was a practice that developed over many generations in the first few centuries CE, becoming an ideal form coincident with the consolidation of political power by the Catholic Church. The nature and structure of this distinctive set of rituals took some time to develop, but once established they became an ideal that was manifest around the globe. The Mass itself encodes a series of metaphysical religious beliefs, materialized in rituals that are replicated in different social contexts. Masses are performed in large cathedrals, modest churches, small chapels, and grottoes; on ships, in forts, and in just about any conceivable space where the faithful can congregate. Mass is performed in scores of languages, by cultures in varied dress and with varied customs. Yet to any but the most culturally uninitiated, there is no doubt when one is witnessing this set of Christian rituals. The key feature is that in spite of its variation, it is immediately recognizable to all participants. This ritual is modified by culture, history, and time. But its core features remain, and its core message is transmittable via visual performance among virtually any audience.

An example of a secular ritual in our own culture would be food presentation and consumption in nondomestic contexts, such as dining in a fine restaurant. This seemingly prosaic activity is actually replete with encoded norms and social expectations. There is a prescribed sequence of food categories, preceded and followed by rituals involving drink and in many countries up to today a round of tobacco. Plates and utensils have their prescribed positions on the table. Different wines have their own glass shapes; desserts and salads have their own implements and so forth. Food rituals differ widely across societies. However, as a structural principle, the ritualized festive meal is found in every society and is in fact one of the most important quotidian social behaviors encoded in norms by virtually all cultures.

The point of these two modern examples is to emphasize again that one can ritualize almost any social activity. In the absence of coercion, rules emerge from social practice, and these become encoded in the particular group. These rules regulate our behavior: whether in a fine restaurant or in a humble Christian Mass, any member of the society that has enculturated these rules knows how to behave. These practices change through time. Sometimes the change is directed from above in hierarchical societies, sometimes not. In small-scale or intermediate societies, it is almost never by fiat of a higher authority; rather the change occurs as a kind of consensus shift among the individual actors. These shifts are historically traceable and tend to be quite conservative in the absence of major cultural disruptions.

Religion

Religion refers to a systematic *body of metaphysical beliefs* shared by a group of people. In the words of Lars Fogelin (2007:57), distilling a Geertzian view, it consists of the "belief and the meaning of symbols – the manner in which belief serves to instill in people a sense of where they belong in the universe." In the West, religion is mainly focused on "the revealed knowledge of the divine" and universal truths, to use Firth's apt words (1970:33). In this Western sense of revealed truth, religion is not a prominent feature of stateless societies. Rather, what we understand to be "tribal religions" are precisely that – religions for and about a particular people. These are the phenomena that most traditional anthropologists used to study. In this sense, the term "religion" is inadequate to capture the essence of particular sets of ritual practices and beliefs of the vast array of cultures on the planet.

Religion as understood in this sense is some kind of universalizing set of beliefs that transcend individual cultures. These kinds of religions developed in tandem with transnational states – the type that developed in the Iron Age Mediterranean and Asia and in the later Post-Classic and Late Horizons in the Americas. These universalizing religions, at a superficial level, were useful ideological tools adopted by states and empires that had to incorporate a vast array of different peoples with different languages and norms. Universal religions fit the needs of an expanding state entity, creating an overarching belief system under which all tribal religions could be safely subsumed.

The definition of concepts like religion in social science is usually affected by the specific discipline in which it is used. Sociologists and economists tend to have short historical frames of reference. Historians have longer frames, and archaeologists have the longest time frames of all. In this light, the concepts proposed by the eminent archaeologist Colin Renfrew (1983, 1994) are applicable here. His formulation explicitly avoids an ethnocentric reliance on the great state religions such as the Abrahamic ones as well as Buddhism, Hinduism, and so forth. Professional priestly classes maintain these religions. They seek logical consistency and a very broad appeal across many peoples. In contrast, as Renfrew says, the work of cultural anthropologists has much to teach us about smaller, tribal societies who have religions that are not "codified (and thus literate), authoritative systems of belief, operating often in an urban context" (Renfrew 1994:47).

In this light, Renfrew notes that by definition all religions provide "answers to profound existential questions." He argues, rightfully in my opinion, that religion offers a guide to influencing the world "favourable to our own aspirations" (Renfrew 1994:49). Religion is also a shared belief system that requires, by definition, a community of adherents. He notes that the earlier functionalist views that saw religion as merely functioning to maintain the social system or make it more efficient is correct but inadequate. Rather, he argues that religion cannot be seen merely as epiphenomenal or secondary, as in neo-Marxist formulations, nor purely functional in the theory of classical structural-functionalism. Rather as Renfrew says, religion has an important role in the explanation of cultural change. This latter point is particularly important for the discussion here insofar as totalizing belief systems can be a powerful way to promote co-operation on a massive scale (see Henrich et al. 2010c; Norenzayan and Shariff 2008).

Magic

In the tradition of Malinowski and Benedict (see H. Geertz 1975:78; Hammond 1970), magic is defined as a *body of practices* that use and manipulate supernatural power against individuals or groups and/or that are used to protect oneself against other social and natural forces. Whereas ritual is patterned, magic can be understood as a tool kit of behaviors available by practitioners to achieve social goals. From an anthropological game theory perspective, the imposition of taboo backed by magic on an economic activity, for instance, can substantially enhance the value of a public good and prevent free-rider threats. There are many examples in the ethnographic and historical literature where the imposition of a taboo serves to keep a resource available for the group. Firth recounts how in Tikopia, chiefs impose taboo on coconuts to allow them to grow to maturity. He relates a story of how a group of men agreed to extend the taboo to areca nuts as well. "The result was a considerable increase of the wealth of the community in these important fruits" (Firth 1965:208). As we will see below, this is a constant theme in discussions of chiefs in stateless societies.

Ethnographic and Historical Examples: How Co-operative Social Structures Work on the Ground

For Westerners, the fact that ritual was central to economic behaviors was particularly striking because in our world the "sacred" and the "profane" were considered to be separate fields of belief and behavior. John Lubbock's glaringly ethnocentric view of non-Western religions, as noted above, held that they were profoundly different than Western religion. And why was it different? Because, according to his views, non-Western people used religion for practical purposes and not for contemplating the metaphysical: "[lower savage religion] is not only different, but even opposite [from Western religion]. Thus, it is an affair of this world, not of the next" (Lubbock 1882:202).

As soon as sympathetic ethnographers began recording non-Western peoples' behavior, they noticed how significant ritual was to the economy in stateless societies. For instance, the great Raymond Firth states unequivocally that ritual was central to the Maori way of life:

> Magic, in fact, in one shape or another, permeates all the economic life of the native. Every craft has its spells and incantations, its rites and omens, its regulations of supernatural import. Magical ideas seem to be ubiquitous in their pervasiveness; at every turn the craftsman is hemmed in by restrictions and prohibitions of quite a different order to the technical rules with their obvious practical utility. (Firth 1929:234)

Malinowski, of course, saw ritual and religion as a primary means of social integration. He saw this as a very positive thing, even superior perhaps to religion as a metaphysical distraction as in the West (to Lubbock's posthumous horror, one can well imagine). He writes about the Trobriand Islanders: "Thus gardening, the most important of these pursuits in Kiriwina, is closely bound up and regulated by elaborate systems of magic" (Malinowski 1918:90). In *Argonauts* he notes that the chief and the garden magician, through "rites and spells . . . initiate each stage of work and each new stage of development of plant life" (Malinowski 1966[1922]:59 and see Malinowski 1921:5). In this and other similar observations, he shows how ritual regulates work schedules.

In this light, Malinowski makes the distinction between everyday magic and what he calls systematic magic. He describes the latter:

> Such magic consists of a connected and consecutive body of incantations and concomitant rites, no one of which can be torn out of its sequence and performed by itself. *They have to be carried out one after the other in a determined order*, and the more important of them, at least, can never be omitted, once the series has been started. *Such a series is always closely connected with some activity*, such as the building of a canoe or an overseas Kula voyage, a fishing expedition or the making and harvesting of a garden. (Malinowski 1966[1922]:413, emphasis added)

He goes on to note: "The association between the practical activity and its magical concomitant is very intimate. The stages and acts of the first, and the rites and spells of the latter, correspond to each other one by one. Certain rites have to be done in order to inaugurate certain activities; others have to be performed at the end of the practical work; others again are part and parcel of the activity" (Malinowski 1966[1922]:413). In this observation we see the illustration of this phenomenon – the use of ritual to schedule labor – in its clearest sense. Every act in a complex labor process has an associated rite embedded in a larger narrative that is rigidly adhered to by the community. As Malinowski says on the same page: "The consecutive progress of work and of magic are inseparable . . . work needs magic, and magic has only meaning as an indispensable ingredient of work."

He ends this extremely important section of his book with the following brilliant summary of his theory of the relationship between ritual and the economy:

> Thus we see that systematic magic consists in a body of rites and spells associated with one enterprise, directed towards one aim, and progressing in a consecutive series of performances which have to be carried out in their proper place. The point – the proper understanding of what is meant by systematic magic – is of the greatest theoretical importance because it reveals the nature of *the relation between magical and practical activities, and shows how deeply the two are connected with one another.* (1966[1922]:414; emphasis added)

These observations are hardly confined to Oceania. In premodern society, as in our own, ritual structured much of human behavior. For instance, Neil Price (2002) discusses the social role that magic played and the high degree to which war and aggression in general were ritualized in medieval Scandinavian society. Irving Goldman (1937a:169) writes about the Ifugao of the Philippines: "Religion pervades all activities of any importance," and about a South African culture: "Behind the orderly pattern of Bathonga life runs the thread of religion" (Goldman 1937b:373). Discussing the function of physical structures in stateless societies from an archaeological perspective, Michael Adler and Richard

Wilshusen (1990:133) note, "By codifying ways of doing things, ritual can more clearly delineate what information and activity the public can expect at the facility."

Sabine Luning (2005:444) writes about a stateless society in Burkina Faso: "In Maane, daily activities and the economic usage of particular places are indistinguishable from ritual practices." As with so many other examples from similar societies around the world, there is a clear ritual cycle that guides agricultural production. It is highly ritualized, and this ritual is organized by leaders referred to as chiefs:

> At the centre of the "first fruit" ceremonies are the chief of Maane, his predecessors and his palace. The annual rituals of the chief coincide with specified stages in the growth of sorghum. Starting in September, when the sorghum "becomes pregnant," the chief performs the first of three rituals, scheduled for three consecutive appearances of the new moon. These rituals are considered to help the sorghum in its "birth." (Luning 2005:457)

Darrell Posey tells us that the Amazonian people known as the Kayapó "have an elaborate and complex ceremonial repertoire. Ceremony is a raison d'être for the Kayapó, for they believe that without the performance of the prescribed rituals, the world would collapse: crops would not grow, children would not be born, the sun and moon would cease to travel across the sky" (Posey 1982:93). While the above observation strikes one as a bit on the romantic side playing into nineteenth-century stereotypes, it does emphasize that ritual is extremely important to this stateless society.

The list of potential citations in the ethnography and history of stateless societies is virtually endless. So there is always a danger that we fall into the same trap as the nineteenth-century travelers and early-twentieth-century anthropologists who saw religion everywhere in non-Western societies. This was part of the creation of the "primitive," as I mentioned above, a subtly racist attempt to scale peoples into different levels of cultural sophistication.

What is absolutely necessary to understand is that these kinds of rituals were useful for people in stateless societies because this was a major component of how their world was governed. Over and over we read in the histories and ethnographies of the practical purpose of religion or religious rites. The ethnographer John Provinse worked in Borneo in 1928. In 1937 he published an article on the economic life of the people. He writes:

> Numerous as these taboos and restrictions may be, powerful as they may appear on paper, the facts of their existence seem to have been overworked and their effects on the economic life to have been overstated. . . . Anyone who has watched a loin-clothed Dyak, armed with a small axe, attack a tree six feet in diameter and after an hour of perspiring labor, see him scurry for safety out of the way of the falling trunk, cannot come away from the scene with the feeling that here is a man whose every movement is regulated by some magical influence. (Provinse 1937:94)

We see therefore, that ritual, magic, and taboo is an effective way to organize complex labor tasks in noncoercive contexts. Such practices promote co-operation. This therefore begs the question: why co-operate?

Reward: The Payoff for Co-operation

In this section I explain how and why groups of people in stateless societies are willing to co-operate in the absence of coercion. This section draws off a large empirical database to illustrate how people in construct co-operative social organizations and how they maintain them through rewards and punishments. Put differently: Where does the payoff for co-operating come from in stateless societies? In economic terms: How do people receive more per capita resources via group participation in productive activities without a corresponding increase in costs? Or, more precisely: How can individuals increase their resources disproportionately to the increase in labor inputs or other costs associated with acquiring those resources? What kinds of social strategies have non-Western peoples created to maintain co-operation? Finally, what other kinds of rewards and punishments do stateless societies employ to maintain group co-operation?

Some Basics: Production and Exchange

The concepts of economic "production" and "exchange" derive out of nineteenth-century political economic theory. They are useful concepts for state economies and particularly for Western, market-driven ones. They remain central to economic anthropology. I suggest that for stateless societies we conceive of production as the "creation and acquisition of resources with utility," both material and intangible (though we retain the name "production" for simplicity).

The exchange of useful resources between autonomous social groups in a stateless society is conducted through various forms of barter – exchanging like for like equivalencies without money. The forms of barter exchange in stateless societies include periodic barter markets or fairs, potlatching, competitive feasts, various forms of Kula exchange, and raiding. Barter is usually conducted between different social groups, while internal exchange within groups is mediated through various mechanisms of deferred reciprocity. However, in complex systems like the Kula ring, trade partners or "friends" engage in vigorous barter called "gimwali."

Raiding is very prominent in the ethnographic literature and constitutes a mechanism of both "production", via acquisition, and exchange. These forms are not exclusive – trading and raiding go hand in hand with feasts and Kula-like trade (and see Junker 1999; Spencer and Redmond 1992). People have created a rich array of social strategies that combine raiding and exchange over time and across the cultural landscape.[5]

For these mechanisms to be successful in all of these cases, it is imperative to elicit co-operation between related and unrelated individuals across space and through time. We will see that there are huge payoffs from economic co-operation in stateless societies, but there are also costs, both economic and social. The rewards must be fairly distributed and the costs must be fairly mitigated or the co-operative organization will revert to the minimum level of co-operation at the household level.

Production is of course only half the equation in any economic analysis. The distribution and consumption of resources is the other half. Exchanging resources obtained by endogenous labor or from raiding for objects of greater value likewise increases wealth. The mechanism of reciprocity is the primary means by which goods and services are exchanged between households. It is the exchange of an equal amount of wealth or "like value for like value." It involves a series of symmetrical exchanges between individuals or groups. It may be socially mediated – a sack of potatoes to my brother-in-law for a pound of meat – or it may be an exchange between nonrelated partners, such as in the famous case of the Kula as first described by Malinowski (1921).

Reciprocity is the normative glue that keeps humans co-operating for mutual benefit. Reciprocity may be immediate, such as a direct exchange in a periodic fair or in commensal feasts (Dietler 1990; Hayden 1996), or it can be deferred for days, weeks, years, or even decades. Deferred reciprocity, in fact, is a fundamental feature of the Kula (Leach 1983:3) and is the most prominent of many "tribal," nonranked, or moderately ranked political economies in the ethnographic literature. Reciprocity can be indirect "down-the-line" trade, where individuals conduct a series of reciprocal trades that link large distances though "friends" or exchange relationships established through a variety of sociological mechanisms.

As Mary Douglas notes, "Right across the globe and as far back as we can go in the history of human civilization, the major transfer of goods has been by cycles of obligatory returns of gifts" (Douglas 1954:viii). She goes on to say that the system is quite simple in the sense that every gift has to be returned in a manner that "sets up a perpetual cycle of exchanges within and between generations" (Douglas 1954:viii). The mechanism of reciprocity, either deferred or immediate, makes up the bulk of exchange in stateless societies.[6]

Reciprocity in particular has been extensively theorized in economic anthropology. Simple reciprocity is effectively the exclusive mode of exchange in mobile hunter-gatherer-forager societies. In complex hunter-gatherer and in settled village societies, additional resources are created by economies of scale and the corresponding more efficient creation of resources through co-operation between households and unrelated individuals. Reciprocity is the epitome of fairness. In anthropological game theory terms, it is a social norm that ensures group co-operation. The norms of reciprocity demand fairness. Household exchange embodies the principle of reciprocity in its most salient way. We can cite many examples in the ethnographic and historical record of how this fundamental and powerful norm works. What is clear, however, is that fairness is at the center of the principle, and if people do not behave properly, the dyadic relationship between individuals or households ends. For instance: "The Siangs . . . are notorious for their hospitality and visitors from other villages are readily accepted and provided for, sometimes for considerable periods of time. But if a man after considerable time makes no attempt or offer to reciprocate in some way for the hospitality afforded him, he is asked to move on" (Provinse 1937:91).

Reciprocity between different groups that do not share in the production of resources effectively is a kind of barter. These reciprocal exchanges must be perceived as fair, and bartering requires norms of behavior, including cessation of hostilities, to work efficiently. Barter is an exchange of perceived equal value. As we will see, the modern concept "marginal utility" allows us to understand even the logic of the simplest kinds of stateless barter.

Redistribution is a form of deferred reciprocity. The redistribution of the products of specialized labor organization turns out to be a fundamental problem in maintaining co-operation. This is due to the simple fact that ownership of the resources is self-evident in household-level production but not in specialized production. When one works in a specialized group, the question of "who gets what when" is at the center of free-rider and fairness issues. I will likewise review redistribution and reciprocity from an anthropological game theory perspective.

Economies of Scale: More Resources for Less Work

To anticipate the major point of this chapter, the biggest payoff for working in a co-operative group is that people will acquire more resources for equal or even less labor per capita. However, such an arrangement creates one large cost: who owns the resources obtained and/or created the group? This is why it is essential to ritualize this co-operative organization, providing norms and rules that incorporate all that we have learned from anthropological game theory. In cases of lengthy co-operative ventures, ritual schedules the key periods when rewards accrue to the individuals. Ritual schedules likewise must conform to norms of fairness in the distribution of the product.

It is always assumed in classic evolutionary and political economic models in archaeology and anthropology that to increase surplus production and thereby create the economic conditions for rank and hierarchy to develop, there must be an increase in the total amount of per capita labor from the vast majority of the population, excepting the elite. In this view, hierarchy is costly to the bulk of the population, and even for the elite there are costs to maintaining the social and political structure conducive to such exploitative arrangements. As briefly mentioned above, Sahlins argued that there are only two ways to create surplus. One is getting people to work more or get more people to work. This position is similar to that of Ofer Bar-Yosef and Anna Belfer-Cohen for population pressure in the shift to agricultural lifeways. Unlike the canonical position articulated by Sahlins, however, this is a form of nonsocial coercion forcing changes in the economy: "Hunter-gatherers maintain their population size below the level of the mean carrying capacity of the region.... If they fail to do so, they must move to a neighbouring territory, change their subsistence base, or die" (Bar-Yosef and Belfer-Cohen 1991:182). In either case, the choice here is a harsh one. People facing stress must either subordinate themselves to a coercive elite, alter their subsistence practices to something less desirable, or die off. This is the classic view of cultural evolution, powered by some sort of stress, either cultural or environmental.

There is an additional way to have an increase in economic surplus without an increase in per capita labor input or by having more people work. It is a rather basic concept in economics, having been developed as a theoretical construct more than 200 years ago. In economic terms, this third way is known as an *economy of scale*. In standard economics, an economy of scale is one in which the unit cost (generally labor, materials, and transport) to produce a good goes down as the quantity of units produced goes up. In economies of scale, total output of any good can increase without a concomitant increase in total labor time. This phenomenon works by reducing redundant tasks that increases labor efficiency.

To most anthropologists, such economic language is dismissed outright because it is associated with analyses of capitalism and modern business practices. There is an implicit assumption that such concepts are of little value in understanding pre- and noncapitalist economies. In many cases, this is true, particularly when we see labor being substituted with other forms of energy. It is true that many of the concepts developed to explain capitalism are of little use in explaining non-Western economies. Examples here include "interest," "income distributions," "price analysis," "inflation," and a whole host of other microeconomic concepts.

However, some of the basic concepts in economics are not restricted to capitalism and in fact can be useful in the analysis of small-scale societies. Marxist analysts, for instance, routinely used the concept of "rent" to explain feudal economies (e.g., Hindess and Hirst 1975). As we will see in the next chapter, theoretical tools such as "utility," "marginal utility," and "comparative advantage" are useful in explaining the nonmarket-based barter exchange typical of stateless societies.

The rudimentary economy of scale fits in well with a small but growing theoretical and empirical literature that demonstrates that overcoming collective actions problems involves a reduction of transaction costs (von Rueden et al. 2014). Michael Alvard and David Nolin (2002:533) for instance, provide an excellent empirical case study in which co-operative labor achieves greater per capita returns for all involved: "Return rates from cooperative whale hunting are greater per capita than those from solitary fishing." They go on to support a key theoretical principle in this book: "Coordination is maintained through complex norms that reduce transaction costs and provide assurances of satisfactory payoffs to participants" (Alvard and Nolin 2002:533). In a preindustrial, premodern society, an increase in efficiency (cutting transactions costs) resulting in more resources for individuals without a corresponding increase in labor is the best way to achieve payoff for all participants in a co-operative effort.

Heightened production in noncoercive, small-group contexts therefore occurs through the creation of rudimentary economies of scale based upon task specialization by individual households in the group. As we will see, economies of scale linking multiple households in productive co-operating groups are very efficient, providing more resources per capita than is possible working in households alone. At the most fundamental economic level, the creation of efficient economies of scale is the material basis of co-operation in stateless societies. The surplus that can be created from this more complex labor organization is the reward to each individual for co-operation. Without this surplus, there would be no way to create and maintain co-operative groups in the absence of force. More resources for the group means that exchange can intensify, thereby bringing in more resources for the individuals in the group.

The Two Types of Economies of Scale in Stateless Societies

Economy of Scale Type 1

The first kind of economy of scale in a small group is a linear phenomenon in which the return of resources is proportional to the size of the group and the degree of specialization.[7] The more that people successfully work together, the more resources that each one receives in the payoff. This linear economy of scale occurs when labor organizations create economic efficiencies from specialized production mainly by eliminating redundant tasks and other transaction costs in the broad sense defined by Douglass C. North.[8] These transaction costs are those associated with "measuring and enforcing agreements" and can be easily understood in game theory terms (North 1990:362). Obviously, a group that has strong norms of co-operation and fairness substantially reduces the costs of enforcing agreements among members.

A Type 1 economy of scale therefore occurs when production costs go down proportionally as the scale (number of specialized laborers) goes up. Each new addition of a specialized laborer results in a proportionally higher output. That is, these kinds of co-operative labor organizations produce more resources for the same amount of labor than groups in which the household is the highest level of co-operative labor organization.

In small societies, we see this phenomenon in simple economic task co-operation: Individual laborers carry out specialized, organized tasks to achieve an economic output instead of creating the product individually or at the household level. As early as the late eighteenth century, political economist Adam Smith (1937 [1776]) outlined the basic phenomenon of economies of scale in simple systems. He provided the now-classic argument that economic specialization (division of labor) by workers properly organized will produce far more in the same amount of time than individual laborers could produce on their own. Smith used the example of the production of metal pins, one of the simplest commodities he could find to illustrate the phenomenon in the late-eighteenth-century industrialized world. Making pins, in fact, is less complicated (in the sense of requiring minimal technical knowledge) than making many objects in preindustrial societies, such as the fine ware pottery, fancy canoes, and metal objects found throughout the ancient world. Pin manufacture in eighteenth-century England is therefore a good analogy for many kinds of labor organization in premodern economies, such as those discussed here.

Smith identified about 18 distinct tasks associated with pin production. In one factory 10 workers produced 48,000 pins in a day. Individually, these workers could have produced only a fraction of this amount in the same time. Increasing the number of individual, nonspecialized workers (those who produce the entire pin themselves) will increase production arithmetically. Increasing the number of specialized workers (those with just one task, no matter how technically simple) will increase production at a much greater rate. That is, this slightly more complex organization will result in greater productivity at the same level of labor input and without a concomitant change in technology. This phenomenon in which a specialized work organization produces more than the sum of the individuals working alone is what Smith called the productive powers of labor and represents gains through efficiency of task specialization.[9]

In modern economics, such an efficiency through specialization is defined as follows: the cost of one produced unit decreases as the capacity to produce the unit increases. In premodern economies, the same general phenomenon is evident, albeit in a more rudimentary form. Economic efficiencies can be achieved when individuals task-specialize and take advantage of situations where a marginal increase in labor cost produces a disproportionately large increase in output.

I already offered the example of pottery manufacturing in the introductory chapter. In breaking down the costs and benefits, we see that there are a number of distinct tasks associated with this production process, including the transport of clay, fuel, tempering materials, pigments, and so forth. Potters also have to build and maintain ovens and then manufacture the pots. If each household produces its own pottery independently, each household has to dedicate labor to each separate task. Obtaining pigment or temper, for instance, may require substantial labor expenditure in travel if the sources are not close to the manufacturing site. The actual transport costs of the raw materials for pottery manufacture can therefore be quite high, even if the investment in carrying them is minimal because of their light weight. The largest costs are in traveling to the area and obtaining the materials.

However, if the five households combine into a specialized labor organization, they can substantially increase production by reducing redundant transaction costs in the separate tasks. For instance, if one person is responsible for acquiring pigment or tempering materials, he or she can, with a marginal increase in actual cost, carry back materials to the manufacturing area for all five households. In other words, there is little difference in cost between walking 6 km and carrying back 2 kg of pigment for one household and traveling the same distance and carrying back 10 kg for five. What you decrease is the cost of four others walking 6 km. Likewise, there is an obvious increase in productive efficiency if the five households jointly use and obtain fuel for one oven through specialization, instead of using five. Again, the marginal cost increase to obtain fuel for five households is substantially less than the increase in productivity associated with sharing one oven and having one individual responsible for obtaining the fuel and perhaps a second one to oversee the firing.

This phenomenon works for any economic activity in which there are a number of distinct tasks, including the preparation of special foodstuffs, alcoholic beverages, and artisan goods. In short, surplus can be increased in an economy of this nature not by getting people to work more but by getting them to work *differently* as specialized producers. Again, it is critical to emphasize that these efficiencies are technically simple – there is nothing new to invent technically; what is new is the *organization of* the labor not the *nature or intensity* of the individual laborer.

Even a couple of dozen people working together can produce much more efficiently than an individual household or family group. Once people in a society can create a co-operative relationship between nonfamily members over a consistent period of time, the economic advantages accrue dramatically. These advantages can be used as the payoff to individuals in the co-operative group, thereby reinforcing the viability of the group in a competitive environment.

What Malinowski calls "socially organized labor" (1966[1922]:157–159) is effectively a kind of Type 1 economy of scale, in which redundant costs are reduced by specialization and coordinated work. He uses many examples, including that of canoe building. He observes, "We have seen in canoe-building a number of men engaged in performing each a definite and difficult task.... Certainly the minute tasks of lashing, caulking and painting, as well as sail-making, were done by communal labour as opposed to individual" (Malinowski 1966[1922]:158). Malinowski reinforces the fact that some kind of noncoercive authority is necessary. He says that there must be a chief, the head of a "well developed social apparatus," who coordinates this process. This "social mechanism must be associated and permeated with economic elements" (Malinowski 1966[1922]:158). Malinowski describes extremely well a noncoercive social environment in which chiefs are given ad hoc authority to represent the group and to coordinate complex tasks. What he calls "communal labor" is also effectively an economy of scale utilizing specialized labor tasks.

John Burton (1984) offers a fascinating insight into the need for co-operative behavior among a large number of people in some highland groups in Papua New Guinea. In one case he studied stone axe makers. He describes vast quantities of rock quarried at a number of mines, all of which were relatively close to the villages. These axes, in turn, were traded over an area of 250 km in all directions from the community. He notes that this was a classic stateless society: "Nowhere were the operations directed by a central authority; they were co-operative ventures in so-called egalitarian, tribal societies" (Burton 1984: 234). The particular group that controlled these mines were settled horticulturalists relying on yams, other cultigens, and pigs. This group, the Tungei, numbered around 2,000 people distributed over several villages. Burton studied the frequency of mining operations and the organization of labor. He found that the men who did the work came not from a single men's house or clan but from clans from several villages. He notes that the normal size of the mining group was smaller, organized by the men's house at intervals between three and five years:

> For the duration of an expedition men set aside personal ambition to work co-operatively.... It was therefore important that equality be achieved in the final shareout of stone. Men certainly differed in strength and aptitude for quarry work, but this did not lead to a corresponding ability to collect a larger share of axe stone. Nor was an expedition sponsored by one big-man and its production owned by him. A respected man in each sub-clan was informally appointed as a tui mok tokm wu, an "axe share hit man"; his job was to apportion the quarried stone in a fair manner. This could have been from one day's work or from several; it depended how long it took to knock down a piece of the rockface. Each man took his share and put it in a palteng, a wooden stall or crate. (Burton 1984:243)

Burton writes, "Social forces also disciplined the workforce into making a constructively co-operative effort." His informants said that the success of this work was due to "ritual purity and the correct axe-making magic" (Burton 1984:240). This supports, yet again, the argument that ritual structures economic activities that are complex and risky. Burton notes that the Tungei described the mining activity as dangerous and that ritual specialists were a critical component of the labor. A series of rites were necessary for the proper conduct of the work. He notes, "The end of an expedition was marked by what can be termed a 'ritual fight,' an attack of women upon men" (Burton 1984:242). In Chapter 2, it was noted that there was no coercion by any leaders in this society, but they certainly had the Big Men–type of leadership common in stateless societies in this area (Burton 1984:243).[10]

The mining work could be done individually. This is important. As mentioned above, the normal labor organization was by individual households or by a small number of households based on the men's house organization. Yet the Tungei chose not to work as individual men's houses. They created a specialized labor organization that was more effective. Burton reports: "Lines of men in basket chains, *grenj kan*, snaked out of each pit passing baskets of spoil from one to another. The spoil was simply thrown onto a talus slope which formed below the sites, a distance of about 30 m from each face. The problem of moving large amounts of rock quickly was therefore easily solved" (Burton 1984:241). Burton goes on with another very important observation:

> Another point is that the relatively skilled – but none the less arduous – task of hammering at the fracture planes on the rockface could only be done by a few men at a time. In contrast to this, the basket chain found useful employment for unskilled labour and could involve every clansman in the quarrying. *Putting more men to work meant that both the overall output of the quarries was significantly increased and more men had a claim to receive axe stone.* (Burton 1984:242, emphasis added)

This quote, of course, is a perfect description of a classic rudimentary economy of scale. In fact, Burton recognized that the Tungei were more successful than other groups precisely because of the ritual sanction, coordination, and higher returns for each individual: "With the benefit of hindsight we can say that *their cult-beliefs were ideally suited to the job in hand*; *though many similarly sized groups owned quarries elsewhere in the highlands, none succeeded in mobilising men in the same numbers for an extended period*" (Burton 1984:243, emphasis added). In effect, he articulates quite precisely a model of ritualized labor organization and strategy selection in which those who organized multiple men's groups were able to produce more resources for the same amount of labor. From this advantageous position, they therefore dominated the trade in axes over a very large region.

Economy of Scale Type 2: Density Dependent

A second kind of economy of scale is a density-dependent one. That is, there is a minimum number of people required to perform a task. This distinction is subtle but is profoundly important for understanding the long-term dynamics of the evolution of co-operation at a deep historical time scale. This is simply because certain tasks require minimum numbers of individuals far greater than what are available from any household unit. In some cases, the minimum number of people exceeds a kin group, and co-operation with nonrelatives becomes a necessity.

A Type 2 economy of scale therefore allows people organized above a household level to conduct activities that are not possibly carried out by an individual or a few households alone. Unlike the first kind of specialized organization, which proportionally increases production through organization, the second kind is characterized by a nonlinear relationship between specialized laborers and production. When the minimum number is reached, a threshold is broken, and new possibilities open up for the group.

The organization of a trading party through potentially hostile territory is a good example. One or two households that want to acquire products from such an area would not be able to provide enough individuals to travel safely. A minimum number of individuals is necessary to exchange safely with other groups and return home. Small groups are vulnerable to attack, whereas sufficiently large groups can defend themselves. Likewise, raiding, certain labor-intensive agricultural constructions, and certain kinds of water transport are all density-dependent tasks that can be carried out only with a labor organization above the household level. Malinowski, in discussing Trobriand canoe-making work parties, notes that "the joining of forces is almost indispensable" in pulling large logs (Malinowski 1966[1922]:159). Acquiring large logs is necessary for building large Kula canoes, which are in turn are required for chiefs to maintain their rank and to host the social events that are the basis of that rank. In other words, this is a threshold kind of labor task that is virtually impossible without a minimum number of workers – more than just a few households.

Kalervo Oberg (1943:579) recognized this principle in house-building behaviors in Tlingit society: "The building and maintenance of the house, which had to be made from heavy cedar planks with stone tools, required a long period of effort. . . . Not only were the risks involved in individual acquisition great but many of the activities could scarcely be carried on at all by a simple family." Oberg compared three societies including the Tlingit. He argued that in some instances, co-operation beyond the household was unnecessary, whereas in cases such as the Northwest Coast of North America, the ecological and social conditions required larger co-operative units:

> The relative extent of cooperation present in these economies, then, was dependent upon the type and degree of economic interdependence arising out of the situation. Collective action was at a minimum among the Bairu, as hoe culture under the particular condition in Ankole called for no greater cooperation. Among the Tlingit conditions demanded wider collective action. (Oberg 1943:580)

From an anthropological game theory perspective, we can see how these rudimentary economies of scale can create surplus (owned by the group) while still maintaining norms of fairness. Basically, one person or household can engage in deferred reciprocal exchange with others, who are then obligated to provide labor at a later date. One can then call in those debts in a prearranged place and time. By getting people to return the similar amount of labor they received, but working in a specialized labor group, a surplus is generated in a way that does not violate norms of fairness.

Simply put, getting lots of people to work together is vastly more economically efficient than if people work alone. The material benefits of co-operative labor, particularly through labor specialization, are tangible and immediately evident. Ethnography and history, as well as our own experiences, demonstrate that working in groups beyond the household is psychologically satisfying, and people in stateless societies look to work feasts with great anticipation. The role of human gregariousness and the need for social interaction cannot be easily measured, but it is a principal motivator for co-operation as well. In short, working together in coordinated ways provides both material and nonmaterial payoffs for everyone participating in a noncoercive environment. This then begs the question: If this is such a positive form of social and economic organization, why did it not develop until the Holocene?

The Big Trade-off: Loss of Autonomy for Higher Material Payoff

If rudimentary economies of scale were so great, they would have developed long ago in the Paleolithic among our species. There is obviously a cost to economic co-operation in these rudimentary economies of scale – this is the loss of autonomy by the primary producers over the goods or other resources that they make (or take in the case of raiding). In effect, people who work in specialized labor organizations are able to produce or acquire substantially more goods than those working strictly as individuals or as households due to the reduced transaction costs typical of economies of scale. But the individuals in the group have to give up some autonomy over the labor process and most importantly, the disposition of those goods. Individuals in a specialized labor organization cannot lay exclusive claim to resources that they shared in creating. The bounty that individuals reap from increased per capita production is offset by the fact that the ability to use those resources shifts from household-level decision making to a larger group. This is a potentially very serious problem at the core of collective action.

The question facing any society, therefore, in such a circumstance is how to prevent defection. Defection from the group will happen if there is a perceived unfairness in the amount of labor by required by individuals and in the perceived unfair disposition of things of value. Failure to keep people co-operating results in political collapse of that group. Ethnography and history teach us that chiefly societies are notoriously politically unstable, a result of managerial leaders' inability to maintain their factions. Disputes between households and villages over resource distribution are legion in the ethnographic record.

To keep the group working together in a specialized labor organization, individual participants must therefore be guaranteed that they will receive their fair share of their production at the end of what can be a lengthy labor process. This is true if it is endogenous production or co-operation in raiding. Firth (1967:306) talks about the essential role of the chief in "the apportionment of a common product among the members of a working party." It has to be fair. It has to be based on notions of reciprocity.

Ritually sanctioned redistribution is therefore an essential tool to maintain the group. One of the necessary conditions for a successful economic organization in ranked, nonstate societies, therefore, is an ideology of reciprocity that guarantees equitable redistribution of resources of value. Likewise, ritually sanctioned work schedules is a way to maintain fairness in the productive process. In other words, these ideologies of fairness keep productive groups together and sanctify the allocation systems mediated or controlled by the leaders.

The Household and Kin Group as Basic Units of Co-operation in Stateless Societies

The approach used here begins with the anthropological game theory assumption that humans are intelligent decision makers who will create or imitate successful strategies to deal with the exigencies of life. From a comparative perspective, people seeking to solve similar problems will come up with similar solutions under similar circumstances. In some critical areas of making a living and defending yourself, there are very few good solutions, and there usually is only one optimal or successful one. Therefore, the development of common strategies in distant peoples in many cases is a product of strategy selection of the medium or long term. It is not a coincidence, for instance, that agricultural practices in semiarid to wet areas are very similar around the world. Agricultural terracing and irrigation technologies in sixth-century BCE Greece are very similar to those in eighteenth-century CE Bali, tenth-century CE highland Peru, and twelfth-century CE Thailand. This is because soil chemistry and the physics of water and soil movement are invariant; there is therefore an optimal way to build terraces and irrigation canals on sloping land. It is no surprise that after a relatively brief archaeological time frame, farmers all around the world developed techniques that approached this optimum. The same can be said for the building of defensive sites (hills encircled by walls that have parapets or are wide enough to walk on are optimal), military maneuvers, roads, and other features. There are a few suboptimal ways of constructing these, and there is one optimal way. In a highly competitive environment, people's strategies in dealing with these problems tend to converge, independently, toward the optimal ones.

The same is true for social organization. We can view the means by which people construct co-operation between themselves as a social technology as important as a material one to achieve economic and social goals. These are the ways in which people construct co-operative social organizations to deal with the exigencies of social life. Some social organizations are simply better than others in achieving the goals of surviving and thriving. We will see, for instance, that social models of co-operation structured by ritual have emerged around the world in various cultures. There is substantial variation in the ways in which these organizations are expressed, but beneath the surface there are substantially similar ways to order co-operation for economic and political ends. We find two levels of organization: the household and the suprahousehold groups.

Household Size and Composition

What we find is that the household, defined as a coresidential group of largely related people who co-operate and share the products of their labor, is virtually always the basic unit of production and exchange of subsistence goods in most premodern societies.[11] This economic anthropological definition of the household is different from sociological or linguistic studies, in which kinship is the focus of the analysis. In these latter cases, we can more properly speak of "families" of people directly related to each other. Nuclear and/or extended family members can live in different households or in different villages given exogamy and other social practices. From the perspective of economic production, exchange, sharing, and trade, the coresidential group is key.[12] In this sense, two siblings living in different villages with different families are not members of the same household because they do not work together for subsistence, though larger kin structures, as described below, will facilitate exchange in many circumstances.

What do we mean by the "basic" unit of production? Again from an economic anthropological perspective, this refers to the smallest social unit of work and sharing in premodern societies (Stanish 1992). In classical economic theory, based on *Homo economicus*, the smallest unit is the individual agent, the rational person. In classic Marxist theory a couple of generations ago, the smallest unit for precapitalist tribal societies was the community that worked together and shared resources with each. A substantial Marxist literature, from Engels (1972) through Sahlins (1972) to Immanuel Wallerstein (1984:18) and Richard Lee (1990), argues that some form of primitive communism characterized stateless societies. In this model, the basic coresidential unit of production would be the multifamily unit of scores of people. The archaeological and ethnographic evidence finds this kind of organization only in a few rare instances. However, the vast bulk of the data suggests that the household, composed of both real and fictive kin, is the largest and most basic unit of production and sharing.

Examples include the Iroquois and some British Columbian groups of North America and possibly the LBK cultures of Neolithic Europe, in which longhouses held multiple families. However, even in these instances, it is likely that the basic unit of sharing was the household, which simply shared living space in the longhouses. There is reason to believe that longhouses simply were "enclosed villages" where individual households were the basic units of co-operation in labor, land, and sharing. Michael Dove (1985) provides a fascinating description of the longhouses of the swidden agriculturalists in Indonesia known as the Kantu.' He studied one longhouse that held 115 people in 12 households (Dove 1985:12).[13] He writes:

> The term bilek refers to each separate, walled off living unit within the longhouse . . . as well as to the separate families that inhabit each unit. . . . Each household builds and maintains its own living quarters as well as the other portions of the longhouse that are aligned with the quarters. . . . The household is partially defined, therefore, as the group that makes common residence in one of the longhouse's living units. One household may nyarang "share" the longhouse quarters of another, but this is permitted only as a short term expedient. (Dove 1985:14)

Dove goes on to observe that the household in Kantu' society is not only the basic residential unit but is also the basic unit of production and consumption. He writes: "It is a tenet of Kantu' law that goods must be shared and consumed within the household with absolute equality, regardless of which members of the household were directly responsible for producing them. . . . Nor does any unit larger than the household, such as the longhouse, function as a unit of consumption" (Dove 1985:14). Kantu' households, therefore, are the basic unit of co-operation in economic matters. While they live under the same roof, Kantu' households conduct their own agricultural labor, own their own swidden lands, share within their own households, and interact with other households and quasi-autonomous entities.

John Provinse observed this same organizational principle among the Siang people in central Borneo in the early part of the twentieth century:

> The family is the basic unit of the social organization, kinship is the most important social and economic tie. . . . Each family, that is, a man, his wife, and their unmarried children, has its own ricefield. The fields are not necessarily located in the same general region, nor in the same direction from the village, though in former years when head-hunting and raiding parties had constantly to be taken into consideration, it was usual for all the fields to be very close together if not actually adjoining. At present, two or more

families, sometimes as many as ten or twelve, often go together in the preparation of the same area, cooperating through all the different stages of felling the trees, burning, planting, watching, and harvesting. (Provinse 1937:81–83)

The Siang at the time of this ethnography had developed a very clear system of reciprocal exchange in labor and prestations via two customary methods. Provinse noted that it was not uncommon to find from 10 to more than 20 people working in a single garden plot that belonged to a household. He describes the social mechanism of the *hando* – a labor exchange arranged by family heads to "club together for the purposes of joining their combined efforts in developing their fields" (Provinse 1937:84). This is a common practice seen around the world. A second form of reciprocity, called *haweh*, allows someone to get out of returning labor by substituting it with a feast at his rice field. As Provinse describes it: "It is a festive work gathering, not unlike a Western corn-husking or barn-raising, and the owner of the field provides not only breakfast for all those who come, but at noon, in addition to bare necessities, provides pig or chicken, and usually tuak, the native rice wine" (Provinse 1937:86). What is clear from this ethnography is that fairness in the allocation of labor and products permeates the social organization of work in this group.

Charles Hose and William McDougall (1912[2001]:70) reaffirm this observation in Borneo from a study in the late nineteenth century: "Each family [of the longhouse] cultivates its own patch of land, selecting it by arrangement with other families, and works as large an area as the strength and number of the roomhold permits." They also noted that each family owned its own boat and that the community possessed larger boats for long journeys.

Raymond Firth observes similar patterns among the Maori. He notes that at certain prescribed times of the year in the ritual cycle people would communally work the land. Firth (1929:345) writes that "each family, however, had its own plot in the communal field, and these were marked off by paths or in some cases by stones at the corners or set in rows along the sides." This is a very typical pattern, found in small-scale stateless societies where economic co-operation in activities that require lots of laborers above the household level is coordinated by managerial leaders at the clan or other kinship group level; but it is the household level that controls the use and distribution of the product. As discussed below, among the Maori, as in many other stateless societies, it is the chief or the equivalent who hosts the feasts for communal activities.

Oberg notes a similar economic structure of the Bairu of Uganda: "Everywhere in Ankole we find the family as the essential producing and consuming unit. The home of the family is the *eka* or homestead. Each wife has her own hut, gardens, and granary. If a man has more than one wife, the huts are arranged in a circle. The co-wives do not cooperate in economic activities. They grow their own food-stuffs and manage their own household affairs" (Oberg 1943:574).

Bruce G. Miller and Daniel L. Boxberger note that among the historic Puget Sound chiefdoms, the local group was composed of a few households or even one household. In this case, "household" refers to a larger unit than the nuclear family. They write, "Although an individual's immediate allegiance was to their household, this did not mean that members of a household necessarily cooperated in resource procurement activities. The nuclear family was the basic unit in the acquisition, preparation, and consumption of food" (Miller and Boxberger 1994:271). In other words, we see yet again that the kin-based household is the basic economic unit. Miller and Boxberger go on to note that households could form ad hoc alliances with other households or even with people from outside the village.

Some additional ethnographic data are useful to understanding the size of these household units. Jack Goody (1972) compiled mean family sizes in traditional (non-Western) societies in Africa and Asia.[14] These family sizes correspond to the group of people who produce and share goods and services. In general, these family sizes represent the minimum coresidential work units (households) in these societies. His results are quite telling. Households vary cross-culturally, of course, but there is a fairly rigid upper limit of family size in his survey. The mean sizes range from 1.8 to 11.9 people. Other ethnographies suggest more flexible numbers on the higher side.

Tsimshian people from northern British Columbia live in "lineages" or "houses." This is the basic coresidential and economic unit and can be up to 30 individuals (Rosman and Rubel 1971:11–12). The lineage is under the direction of the chief. The lineage includes the chief and his immediate family, including unmarried children, plus married maternal nephews and their nuclear families.[15] Oberg (1943:578) confirms this, writing that the economic unit was the house, consisting of 5 to 10 men, plus wives and children. Paul Roscoe (2009:79) notes that in the New Guinea highlands, "the 'basic' or 'primary' 'economic unit,' 'the smallest productive and consumptive unit,' the 'self-sufficient' economic unit was said to be the 'nuclear family,' 'family,' or 'a woman and her husband.'" The list goes on and on in which the household or family unit is the basic economic group in small-scale, stateless societies.

What do these data mean? The empirical evidence, both ethnographic and archaeological from several continents, shows that in the substantial majority of instances, the coresidential household appears to be the fundamental unit of economic production and sharing in stateless societies. Labor parties for communal activities are also recruited through households. Wealth differences in stateless societies are rarely expressed in individual terms but rather are usually defined by the household. In short, households socially mediate the vast bulk of subsistence-level economic activities. Complexity in stateless societies is built upon the foundation of household production.

Suprahousehold Kin-Based Ritual and Co-operative Units

We likewise find that some kind of larger kin-based organization of multiple households is the primary organization for most co-operation for activities beyond the household, including the hosting of feasts, trade in surplus goods or preciosities, holding title to land, and raiding. This is a source of some confusion in the anthropological literature. Villages, for instance, will co-operate to achieve common goals, but the household is the basic unit of production in any economy of scale. By this, I mean that it is not the individual but the household that co-operates as a component of the co-operative group. It is the household that provides the labor, that withdraws its labor, and that receives the resources. The household is the unit of account in the group as well. Households are bound together by much stronger normative rules than are larger, suprahousehold organizations. The latter organizations break down all the time; one almost never reads of household organization collapsing.

The household, composed of the nuclear and/or extended family, seems to be the almost universal *economic* unit of production and sharing. However, in a stateless society lacking coercive, hierarchical social units or classes, the coordination of community-wide or larger political and ceremonial activities is vested in these suprahousehold units. John Monaghan (1990:768) writes about an agricultural community in the Mixtec region of Mexico: "The basic social unit in Nuyoo is the household, and the community is internally constituted by the relations households maintain with one another." Maintaining that co-operative relationship between households is a collective action problem.

A review of the ethnographic literature indicates that these suprahousehold organizations are almost universally organized by kinship and are responsible for many activities, most importantly those involving ritual and community feasting. Social anthropologists refer to these kin-based organizations by various names. The terms include "lineage," "clan," "phratry," "moieties,"[16] and "gens." Within the diverse societies around the world, there are countless names.

These larger kin-based units oftentimes control land use, mediate disputes between households, and in general provide a framework in which multiple households can co-operate on activities involving higher than normal risk and complexity. The larger kin group is hugely important in promoting multihousehold co-operative activities. These activities are usually permeated by ritual. In the example on the Kantu' people discussed by Dove, above, we find that in spite of the role of the household as the basic economic unit, the longhouse functions as the primary ritual body (Dove 1985:23). Kantu' longhouses are also the major jural unit, where disputes are regulated, taboo is enforced, and punishment is directed. Among the rules that longhouse members must obey are the following: A household will be ejected from the longhouse if they "do not join in the longhouse MOOTS … do not join the others in clearing the trails … do not follow [taboos on swidden labor and] do not work in the swiddens with the others (e.g., when the others have proposed an interhousehold work party]" (Dove 1985:26). Notice how Dove points out that it is the *household* that is ejected from the longhouse, not the individual person.

Andean societies, both historical and prehistoric, were organized around households within a larger kin-based organization called the *ayllu. Ayllu* structured communal projects such as canal cleaning. Even in these instances, however, *ayllu* leaders recruited workers from households. Households in imperial contexts, such as the Inca state, as well as in decentralized ones at the village level, were the unit of account in labor. Extended households in general were the unit of work and sharing for most commodities, though ritual activities occasionally involved the entire *ayllu*.

Henry Lewis Morgan's (1962[1851]) classic description of Iroquois social and political organization of course included the gens or clan, phratry, tribe, and confederacy (see Fenton 1965:253). The longhouse, composed of several matrilineal extended families, was most likely similar to other longhouse organizations in agricultural societies around the world in that it housed a number of nuclear families. These individual families were most likely the basic economic unit of work and sharing. Samuel de Champlain wrote that the longhouses contained nuclear families (Johansen and Mann 2000:200). Morgan confirmed that longhouses were between 50 and 130 feet in length and were composed of separate apartments. Each apartment housed two families, who shared a fire.

It is significant that Morgan's list of the 10 functions of the gens did not include any economic activities; the functions were exclusively social, political, and ceremonial. These functions included electing and deposing leaders, exogamy, inheritance procedures, vendettas, names, adoptions, burials, and other ceremonies (Fenton 1965:254). We see the gens acting like the *ayllu* and other kin-based units to coordinate political and social functions between autonomous households. When the gens or ayllu did act as coordinators of economic activity, the collective action problem kicked in and ritual served to structure the labor.

The Hopi of the southwestern United States is another example. Michelle Hegmon (1991:314; and see Hegmon 1989) notes, "The matrilineage is the core of the kinship structure; it controls land and is the primary unit in ritual and inheritance … [while] the household is the basic unit of economic activity."[17]

I agree with Roscoe, who argues that group size is optimized for different tasks in what he calls a "modular" organization. He suggests that smaller groups, "each optimally sized to serve one outcome (e.g. reproduction), are nested so that they can simultaneously co-operate as a larger group optimally sized to serve another (e.g. defense)" (Roscoe 2009:71). In this sense, "reproduction" includes both biological and economic reproduction, and larger groups function for activities such as defense, multihousehold festivals, and other tasks.

We can envision millennia of experimentation (see Chapter 6) in which people tried different kinds of social organizations, ranging from large, communal structures to production at the household level. In fact, the data show such variation in the archaeological record. Over time, certain kinds of organizations were more effective than others. The household as a co-operative, coresidential labor unit appears to have evolved at the expense of other kinds of organizations. Likewise, kin-based organizations of much greater variety developed as the most effective way to politically co-operate at a larger scale in the absence of state organizations. In other words, the evolution of complexity can be seen as a process in which more and more socially autonomous households work together as units to co-operate for mutually beneficial economic activities. Ritualized co-operation is the means by which these organizations survive and thrive.

Ritualized Co-operation in Stateless Societies

The fundamental economic principles of production and social reproduction in complex stateless societies can be conveniently reduced to a simple aphorism: "Make it or take it, and then exchange it to greater advantage." Households and groups of households can engage in specialized production to create valued resources. The surplus created in these activities can be exchanged on a local or regional level with people in other villages via various forms of barter. "Taking it" refers to raiding. Raiding between villages is a common phenomenon in stateless societies, so common in fact that it can be considered is a form of economic exchange. Ritualized raiding in particular serves an economic as well as social and political function. Raiding and trading usually occur simultaneously over a landscape and constitute the principal means by which valuables circulate in any landscape.[18] Finally, in the absence of money, markets, and formal means to store wealth, successful leadership in complex stateless societies strategically exchanges resources to keep factions together and to keep specialized labor organizations functioning.

In short, people in complex stateless societies (beyond the simple reciprocity of the hunter-gatherer band) create new resources (surplus) through specialization and intense co-operation. Surplus resources are internally created through rudimentary economies of scale. They are externally obtained by raiding, a form of intense in-group co-operation. Booty from raiding in this economic sense is a form of surplus. The resources are strategically used within groups with redistributive feasts and other ceremonies as a payoff to households for their co-operation. The resources are likewise used by households between groups in potlatching events to enhance prestige and to compete for status in a nonviolent manner. Proper ritually sanctioned redistribution is essential to maintain fairness in complex group dynamics. Most important, leaders must be fair. Firth, in his extensive discussion of Maori principles of the "apportionment of communal products," notes that it is critical to have the person responsible for the distribution be someone who will not favor his relatives or others. This person was either the headman of the village or the kin group, but it was essential that the person be fair (Firth 1929:279). Finally, virtually all trade in stateless societies is conducted in barter markets, usually in periodic festive fairs. Essentially, ethnography, history, and archaeology show us that people in stateless societies both acquire and produce resources in a sophisticated and complex web of rule-bound, ritualized behavior. Let's now take a look at that ethnographic record.

Ritualized Production: Make It

> Is magic, from the economic point of view, a mere waste of time? By no means. . . . It will be seen clearly that magic puts order and sequence into the various activities, and that it and its associated ceremonial are instrumental in securing the co-operation of the community, and the organisation of communal labour. . . . The belief that the magician is a man endowed with special powers . . . makes him a natural leader whose command is obeyed, who can fix dates, apportion work, and keep the worker up to the mark. – Malinowski 1966[1922]:116)

> All human activities [among the Maori], whether agriculture, military, hunting, fishing, fowling, building, travelling or human relationships were governed by elaborate ceremonials. – Winiata (1956:228)

As I mentioned in the opening chapters, the early ethnographers were impressed with the seemingly ubiquitous ritual and religious behaviors that they observed in stateless societies. Firth, for instance, clearly saw firsthand how ritual, magic, and taboo were essential to economic co-operation (Figure 5.2). He writes about the Maori, "The system of magic, by its prohibitions and imperatives, definitely regulated the conduct of people of the community towards their economic environment" (Firth 1929:252). In a section of his book titled "Magic of Production," Firth emphatically says, "In all economic enterprises of any moment, magic had its assured place in the scheme of events" (Firth 1929:253).

Likewise, Malinowski's extensive ethnographic research highlights in great detail the role of communal labor in building complex products such as canoes: "Technical difficulties face them, which require knowledge, and can only be overcome by a continuous, systematic effort, and at certain stages must be met by means of communal labour" (Malinowski 1966[1922]:13). In fact, the

Work	**Magic**
Preparation of the ground	
Clearing of brush	
Digging over soil	Digging chants, sometimes
Breaking up ground	with magical influence
Pulverising soil	
Gravelling soil	
Loosening soil	
Throwing into hillocks	
Planting	**Magic of planting**
Distribution of seed-tubers	Preparation of mara tautane
Planting of seed-tubers	Magic of weaving sacred basket
Covering of seed-tubers	Magic of seed distribution
	Magic before planting
	Ritual chant during planting
	Ritual chante during seed covering
	Ritual change at completion
	Lifting of taboo
	Ritual feast
Cultivating crops	**Magic of growth**
Tilling the soil	Magic for rain
Dispersal of rain water	Magic against frosts
Weeding	Magic against pests
Tending of runners	Magic to promote growth
Erecting breakwinds	Ritual offering of food to Pleiades
	Magic for broken tubers
Harvest	**Magic of Harvest**
Inspection of crop	Lifting of taboo from crop
Preparation of storage pits	Offering of first fruits
Lifting of crop	Ceremonial binding of shoots
Sorting and storing of crop	Digging of hillock
	Burying of tubers
	Unearthing of buried tubers
	Magic of storing

Figure 5.2. Ritual cycle and agriculture among the Maori. From Firth (1929:265).

actual amount of labor time spent building a canoe is low, just several days. But if labor were not coordinated in a group, the total labor hours would be substantially greater. The same is true for making a canoe sail. Malinowski says that communal labor "with a number of people, [allows] the tedious and complicated work . . . [to be] performed in a relatively short time" (Malinowski 1966[1922]:141). Malinowski (1920a) describes the building of the larger canoes as a "big tribal affair," meaning that it is a communal production involving many people and at times the entire village. Magic permeates all the productive steps. As Malinowski says, "The magic is always interwoven with the technical operations and is to the native mind absolutely indispensable to the successful accomplishment of the task" (1920a:102). In fact, he refers to the expert in charge of the building project as the "magician-constructor." The canoe owner pays for the work with gifts provided at feasts and other ceremonies. In one of Malinowski's observations, we see the clear connection between ritual feasting and a norm to maintain group co-operation, in this case a means to obligate a crew to the canoe owner: "They have a ceremonial distribution of food, which imposes an obligation on the *usagelu* (members of the crews) towards the *toliwaga* (owners of canoes) to carry out the expedition even in the face of contrary winds and bad, weather" (Malinowski 1920a:103).

Raymond Firth describes how Maori chiefs effectively regulated task-specific, co-operative labor with all sorts of ritual tools: "In communal tasks of any kind the chief held a position of command in the work . . . his authority was often exercised as the initiator of economic enterprises" (Firth 1929:236). He goes on to say, "The tapu regulated their behaviour and enabled them to perform many tasks which otherwise they could not have done. Through it their food, fish, and forests were preserved. Notwithstanding, they certainly never liked it," (Firth 1929:248–249)." This is the use of what Firth (1929:254) called the "magic of protection." Streams, fish, forest products, food plants, eels, certain pigs, ducks, flax, and red ochre sources were all protected by this magic and taboo. From his direct observations of the Maori over many years, he noted that the "tapu also assisted the maintenance of law and order in the community, and, inter alia, was often useful in facilitating the protection of private property" (Firth 1929:248). Again, he emphasizes the role of magic as well as taboo in economic activities among the Maori: "To guard against careless behaviour a whole-system of stringent magical regulations was in force, regulating the attitude of man towards the raw materials of production" (Firth 1929:256). As he sums it up in a classic functionalist formulation: "Magic is a very useful reinforcement for rational rules" (Firth 1929:263).

Firth notes that the Maori were particularly pragmatic with taboo as well in that once an object or possession was created, it was possible to lift the taboo to make it available for use by the community. What he describes as *whakanoa* rites served to remove sacred qualities from things such as a carved meeting-house. There were a whole series of taboos and rituals bound up with constructing such a building, because it was a complex task that required substantial co-operation over a considerable period of time. But once it was complete, those rules were removed and the house, while still sacred in certain senses, was now available to anyone who was part of the group.

Malinowski's description of the garden magician's role in keeping lazy individuals from free-riding or damaging public goods reads like a textbook case of irrational prosocial behavior in collective action theory: "Thus, the neglect of one careless individual might result in a damage to all, for bush pigs or wallabies might find their way in and destroy the new crops. If this happens, the garden magician gets up in front of his house in the evening and harangues the village, often mentioning the culprit by name and heaping blame on him – a proceeding which seldom fails to take effect" (Malinowski 1921:6).

For both Malinowski and Firth, managerial leaders possessing ad hoc ritual power were necessary for the effective functioning of the social group in economic activities:

> It is clear that this differentiation of tasks, co-ordinated to a general purpose, requires a well developed social apparatus to back it up, and that on the other hand, this social mechanism must be associated and permeated with economic elements. There must be a chief, regarded as representative of a group; he must have certain formal rights and privileges, and a certain amount of authority and also he must dispose of part of the wealth of the community. There must also be a man or men with knowledge sufficient to direct and co-ordinate the technical operations. (Malinowski 1966[1922]:158)

As discussed above and necessary to repeat here, ritual cannot be reduced to a simple set of functions. It has many meanings and uses, some of which are practical, some of which are not. But as Firth says about the rituals surrounding the sacred yam in Tikopia: "The opening rite of the work of the sacred yam consists in the renewal of them, and thus religion and practical aim are fused together" (Firth 1967:142). Indeed, it is this fusing of the religious and the practical that makes the former so powerful. Ritual provides the social and political framework to maintain co-operation in activities that people know to be in their long-term self-interest, but this co-operation is constantly in danger of falling apart due to free-riders and perceptions of unfairness.

Ritual behavior in which all group members implicitly and explicitly know a strict sequence of rites is ideally suited for the scheduling of labor, the equitable distribution of goods, and overcoming many of the problems inherent in co-operative group dynamics. The ethnographic work of Malinowski in the Trobriands is full of references to the role of ritual and magic in regulating labor and scheduling the timing of redistribution. There are, in fact, so many references to the role of ritual leaders in scheduling labor in the Trobriand Islands that is beyond doubt that the timing of agricultural production was seen as central to their duties. As Malinowski forcefully tells us: "The magician inaugurates successfully all the various stages which follow one another . . . [in planting and harvesting]" (Malinowski 1966[1922] :59).

Malinowski points out that ritual is also critical in managing nonagricultural labor tasks, such as shark and mullet fishing. He writes: "Thus fishing, an activity of great economic importance and favorite sport of the Trobriands, ranges from a purely economic pursuit to almost a magico-religious ceremony. In fact, the mullet fishing in Labai is surrounded with more stringent taboos and is bound up with tradition and ceremony than any other social activity in the Trobriands" (Malinowski 1918:92). The same is true for shark fishing, which involves numerous rituals, taboos, and ritual-specialist control.

The same can be seen in Borneo among the Dyaks as described by Roth (1968[1896]:411–416). Here we see that series of feasts are associated with the agricultural cycle. These include Nyipaän, which is the harvesting of the first fruits. It is a two-day feast that includes processions and dancing. As Roth (1968[1896]:412) reports, after the feast is over, "Nothing but work, work, work, is thought of for a fortnight or three weeks." Roth goes on to report that the next feast, that of Man Sawa, provides a respite from the hard agricultural work. This is also a harvest feast. It lasts for four days and also includes singing, dancing, and eating. The final harvest festival is called Nyishupen and is scheduled after the food crops have been stored. This is an eight-day-long feast and is quite elaborate, involving fixing up the village and elaborate dress.

Firth's work among the Maori likewise reveals a similar level of ritual control over fishing, farming, and crafts. "For despite the supernaturalism which forms its very essence, magic in a primitive community is a distinctly practical matter. It is called into requisition for all kinds of commonplace affairs, even in matters of work and trade" (Firth 1929:234).

One of the most detailed analyses of such ceremony is perhaps that of Firth in Tikopia. In *The Work of the Gods*, Firth describes the ceremonies surrounding the yam, a highly prized food. As the most desired crop it is not surprising that the yam was surrounded with the most ritual. The island of Tikopia is small measuring only 10 km^2. It was divided into several chiefdoms, with each chief ritually controlling one of the principal foodstuffs: taro, coconut, breadfruit, and yam. The first yam ceremony took place to consecrate the newly dug crop. There was then a highly scripted set of ceremonies that began with the plucking of the *repa*, or leaf oven covers. Once this was finished, the workers waited for the chief to arrive. The baskets for the kava feast were opened, libations were poured, and significantly there was a distribution of food to the people of the chief's group. The chief himself dressed elaborately and went to the temple to continue with more incantations.

The following day, the chief returned to his royal seat, and bundles of yams were taken to the oven house. Libations were poured, food was again distributed, the chief's genealogy was recited, and the community was then allowed to freely use the crop. Firth notes that similar though smaller rites were performed by lesser chiefs in other clans. At this point, the hot food ceremony was readied in the oven. This was a competitive rite called the *kai vera*, in which the eldest son of the chief took a central role. The oven was uncovered. Food was removed and placed in a special dish that was taken to the temple. Each person, except the chief, was given a large leaf, and food from a basket was given to each man, who had to eat as quickly as possible. The following day, there were more libations, incantations, hearty meals, pierced coconuts, and so forth along a path.

The ritual the next day was called by Firth "sanctifying the seed tubers." All the tubers were removed and eaten, except for a few that were planted on a sacred hill, which was a place of great ceremony during planting. These residual yams were used as seed for next year. They were considered highly sacred and surrounded with taboo (*tapu*). Firth writes: "The ceremonies connected with the sacred yam are almost inseparable in some aspects from those relating to the renovation and re-furnishing of the temple of Kafika [the chief], in that theoretically the same god controls both, and practically their rites take place in the same place and overlap or even merge on certain days" (Firth 1959:164).

Similar and separate rites were conducted for taro, coconut, and breadfruit. The rituals involved elaborate preparations that were timed to coincide with the agricultural labor cycle. The feasts involved with the rituals brought the entire village together. As a result, the labor force was guaranteed to be together during critical periods when co-operation was necessary. The feasts also provided the opportunity to determine who was working well and who was avoiding their responsibilities. These public and community events allowed chiefs and religious specialists to criticize those who were not pulling their weight and to assess the needs of the agricultural labor.

Patrick Kirch (1994) describes the ritual cycle in the South Pacific island of Futuna in western Polynesia. The Futuna rituals were recorded by a French missionary in the mid-nineteenth century and therefore represent pre-Christian traditions. Kirch notes that the Futuna cycle is very similar to the Tikopian one, suggesting that both derive from an ancestral Polynesian society. Like the Tikopian rituals, the Futuna cycle follows the agricultural seasons and ecological patterns. There were numerous ceremonies throughout the year; most of them involved redistributions of foods. Most were held in the open village plaza. In one village, Kirch describes this plaza, or *malae*, as a rectangular space, about 1,000 m^2. At one end of the *malae* was a stone alignment of coral slabs that marked the edge of the ritual space.

The historical document described by Kirch lists 29 and 22 ceremonial events in the monsoon ritual seasons for 1838 and 1839, respectively. Most (if not all) of the Futuna ritual ceremonies involved distributions of food. Kava ceremonies involved presentations of food as well as of the famous mildly narcotic drink that is a staple of Polynesian culture (Collocott 1927). Given the number of ceremonies, the actual quantity of food brought in represented a considerable portion of the diet. This is not to say that chiefs were controlling and redistributing the food on a massive basis but that a large share of production was stimulated by these ceremonial events. Production of basic foodstuffs was at the household level, but ritual obligations scheduled substantial redistributions of food between the households. Household labor was indirectly affected by ritual obligations.

Kirch also describes the work required to manufacture turmeric dye. In fact, this complex task had its own ritual season. Kirch (1994:357) writes, "The principal focus of the trade-wind ritual season seems to have been the extraction and manufacture of turmeric dye . . . a substance of great importance throughout Polynesia." Prayers opened this ceremony, and it is noteworthy that the usual kava ceremony did not take place. This was a monthlong project. The missionary says that two valleys (most likely representing the people of these valleys in the chiefdom) gathered at one village to make filtration bags. The work was accompanied by prayers at the end of the work and was celebrated by a kava ceremony and a feast.

The most important ritual season coincided with the first yam and breadfruit harvests (Kirch 1994:269). A council of chiefs decided when the feast was to begin. Preparations for this feast included the bringing of "a great quantity of taro," and according to the missionary, on the following day, "people arrived [from the other side of the island] in large numbers and there was a distribution of great numbers of food, raw and cooked taro, pigs, fish, bananas, coconuts, etc." (Kirch 1994:270).

The Tsembaba of Papua New Guinea (Rappaport 1967) lived in a very small-scale society, with no authorities who can be called chiefs let alone Big Men. Rappaport notes that without any political authorities with coercive power, "the ritual cycle in Tsembaba society likewise provides a means for mobilizing allies when warfare may be undertaken" (Rappaport 1967:18). The actions of the individuals in these rituals are highly nuanced and communicate nonverbal information that all key actors understand. This characteristic is perhaps one of the most important attributes of rituals such as these.

Rappaport also describes how a ritualized dance ceremony determines whether a large enough group of men will mobilize to go on a raid: "The massed dancing of the visitors at a *kaiko* [pig ceremony] entertainment communicates to the hosts . . . information concerning the amount of support they may expect from visitors in the bellicose enterprises that they are likely to embark upon soon after the termination of the pig festival" (Rappaport 1967:27). In other words, the ritual dance encodes messages that can be understood by people seeking to organize an activity, in this case a raid. Words need not be passed, and men can choose to not participate by mere inaction or by behaving in a certain way that communicates their refusal to join. If indeed there are enough men willing to participate in the raid, the dance can be viewed as one of the first rituals that organize this social activity.

Ritual also provides a mechanism for redistributing local pig surpluses (Rappaport 1967:17). The dancing ground of the pig festival the next day becomes the center of a marketplace, where axes, bird plumes, shells, baby pigs, and in earlier times salt are exchanged (Rappaport 1967:27). The pig festival is also the place for a huge redistribution of salted pigs to allies in the previous raid.

An ethnography conducted in 1915 by Ivor Evans among the Dusun of northern Borneo likewise emphasizes the importance of ritual and taboo in their daily lives:

> The Dusun's conduct, needless to say, is largely regulated by his belief in spirits and omens, and, in addition to the influence exercised by these, he is further hampered by the prohibitions and tabus which restrain his actions in many of his daily occupations. There are war tabus and birth tabus, house tabus and tabus for the padi-field, tabus to be observed in speaking and tabu days when no work must be done. (Evans 1922:153)[19]

Even in more complex sociopolitical contexts, we see how rituals regulate or at least schedule and facilitate communal activities such as canal cleaning. In his study of an indigenous community in Ayacucho in the heart of the Andean highlands, William Mitchell (1976:36) tells us, "*Yurqa Aspiy* ('cleaning of the irrigation canals') or *Yurqa Ruwuy*('working of the irrigation canals') are Quechua terms used to designate the cleaning and ritual celebration of the irrigation system." Mitchell says that this is one of the most important festivals in the town and is most likely one of the most important throughout the highlands. Each household that uses the system must supply a worker and the local sections in the region put on the festival. The seasonal work is celebrated with feasting, music, and dance. In this sense, the festival serves to mark the beginning of rainy-season work on the canals.

Tsuyoshi Kato (1988) describes similar rituals associated with small-scale rice cultivators in the Malaysian state of Negeri Sembilan (Figure 5.3). As with virtually all agriculturalists who have such rituals, the calendar follows the planting season. The initial ritual is called *minta doa di hulu sungai*, which means "to pray at the headwaters" (Kato 1988:114). Kato notes that the person who decides the schedule for this ritual is called a *pawang*, or "magician." He goes on to describe an earlier ethnography, noting that several men in the area claimed to be *pawang* but none of them were universally recognized as

Month	Agricultural activities	Agricultural rituals
June		Minta Dos di Hulu Sungai
	Land preparation	
July		
August	Sowing	Tanam Setabun
		Ambil Hari Obah
September		
	Transplanting	
October		
	Weeding	
November		
December		Various taboos
January		
February	Harvesting	
		Tekachi Berpuar Ambil Semangat Padi
March	Threshing Winnowing	
April	Storing	

Kato 1998:128

Figure 5.3. Ritual cycle and agriculture among small-scale rice cultivators in Malaysia. From Kato (1988:128).

having the skills. Kato also describes one or two of these men as having "sizable followings." In other words, this organization fits very well the model of managerial leadership that helps coordinate labor and simultaneously competes for followers. In fact, Kato (1988:115) notes that the *pawang*"was more concerned with the welfare of the luak [region] as a whole." These *pawangs* likewise had other special ritual duties, including "to preside over the community-wide agricultural rituals."

The next major ritual starts with preparation of the fields. Kato's description again is similar to those of so many other such agricultural ritual cycles from around the world in small-scale societies:

> On an appointed day, villagers gathered at the designated keramat, each with cooked food for ritual offerings and a small amount of rice seeds. The pawang burned incense (kemenyan) and chanted incantations asking for enough water and for the protection of rice during the coming planting season. He blessed the rice seeds with incense smoke, *tepung tawar* (magical rice water or rice flour mixed with water) and charms in order to protect them from evil spirits. The seeds were to be brought back to individual households and mixed with other seeds for seedlings. (Kato 1988:116)

It is significant that seeds were collected and redistributed to households in this ritual. Such a practice ensures hardier crops by maintaining genetic diversity.

The third ritual begins with a prayer and is conducted only once every three years. It includes incantations, incense, and the playing of gongs along a procession that stops at each house. Ritual feasts are enjoyed along the way. An element of this ritual includes the fencing of the village, metaphorically, with magical rice water. This can be interpreted as a physical manifestation of a taboo that signals community-wide acceptance of a norm of behavior designed to mark a group and ensure co-operation. Breaking of this taboo would be immediately recognized by all parties and would most likely lead to costly prosocial behaviors on the part of the community members and their leaders.

After these community-wide rituals were completed, two household-based ones were conducted during the first sowing of the seeds and during the transplanting of the seedlings. These rituals included the planting of mixed seeds from the previous ritual. The rituals involved various taboos that had to be followed by participants (Kato 1988:119). More harvest festivals followed. These events included the *tekachi* which came after the fifth month of planting the seedlings. It was performed once every three years and, according to Kato (1988:120), this ritual "combined elements of rituals, games, plays and entertainments."

In this example from Malaysia, we see how a clearly defined set of rituals over the entire agricultural season schedules the work for all community members. There is no state authority involved in this ritual cycle, and there is no interference from outside entities.[20] Apart from giving precise times for the agricultural work, the ritual cycle enriches the lives of all participants and provides priceless social, psychological, and practical benefits for being part of the group. Ritual is the glue that binds this essential, co-operative economic activity.[21]

A similar agricultural ritual cycle is documented by Michael Dove (1985) in Indonesia. As seen in Figure 5.4, the entire phase of the swidden agriculture cycle – nine different activities and time periods – is organized by six

Swidden phase	**Associated ritual phases**
Selecting	Bedurong (using the birds) Nganyung batu umal (bringing the ritual whetstone to the swidden)
Slashing	Ngami' batu umai (bringing the ritual whetstone back from the swidden)
Felling	
Burning	
Planting	A complex of planting rituals
Weeding	Nyingkelan umai (swidden offering)
Guarding	
Harvesting	Matah padi (rice breaking)
Carrying	A complex of rituals Gawa' batu (post harvest year's end feast of the whetstone)

Figure 5.4. Ritual cycle in swidden agriculture in Indonesia. From Dove (1985:28).

different ritual phases that include a number of events, including feasts. In an article about swidden agriculture in Southeast Asia in general, Dove makes a very important point: "Thus, the multi-household work group represents not communal labor but reciprocal labor, since the labor input of each household is always reckoned separately, and since the venue of labor, namely the swiddens, are also always reckoned as the private property of one household or another" (Dove 1983:88). This is a subtle but critical observation. The labor relationships between households effectively involve a very complex series of reciprocal debts that are continually repaid over at least one agricultural cycle. In this social context, it is easy to recognize the importance of feasting and rituals. These serve to reinforce the reciprocal debts between households, reward people who do reciprocate, and punish non-co-operators by excluding them from the events as well as the benefits of co-operation.

Patrick Kirch describes how the feasting cycle and agricultural production are intertwined in many Polynesian societies. He says: "As in many other

Polynesian societies, religious ceremonials in Futuna revolved around an annual ritual cycle that was intimately connected with agricultural production" (Kirch 1994:204).

There are many more examples of small-scale agricultural societies in which rituals and feasts permeate the agricultural calendar. The Kayans of Borneo have a very rich set of taboos that regulated the labor schedule: "The system of taboo is greatly practiced by the Kayans during the times of planting and harvesting the crops, and more especially when the paddy is being stored" (Roth 1968[1896]:402). Roth also relates how no one can cut a paddy or engage in other work until a first-fruits feast is concluded. The feast lasts two days and is quite elaborate. Another feast (Man Sawa) requires a cessation of work and rest. Additional rituals that time the agricultural season follow this feast and provide a framework for scheduling co-operative work in the fields, either by household units or by multiple household labor.

The Principle of Proportional Ritualization in Production

> No native would ever make a yam or taro garden without magic. Yet certain important types of planting . . . are devoid of magic. Fishing, the economic activity only second in importance to agriculture, has in some of its forms a highly developed magic. Thus the dangerous fishing of the shark, the pursuit of the uncertain *kalala* or of the *to'ulam* are smothered in magic. The equally vital, but easy and reliable method of fishing by poison *has no magic whatever.* In the construction of a canoe – an enterprise surrounded with technical difficulties, requiring organized labor, and leading to an ever-dangerous pursuit – *the ritual is complex, deeply associated with the work, and regarded as absolutely indispensable.* In the construction of houses, technically quite as difficult a pursuit, *but involving neither danger, nor chance, nor yet such complex forms of co-operation as the canoe, there is no magic whatsoever.* (Malinowski 1948:115, emphasis added)

This extended quote by Malinowski highlights one of the most important facets of production in stateless, ritualized economies. It is what I call *the principle of proportional ritualization* – the degree to which an activity requires the co-operation of many people, requires complex coordination, and involves risk is the degree to which ritual regulation and taboo will be important to that activity. In the Trobriand Islands, Malinowski noted that some kinds of fishing and gardening were heavily regulated by ritual while other kinds were not. As he points out, communal gardening, the most important and risky in terms of potential economic return, was heavily imbued with magic. Fishing in ponds and lagoons, on the other hand, was not subject to ritual or magic at all. Likewise, working the family plots did not involve chiefly or ritual-specialist intervention, except when a family's fences were untended and let pigs into other people's gardens. There was a clear distinction between labor conducted under ritual schedules and taboos and labor that were not.

Pond fishing involved lone individuals using poisons to bring fish to the surface. The same was true for net fishing by individuals. Malinowski tells us that there was very little risk or investment and no coordination with others. The same was true for agricultural work on individual plots. In effect, ritual regulation corresponded to degree to which co-operation was essential to the success of the task. The following passage reinforces this observation. I believe this is such an important observation that I provide another extended quote:

> In the western villages fishing in the lagoon is an activity which always gives an abundant yield, *without uncertainty and without risk to the fishermen.* This is especially true in the case of fish poisoning among the coral patches, where a man is sure to obtain an abundant catch easily each time he goes out. *Now there is absolutely no magic in connection with fish poisoning*, and very little in connection with the ordinary fishing by means of nets. Yet on the northern shore, both in the *kalala* and shark fishing, we find that the whole proceedings *are absolutely governed by magic*.... Shark fishing is hardly less subject to magico-religious observances, the construction of the special canoes, the opening of the fishing season, and the actual expeditions, being all connected with magical ceremonies and taboos. (Malinowski 1918:90, emphasis added)

A critical observation that supports this concept of proportional ritualization is that the labor involved in all these activities is not technically difficult. In fact, people do these kinds of tasks all the time alone. A specialized knowledge is rarely required. Rather, the main challenge is keeping people co-operating in specialized labor organizations for their perceived self-interest. The group must ensure that potential defectors do not defect. It must ensure that fairness pervades the activities to avoid the rise of antigroup behavior. These conditions are met by ritual that exists outside of any individual but that express the needs and ideology of the group.[22]

The same concept is expressed by Firth in his discussion of Maori taboo: "Objects of importance to the community, such as large canoes, eel weirs, and superior houses always had a certain amount of tapu pertaining to them, whereas small fishing canoes and dwellings of ordinary type were void of such quality. *The intensification of the tapu in the case of things of greater social importance is clear*" (Firth 1929:248, emphasis added). This is a direct confirmation that the degree of importance of an activity or object to the community was proportional to the degree that it was ritualized in some form or another.

We see this principle among the Tlingit of northwest North America as well, as reported by Oberg (1943:577): "While magic was employed in the cure of sickness, in many economic pursuits it was singularly lacking. In this the basic enterprise of salmon fishing I could discern no magic at all. In the highly dangerous occupation of bear hunting, on the other hand, numerous taboos and a course of ritual were involved." Here it is very clear that the co-operative activities involving risk were heavily tabooed while those that were not risky did not have taboo. In short, taboo in the economic sphere served clear practical purposes.

Phyllis Kaberry tells us that in the Sepik district of New Guinea, "The younger men only observe taboos when actually planting the yams, or when they enter the garden to weed and perform sacrifices. The older men assume full responsibility for performing the important magic for the large yams in the large yam gardens and they observe taboos for the period between planting and harvesting" (Kaberry 1941b:356). In this case we see the highly selective use of ritual to regulate agricultural activities. It is a very clear instance of ritual functioning to deal with the everyday issues of group co-operation.

Meggitt describes how prosaic exchange among the Mae Enga is not associated with any ritual, but exchange events that establish political power and authority are heavily ritualized:

> Trade relations are, however, contingent, finite and private transactions between individuals. They lack ceremony, do not involve the extension of credit and are not formally or structurally connected with the networks of public prestations. . . . The prestations, on the other hand, are elements in a system of delayed exchanges which include the notion of credit, are highly elaborated or "ceremonialized" and usually take place with maximum publicity between groups of people (generally relatives, although other non-relatives may sometimes participate). Even though, in the last analysis, individuals are giving and receiving the valuables, the proceedings are mediated through the action of Big Men who represent the groups concerned. (Meggitt 1974:169)

Sean Swezey and Robert Heizer (1977) describe the ritual management of fish by tribes in northern California. Using ethnographic data from the nineteenth century, they define how a complex set of taboos and ceremonies regulate the use of these resources to make "salmon fishing perhaps the most efficient subsistence undertaking in Native California" (Swezey and Heizer 1977:11).

We can find numerous additional examples in virtually any society. In short, *when an activity is risky, complicated, politically charged and/or involves many co-operating individuals, it tends to be heavily ritualized.* When it involves everyday, prosaic activities conducted by individuals or households, ritual tends to be absent.

Ritualized Production: Take It

> In the good old days, Big Men led raids against enemy villages. – Flannery and Marcus (2012:102)

Most modern economists would not recognize stealing as a viable form of exchange. Of course, it is not a socially acceptable thing to do in Western society or in virtually any other society known to ethnographers. However, the distinction that people often make between stealing from your own group and stealing from the other is well known in anthropology, sociology, and philosophy. Putting morality momentarily aside, stealing from other groups is a time-tested mechanism of acquiring resources. In fact, in societies that practice raiding against other groups, we see the evolution of a whole series of norms that validate it as beneficial to the group. Under the appropriate conditions, raiding is a widely acceptable form of co-operative behavior of one group used against others.

It goes without saying that raiding your neighbors is risky, and it by definition involves coercion if not outright violence against others. There is a huge debate about the nature and intensity of conflict in stateless societies.[23] My read of the ethnographic and archaeological records is that raiding is a common though not always a lethal activity in stateless societies. There are as many documented cases of low lethality as there cases of vicious treatment of enemies. All peoples did not universally conduct raiding in all times, but it certainly was an extremely common form of "economic" activity that is readily understandable by anthropological game theory.

Superficially, raiding, war, and other kinds of intergroup conflict would appear to be the opposite of co-operation. Many early scholars would assess the archaeological data, see conflict in the record, and then presume some kind of Hobbesian universe. In reality, organized conflict requires some of the most intensive kind of co-operation within a group. As a density-dependent form of economic activity in stateless societies, raiding requires a fair amount of co-operators.[24] Jung-Kyoo Choi and Samuel Bowles (2007), in an article titled "The coevolution of parochial altruism and war," argue that the creation of agonistic, parochial groups ("us versus them") in an environment characterized by hostilities will promote co-operation (altruism) among group members." They continue "Our game-theoretic analysis and agent-based simulations show that under conditions likely to have been experienced by late Pleistocene and early Holocene humans, neither parochialism nor altruism would have been viable singly, but by promoting group conflict, they could have evolved jointly" (Choi and Bowles 2007:636). Sam Bowles (2008) goes so far as to call conflict "the midwife of altruism."

Likewise, Mathew and Boyd (2011:11375) correctly describe raiding or "warfare" as a "high stakes form of cooperation."[25] Richard Sosis, Howard Kress, and James Boster (2007:234) write, "We argue that costly male rites signal commitment and promote solidarity among males who must organise for warfare." Macfarlan et al. (2014) refer to "lethal coalitionary aggression," a term used in primate studies, as a particularly strong form of co-operation between males in stateless societies (and see Wrangham and Glowacki 2012). "Male-centered, descent-based power groups cause affiliated males to coreside with one another throughout life and organize individuals . . . into cooperative units that compete violently against other similarly organized factional polities in cycles of retaliatory violence and internal warfare" (Macfarlan et al. 2014:16662). Gavrilets (2015:5) takes this idea a bit deeper into our Paleolithic past. He argues, on theoretical grounds, that "between-group conflict can select for increased intelligence and cognitive abilities used to coordinate group activities, potentially overcoming both the high costs of large brains and the collective action problem."

One of the debates in anthropology today focuses on the goals of raiding. On one side, sociobiologists argue that the goals of raiding were status and the capture of women and slaves by young men, effectively increasing men's fitness (see Chagnon 1988, 2012; Macfarlan et al. 2014). Such behavior may also benefit the individual's kin group (Maschner and Reedy-Maschner 1998:21). In this view, men raided exclusively to obtain women and prestige. They point out that raiding was not conducted to conquer and keep territory or other material resources.

On the other side, the so-called materialist position argues that the primary goal of raiding is to obtain resources (Harris 1984). It argues that the bulk of raiding was a response to environmental or demographic stresses. In this view, raiding is at base a response to scarce resources.

Abigail Levine and I noted that from an evolution of co-operation perspective, war and trade have similar effects on stateless societies – increased co-operation at the group level and a net increase in resources. War and trade "effectively lead to high levels of in-group cooperation to achieve material and political ends." Furthermore:

> In a context of sufficiently high regional population densities, conflict emerges as a viable strategy for highly cooperative groups to acquire more resources and protect themselves against potential adversaries. In effect, organized intergroup conflict is a type of *within-group* cooperation, albeit a form of behavior that is used in a decidedly non-cooperative manner *against* other groups. Groups that successfully organise themselves to raid others will acquire external resources and, in the long run, will be at a selective advantage against groups that are less well organized. (Stanish and Levine 2011:13902)

A good body of ethnographic and historical data describes the nature and intensity of raiding in stateless societies. These data tend to support the materialist position a bit more strongly than the sociobiological one, in my view, though there clearly are multiple motivations involved in raiding. The problem here is that the debate is improperly defined. One side focuses on nonmaterial prestige and fitness, while the other side focuses on material resources. Prestige is a very important motivation, as are material resources. A brief look at some data from the historical record indicates that both prestige and material resources are motivators in raiding in stateless societies. It is less a stark dichotomy and more of a continuum of economic utility that vary in different cultural and historical contexts.

Louise Sweet's classic article on camel raiding among the Arabian Bedouin summarizes the observations of many ethnographers and travelers during the nineteenth and early twentieth centuries. She notes, among other things, that raids were continual and were conducted against other tribes and settled villages. Raids were conducted mainly to obtain camels. These raids therefore had the effect of circulating one of the most valuable resources – camels – throughout the region. She notes that the raiding parties avoided the killing of people and that raids did not result in permanently seized territory. Most significantly, she notes, "Raids are necessary in an economic sense . . . for Bedouin survival" (Sweet 1965:1133).

It is fascinating that the Bedouin language distinguishes between raiding, fighting for territory, common thievery, and pillaging of non-Bedouin (Sweet 1965:1133). This linguistic distinction shows a bit more nuanced view of this behavior than we find in Western languages. It also suggests some subtle moral distinctions involving taking other people's resources. The typical raid of other Bedouin is known as "reciprocal" and is carried out with very specific rules of etiquette. The focus is on camels, but in a very successful raid, Bedouin may also take tents and weapons. Other livestock, persons, slaves, or chiefly emblems are occasionally taken as well (Sweet 1965:1141).

Sweet notes that joining a raiding party is voluntary and based upon material greed. Men "join raiding parties in the hopes of acquiring enough camels to become self-sufficient" (Sweet 1965:1136). Raids are formally organized groups of between 14 and 40 men, a kind of Type 2 economy of scale. Here, ad hoc leaders have unquestioned authority, an otherwise rare circumstance in Bedouin society. To maintain norms of fairness, all participants agree on the division of the spoils "according to several customary modes." Magic is also involved: "Prosecution of raids is also surrounded with lucky and unlucky days, good and bad omens" (Sweet 1965:1141).

While camels are used as a medium of exchange for gifts and for social and political obligations, they are never bought and sold between chiefdoms of equal status. Rather, Sweet notes, the only way that groups exchange camels is by "thievery and raid" (Sweet 1965:1136). In other words, raiding is a major economic exchange mechanism in this stateless society. The most valuable resource in this society is arguably camels, and they circulate according to raiding and obligatory gift giving. In Sweet's works, "reciprocal raiding [is] a continuously employed mechanism of intertribal exchange of camels" (Sweet 1965:1142).

Yet it is common for a man to give away all his camels after a successful raid. Sweet notes that successful raiding "is a prime means to gain prestige and influence in the chiefdom." This may result in a man being adopted by his chief and given a daughter as a wife (Sweet 1965:1146). In other words, while camels are the material resource objective in raiding, status enhancement is also a goal. The material and the psychological cannot be easily separated in trying to understand the motivations of raiding in this or any society.

As one reviews camel raiding as described by Sweet, one cannot but help note the numerous instances in which anthropological game theory can be used to understand these behaviors. Some men give away (presumably surplus) camels to others in exchange for nonmaterial prestige. The reciprocal nature of raiding and notions of fairness permeate the custom. Ad hoc leaders emerge at the behest of the group, and their authority ceases after the activity. There is a minimum number of men required to conduct the activity (Type 2 economy of scale), and the redistribution of the "product" is heavily regulated by customary ritual practice. And all of this is done without any centralized leadership. The norms of raiding are clear to all members of the society, and these norms are embedded in a variety of habituated or ritual practice.

Another classic article about raiding in a stateless society – the Chiricahua Apache of the American Southwest and northern Mexico – describes the nature of late-nineteenth-century raiding in this particular cultural context (Opler and Hoijer 1940).[26] Here, raiding for booty was a major economic and prestige-enhancing activity, described as "a regular part of the Chiricahua economy and... the royal road to wealth and distinction" (Opler and Hoijer 1940:621, note 8). Participation was voluntary. Raiding parties had their own vocabulary or sacred language. Elders and religious specialists trained young men in this language and the requisites of being a good warrior. Shamans specialized in raiding and "war" and were critical in the organization and prosecution of these activities.

There are numerous references to raiding for both prestige and resources throughout the vast ethnographic literature. The celebrated ethnographer Robert Lowie (1920:356) writes, "The Plains Indian fought not for territorial aggrandizement nor for the victor's spoils, but above all because fighting was a game worth while because of the social recognition it brought when played according to the rules." But as Newcomb (1950:319) points out, Lowie seemed to reverse himself a bit decades later when he wrote that horses were the primary economic objective of raiding, though "utilitarian urges" were not dominant. The nature and aims of Plains warfare in the historical record is hotly debated between those who stress economic motivations over psychological ones and vice versa. It is safe to say, given the ethnographic record, that both factors were significant.[27]

Malinowski's description of Melanesian culture illustrates how war and trade are deeply intertwined. He tells us: "The coastal populations of the South Sea Islands, with very few exceptions, are, or were before their extinction, expert navigators and traders. Several of them had evolved excellent types of large sea-going canoes, and used to embark in them on distant trade expeditions or raids of war and conquest" (Malinowski 1920b, 1966[1922]:1).

Likewise, his description of "war" is more in line with other ethnographies that emphasize the nonlethal but still politically and economically serious nature of raiding:

> From time to time, in the old days, war used to break out between the two provinces, each of which could muster some twelve villages for the fight. These wars were never very bloody or of long duration, and they were in many ways fought in a competitive, sporting manner, since, unlike with the Dobuans and Southern Massim, there were neither head-hunting nor cannibalistic practices among the Boyowans. Nevertheless, defeat was a serious matter. It meant a temporary destruction of the loser's villages, and exile for a year or two. (Malinowski 1966[1922]:66)

Organized conflict in stateless societies has been documented as early as the late Paleolithic (about 13,000 years ago) in Nubia. Site 117 near Jebel Sahaba has a graveyard with almost 30 people pierced by projectile points (Wendorf and Schild 2004). This is probably the earliest and best evidence we have of organized conflict. The bodies had numerous wounds, some of which had healed, indicating that there was a history of violence. The dead included men, women, and children. Kent Flannery and Joyce Marcus (2012:41) interpret this cemetery as an example of continuous clan-based blood feuds.

We see strong evidence of raiding prior to the establishment of settled village societies in both hemispheres. Raiding is endemic in complex stateless societies as one means of obtaining external resources. Raiding enhances prestige for young people, mainly males, who successfully conduct raids; it also constitutes a means of increasing resources. Most raiding in the ethnographic record is not particularly violent, because strict norms of behavior keep such practices relatively nonlethal. There are some examples in the literature (the Yanomami, for instance) in which violence results in high casualties. In most instances, however, raiding is strictly controlled by ritually sanctioned norms that keeps casualty rates reasonably low in many stateless societies. There are, of course, exceptions to this (see Keeley 1996) and the archaeological record records numerous instances of lethal raiding in virtually every cultural area of the world.

Ritualized Exchange

We know from ethnographic studies that hunter-forager bands carried on exchange in the form of reciprocal sharing within the family group and the larger band. We also know that many objects traveled around the landscape via down-the-line trade. Furthermore, trade between hunter-foragers and settled farming villages is very common in regions where both lifeways coexist (e.g., see Peterson 1978). John Speth and Katherine Spielmann (1983:20) note,

"Hunter-gatherer exchange of meat with horticultural populations in return for carbohydrates is documented for many areas of the world." The ethnographic literature is replete with instances of exchange between mobile groups and settled agriculturalists (e.g., Boyd 1998).

The literature on exchange *between* hunter-gatherer-forager groups is less clear. There are some extreme positions: James Woodburn claims that sharing without expectation of reward is the only way in which people like the Hadza exchange. In an article titled "Sharing is not a form of exchange," he writes: "Exchanging with other Hadza is reprehensible. To barter, to trade or to sell to other Hadza is, even in the 1990s, really not acceptable.... Even mutually beneficial bartering without thought of profit is very difficult.... People should give freely without expectation of return. People should share, not exchange" (Woodburn 1998:54). Of course, even Sahlins saw meat sharing as a form of reciprocal exchange, so Woodburn's position is at odds with most economic anthropological analyses. It seems to be more informed by political ideology rather than theoretical accuracy.

Most scholars of hunter-gatherer-foragers recognized that there was at least sporadic exchange of resources and information in periodic "aggregations" of different groups:

> Perhaps the best hunter-gatherer examples of the coincidence of informational and network mobility are periodic aggregations. David Hurst Thomas (1972), for example, describes fandangos as "clearing houses of information." The ethnographic record is full of similar examples of trade fairs, corroborees, fandangos, ceremonial, and other periodic aggregations that promote the pooling and exchange of resources and information. (Newlander 2012:74–75)

In short, hunter-gatherer-forager peoples also exchange with each other and with their settled neighbors. There is some disagreement about the nature of this exchange. As best as we can deduce, the nature of this exchange is fairly basic. We can suggest that formal exchange commences with the establishment of special places on the landscape, with some kind of monumental constructions where large numbers of unrelated people meet on a regular basis under some kind of predetermined calendar. As we will see in the following chapters, this occurred in the distant past among mobile hunter-foragers – no agriculture required. Simple reciprocal sharing within the family in a hunter-forager band of course requires certain norms of behavior based upon reciprocity and/or culturally determined ideas of fairness. But, as we will see, once special places are created on the landscape to facilitate the exchange of goods and information between distantly related or unrelated people, a more complicated set of formal norms is required, because the interactions are often between nonrelated people who may or may not routinely meet trading partners. In these more complex contexts, the full range of collective action and free-riding problems kicks in. These are the norms of barter exchange, a kind of exchange that occurs most frequently between non- or distantly related people in stateless societies.

Barter Exchange in Complex Stateless Societies

In broadest terms, there are two fundamental ways to exchange for mutual benefit:[28] (1) through the direct exchange of equivalent value, commonly known as barter; (2) through the use of a medium of exchange or money in some kind of market interaction.

In the absence of money, coercive contracts, and policing powers beyond the village, all face-to-face, noncoercive exchange in stateless societies between unrelated people is effectively conducted via barter (e.g. Chapman 1980). Barter can be conducted in just about any place, mundane or special. We see in the ethnographic record extremely complex barter systems, such as the Kula ring. Malinowski characterizes the Kula as a form of "ceremonial barter with deferred payment" or "payments which are ceremonially offered, and must be received and re-paid later on. The exchange is based on a permanent partnership, and the articles have to be roughly equivalent in value" (Malinowski 1966[1922]:187). Direct exchange of like value for like value, barter is the simplest and most powerful form of exchange in stateless societies.

From a more narrow perspective, however, a number of features distinguish these two kinds of exchange. Exchange can be immediate or deferred. A deferred exchange is a debt incurred by the party that it has to repay. Debts can be accumulated and discharged in a variety of ways. Likewise, many kinds of money can be incorporated into barter exchange. And most importantly, most stateless economies incorporate several forms of exchange simultaneously. Stateless economies effectively do not have markets in the modern sense of the term. The economic anthropological literature is somewhat imprecise with terms. We therefore have to take a brief digression to more carefully distinguish between different kinds of markets, marketplaces, and exchange mechanisms in economic anthropology.

A Short Digression about Marketplaces, Fairs, Price-Making Markets, and Currency Markets

The most common exchange mechanism in modern society is *market exchange*, which has dominated the world since the beginning of the modern era and it is clear that forms of market exchange go back millennia in the Western world and Asia.[29] Market exchange is one of the most complicated mechanisms in the economic anthropological literature. The word "market" is old and pervasive in European languages, occurring as early as the eighth century in Old

High German (see the *OED* entry on "markets"). There is a long tradition in economic anthropology, particularly in the substantivist tradition, which sees market exchange as a uniquely Western phenomenon. This is dramatically clear in the title of Karl Polanyi's *The Great Transformation*, which allegedly characterized the shift from noncapitalist to capitalist economies at the beginning of the sixteenth century. Polanyi felt that all trade prior to the modern period in the Western world was not based upon marketing mechanisms (supply-demand pricing in particular). Rather, he saw all trade embedded in social relations and controlled by kin or political institutions. Markets in the modern sense did not develop until the modern period with mercantile capitalism. Polanyi was wrong on this issue,[30] and a huge literature, both empirical and theoretical, demonstrates that marketing mechanisms go back well before the modern period and are found in any number of non-Western contexts.

The modern economic use of the term "market" implies the existence of currencies and supply-demand mechanisms that set price. However, it is important to remember that supply and demand affect all economic exchange, including that of stateless societies that do not have currencies or other forms of money.[31] Furthermore, stateless societies exchange goods both informally between families and also in special places that I refer to as barter markets (described below).

Donald Kurtz raised this issue a generation ago, and it is still a very useful distinction:

> The markets with which anthropologists have traditionally been concerned are the sites and places in which goods and services are traded and exchanged. This is a considerably different concept of the market from that with which economists are concerned. The market of the economist is a specific institution with rules of its own which function to regulate the supply of goods and services in relationship to demand and considerations of price so that presumed infinite human wants may be satisfied with the finite resources at hand. (Kurtz 1974:686)

Markets as Places

In the most common use of the term, a market is simply a *place* where goods and services can be exchanged, a geographically confined place for exchange between buyers and sellers. A typical broad definition in this sense is "an organized public gathering of buyers and sellers of commodities meeting at an appointed place at regular intervals" (Hodder 1965:57). This is similar to the OED definition of "marketplace": "A meeting or gathering together of people for the purchase and sale of provisions or livestock, publicly displayed, at a fixed time and place; the occasion or time of this." Quoting this definition in their review of marketplace trade in Latin America, R. J. Bromley and Richard Symanski elaborate:

> These markets are held in open market squares or plazas, in streets and open spaces, at roads junctions, and in public, municipal market buildings. Locally, markets are referred to by such names as *ferias*, *plazas*, or *mercados* in Spanish-speaking Latin America . . . *feiras* in Brazil, tianguis in Indian areas of central Mexico, catus in some Quechua-speaking Andean areas, and *marches* in French-speaking countries. Generally found in medium to large nucleated settlements, markets are also encountered in very small hamlets, particularly those located at nodal points in communication networks. A settlement may have one or more different market locations. . . . Places with one or more markets are generally referred to as market centers. (Bromley and Symanski 1974:3)

Kenneth Hirth defines markets to include those where exchange occurs with either barter or currencies. He notes, "These exchanges [in markets] may occur as single balanced and reciprocal events or in clusters as they do in a marketplace" (Hirth 1998:451–452). In this perspective, a market is both a place and system where goods and services are exchanged via a number of different mechanisms (and see Hirth 1996). In this sense, any beach in the Trobriand Islands where Kula merchants land and barter goods is a barter marketplace.

Such a definition of a market is quite useful. It effectively covers the exchange of goods in most settled village societies known in the ethnographic and historical record. We also call these institutions fairs. The definition would even cover some meeting areas in mobile societies that periodically meet to exchange goods, information, and so forth in a predetermined place.

Markets as a Set of Rules and/or Strategies

Another tradition in the study of exchange in the premodern world creates distinctions between different kinds of markets. This is a view of the market *as a set of rules and/or strategies for exchange* and not necessarily as a place where exchange takes place. This concept has been developed largely by economists and economic historians, though recently it has been taken up by anthropologists and archaeologists (e.g., Hirth and Pillsbury 2013). This more restrictive definition of "market" centers on the means by which value is established and the nature of transaction costs (in the sense defined by North [1989]).

Price-Making Markets

The strategy-based definition of markets describes what economists call price-fixing markets or price-making markets (North 1977). A price-making market is one in which prices are determined by negotiation between autonomous buyers and sellers without regard to social class or status. Supply and demand are the principal forces behind the negotiation and the ultimate exchange value reached, expressed almost always in a "price" based on some kind of independent unit of exchange or currency. In fact, money or some kind of medium of exchange is virtually a prerequisite for the efficient operation of price-fixing markets.

In price-making markets, competition between individual traders is played out via price negotiation. Price does not necessarily have to be expressed in terms of money, but some kind of unit of value is necessary. Individual sellers maximize advantage by selling more commodities at appropriate prices that fluctuate on a supply-demand basis. Buyers, in contrast, seek to bring the price down as much as possible. In economic terms, the price reaches equilibrium between supply and demand. Merchants enrich themselves by finding the right price to maximize profit, selling to a buyer with little regard to other factors. The principles of price-fixing markets are such a staple of Western economies that they require little explanation.

Price-making markets have never been documented in stateless societies. However, it is important to comprehend price-making markets so as to understand the other type of exchange that represents the vast bulk of all exchange in human history: barter exchange.

The Principles of Barter Exchange

Barter may well be one of the most archaic and universal means by which people acquired needed or desired objects they lacked as well as products they had never seen before. – Chapman (1980:43)

Barter exchange is the most common form of exchange in stateless societies between villages and different groups.[32] A barter exchange is one in which goods and services are traded without the use of currencies. Barter occurs in market *places*. Usually referred to as fairs or festivals, these places are ubiquitous in stateless societies. Over a generation ago, Chapman (1980) forcefully argued that barter was a universal exchange mechanism based upon an extensive review of the ethnographic and historical literature.[33] She defined "barter" as a purely economic transaction of direct exchange without money or intermediaries. Chapman documented barter in a wide variety of ethnographic cases around the world, including among the San, the Trobriand Islanders, the Nambikuara and Xingú of Brazil, people of central Sudan, aboriginal Australians, groups from Venezuela, and people of Suriname. Typical of the ethnographic record is Jeannette Mirsky (1937:60) discussing exchange among the indigenous peoples of Greenland: "There is no organized trade, but bartering is carried on. Bartering can include everything from a wife to a bone dart; it can be a permanent transaction or a temporary exchange." As she describes it, the bartering is a very simple exchange of like value for like value, as determined by the partners themselves using their own culturally specific criteria.

From a game theory perspective, barter exchange is good for all parties when it is done properly. By properly, we mean that the outcomes are fair, are reciprocal, and reach Nash or Pareto equilibria. As mentioned earlier, Nash equilibrium is when no actor's payoff can increase by arbitrarily changing his/her strategies. A Pareto optimum is a zero-sum outcome where no one can increase his or her payoff without negatively impacting another person in the exchange. Proper barter allows for the increased material and social well-being of the exchange partners, and it is one of the greatest promoters of co-operation in human society. A basket of corn for a bushel of wheat or a stone ax for a worked marine shell epitomizes this kind of exchange. Without rules of barter, complex human society could not exist.

There are a seemingly infinite number of variations of exchange equivalencies using goods to establish equivalencies, all without the use of money. As with *all* economic transactions, supply and demand affect the relative equivalencies (Smith 1980:876), but they do so within a larger social context. That social context involves kin groups, political and religious authorities, and the like. These exchanges use customarily agreed-upon equivalencies that also fluctuate, but they fluctuate far less than those seen in price-making markets, which use these equivalencies and supply-demand mechanisms to set prices.

The most efficient barter exchange, in economic terms, is when people trade resources of mutual marginal utility. As we mentioned, utility is the satisfaction one gets in consuming something of value. A marginal utility is the *additional* satisfaction one receives from using an additional unit of a valuable, such as a commodity. A good analogy is chocolate. One chocolate bar has high value for many people. If one has to consume it in a very short time, the second bar might have utility as well but surely the tenth bar has minimal utility to the point that it would make you ill to consume. Likewise, a cup of peanut butter has high utility for most people, while the tenth cup, under the same time constraints, has low marginal utility. Combined, chocolate and peanut butter have even higher utility. So the trading of each other's "tenth unit" results in each party giving away something of little value to themselves and receiving something of great value. This is an ideal interaction in which each party gains with no disadvantage to either. Now, add to this social context the fact that stateless societies often have little ability to store many products. These products have a short use span. In this circumstance, there is great incentive to exchange products with marginal utility to each other so as not to lose the value completely.

The ideal solution for both parties is to exchange resources of respective marginal utility, thereby creating a win-win situation in which all parties achieve Nash equilibrium. Applying this trivial chocolate example to something more serious, like marine resources and mountain goods, we can see how exchange can quickly develop between groups of people. On the coast, fish have increasingly low marginal utility because of their abundance, as do acorns or pine nuts in the highlands, again because of their local abundance. This is the fundamental principle of regional exchange between autonomous groups – trading something that is abundant to you but rare to another and vice versa.

A key point is that in both kinds of market exchanges – price-making markets as well as barter – people act rationally, either economically or socially, to maximize their material and social resources. People seek to get acquire more resources, achieve social status, and vie with one another for advantage in each of these systems. As anthropologists, we understand that different cultures value goods and services in different ways. It is almost self-evident for the anthropologist to note that the utility of goods and services in different societies will be culturally determined. Less obvious perhaps is the observation that what differs between these various kinds of exchange systems are the *rules* of competition and the corresponding *strategies* used by people as they seek to maximize utility, culturally determined as it is. And of course, the two systems are not mutually exclusive: barter and price-making markets frequently coexist, even in today's modern globalized world.

Some Ethnographic Examples of Barter Exchange in Stateless Societies

Barter trade is ubiquitous in stateless and premodern societies. Most barter exchange is external to the social group. That is, people engage in barter with others outside of their co-operative group. Internally within households and clans, exchange was dominated by deferred reciprocity mechanisms. In this section, I briefly summarize some of the many descriptions of barter exchange around the world in premodern contexts.

Barter markets or fairs were common in premodern, precapitalist Europe where we have massive documentation. While these are not stateless societies in the sense used here, they provide some analogs for barter markets in other nonmarket, nonindustrial contexts. In his celebrated *The Economic History of England*, Ephraim Lipson (1937:196) stated that "the bulk of internal trade in Medieval England was conducted in fairs and markets."[34] Like Chapman and other economic anthropologists for non-Western societies, Lipson defines "fairs" as places where "distribution and exchange . . . take place at periodical gatherings" (Lipson 1937:196). Most significantly for this discussion, Lipson notes how important religion was in these fairs. In fact, he agues that the first fairs and markets were more religious institutions than commercial ones. He points out that the fairs started as periodic congregations of the pious around famous religious shrines. Fairs were intimately linked to religious festivals, and in the beginning the two were largely inseparable.

Lipson also notes, "These periodical gatherings became the natural centres for commercial dealings" (Lipson 1937:197). The pilgrims were often merchants. In fact, Lipson's description reads like those of the many fairs in premodern societies from around the world. He relates how the "cloak of religion" guaranteed security for the pilgrims/merchants (Lipson 1937:197). What we see here is nothing less than a kind of ritualized economy in premodern Western society. Fairs/festivals were common throughout the Western and Mediterranean world throughout history and antiquity in the great states and empires. They were a mainstay of Roman Greece (Spawforth 1989). Fairs were common in Muslim Spain in the Middle Ages (Imamuddin 1981:129) and throughout India.

While it is true that a state and church exercised substantial power over people, these local fairs were largely outside of the rudimentary monetary system and state control. In this context, medieval religious fairs were an effective way to conduct gatherings using a ritual calendar and ritualized norms of behavior. We see many examples in which people from different villages or regions meet in a predetermined place to exchange goods and other valuables. Furthermore, we see that fairs were never merely about economic exchange. People met to gossip, find marriage partners (particularly important in exogamous societies), gamble, worship, and generally do what people do when they get together for common purposes.

Fairs and festivals were not confined to ancient states in the West. Meeting in predetermined places is common in stateless societies throughout the ethnographic record. As with the great states, ritual was an overriding feature in most instances. North American indigenous peoples, for instance, had extremely well-developed fair systems well before the arrival of Europeans. These fairs included relatively simple gatherings between two groups in neutral or safe areas. Other fairs were more elaborate and featured long-distance traders and other kinds of middlemen. Henry Dobyns explicitly (1984:23) refers to many of these aboriginal North American phenomena, located on major exchange routes, as "trade centers."[35]

The first European visitors recognized that the Mandans of the middle Missouri River drainage had a vibrant trade center. Dobyns describes it as an economic and cultural phenomenon: "Almost anything could and did change hands at Mandan trade fairs: weapons, tools of many kinds, jewelry, folk tales, dances, and songs. Courting at Mandan fairs led to gene flow between different ethnic groups as young people married exogamously" (Dobyns 1984:24). He also says that the Mandan held surpluses of basic foodstuffs such as maize and beans, which they traded for other goods. Citing several sources, Pekka Hämäläinen (1998:485) writes, "The corn, horses, manufactured goods, and luxuries of the villagers attracted both Indians and Euro-Americans from all directions, creating one of the busiest trading hubs in Native North America." There were also trade centers maintained by the Hidatsa and Arikara, as well as the

Western Comanches. Hämäläinen (1998:489) notes that the Comanches traded with the Pueblo peoples, "bartering hides, dried meat, and tallow for corn, beans, squash, tobacco, and pottery." They also traded with the Spaniards when came on the scene.

Dobyns also identifies trade centers among the Great Lakes Huron and the Northern Panya of the lower Colorado. Regarding the latter, the first descriptions of their trade centers by the Jesuits included "a very large trade in food – watermelons, melons, squash, beans . . . and other grains" (Dobyns 1984:24). This trade also included blankets and sashes, woolen textiles, and course homespun obtained by the Northern Panya from the Hopi and other Pueblo peoples.

Dobyns describes major commodity production of fish by indigenous peoples of the Columbia River drainage. Two of the first Anglos to visit the Dalles area, Meriwether Lewis and William Clark, reported more than 15 tons of dried fish (Dobyns 1984:28). As quoted by W. Raymond Wood, they reported, "Trade climaxed in the annual fall trade fair when Wishram and Wasco hosted numerous groups that traveled to The Dalles. Every ethnic group in the Columbia drainage basin seems to have participated either directly or indirectly in the fair" (Dobyns 1984:28). The trade items were many and varied. These included *Olivella* shell, body paint, hematite, turquoise, bison hides, steatite, sealskins, cactus syrup, and piñon nuts.

One notable example of periodic barter market fairs is found at Pecos Pueblo in what is today north-central New Mexico. The Pecos Pueblo architecture seen today was built in the fourteenth century. There were earlier occupations, but it was in the next century that the town became a major regional center. Other settlements were abandoned, but Pecos thrived at this time, due in part to its position as a major trading center. During the historical period, Pecos was an integral part of the Plains-Pueblo and Pueblo-Spanish relationships, particularly in regard to exchange (Levine and LaBauve 1997). Archaeologists have long noted that Pecos was an integral part of the extensive regional networks prior to European contact (Spielmann 1983). It is therefore highly likely that the descriptions of the barter markets held at Pecos in the earliest documents reflect pre-Hispanic patterns.

The site of Pecos is an impressive late prehistoric and early historic pueblo (Figure 5.5). Below the pueblo is a large plain that rolls on a low slope. Pecos is an ideal example because we have good documentation from the early sources that depict a society that developed prior to the European contacts. Early descriptions tell how this plain filled up with tents from late August to October with hundreds of Apaches who arrived to "trade and traffic" (Kessell 1995:134). The "nomads" brought hides, leather, dried meat, tanned skins, bone and flint tools, salt, and slaves. In return, they bartered maize, other agricultural foodstuffs, blankets, pottery, and turquoise (Kessell 1995:136).

Figure 5.5. Hypothetical reconstruction of Pecos Pueblo in the early historic period. Drawing by Kathryn Killackey.

Joel Janetski writes:

> Trade fairs were held at Taos and Pecos as well as other pueblos and were attended by various nomadic groups (Comanche, Apache, etc.). Here traditional enemies met together under truce to exchange goods, especially foodstuffs, but also objects necessary for ritual activities (Ford, 1983; Simmons, 1979). The Pima held similar fairs in the Phoenix area. . . . The Southwest trade fair pattern intensified and was controlled in part by the Spanish to enhance economic purposes in the 1700s. . . . Fairs were held in the late summer and fall when agricultural products were ready and bison hides and meat were available. (Janetski 2002:347)

The reconstruction in Figure 5.5 illustrates how people would meet at the Pecos site. The tents in the foreground represent the nomadic groups mentioned in the above quote. They would set up camp well outside of the buildings at the site, in the large open area in front of the buildings. The festivities would last for as long as two months. While economic exchange was a central feature of this activity, there were many other social activities that occurred at the site. Even though there were at times hostile relationships between many of the people meeting in this predetermined time of year, the trade fair flourished without conflict.

Precontact peoples in Hawaii also had elaborate barter fairs. Hawaiian society was very complex; the leaders had coercive powers and had a borderline state-level political system. It was not a market society and was certainly not an industrial society; nor was it affected by European expansion at the time of many descriptions of its barter fairs. Captain Cook's diary entry of December

21, 1778, notes, "They understand trading as well as most people." The missionary and early ethnographer William Ellis visited Hilo in the 1820s, being one of the first nonnatives to visit the area. He writes that there were markets or fairs on the banks of the rivers. People from all over the island brought mats, cloth, dried fish, and other goods. His description of the barter market fair is quite fascinating:

> From bank to bank the traders shouted to each other, and arranged the preliminaries of their bargains. From thence the articles were taken down to the before-mentioned rock in the middle of the stream, which in this place is almost covered with large stones. Here they were examined by the parties immediately concerned, in the presence of the collectors, who stood on each side of the rock, and were the general arbiters, in the event of any disputes arising. To them also was committed the preservation of good order during the fair, and they, of course, received a suitable remuneration from the different parties. (Ellis 1826)

Robert Hommon (2013:107–109) reinforces the extensive gift giving and barter exchange that characterized the Hawaiian state. It is quite fascinating that these market fairs were apparently suppressed with the ascension of Prince Liholiho to the kingship (taking the name King Kamehameha II). The period around 1819 was a revolutionary time in Hawaiian society, marked by the abolition of the traditional *kapu*, or taboo, system (Seaton 1974). The overall effect of this revolution was to concentrate power into a more centralized elite class. We can surmise that the market fairs were a source of power for the lesser elite and commoner alike and were therefore a target of an aggressive monarchy. This is a common pattern seen in comparative analysis. As states develop coercive powers, they monopolize important trade goods and suppress the unstructured and difficult-to-control trade fairs where so much wealth is exchanged.

The ubiquity of fairs or regional trading festivals in historical sources in societies that had no contact with each other suggests that they were a very common strategy that evolved in all human groups beyond the hunter-forager band under the right demographic conditions.

Establishing Value in Barter Exchange

By the first decades of the twentieth century, Western economists had effectively and uncritically accepted the idea that supply and demand was the principal if not only means by which value was established in *any* economic system. Of course, having a medium of exchange in a price-making market system was fairly straightforward and allowed for the unambiguous setting of a price. People would negotiate and a price would emerge. Over multiple episodes of negotiation through time, imprecisions would theoretically be averaged out and a price that reflected the demand for the good and the total supply available would emerge.

But another mechanism, known as the labor theory of value, was proposed by Adam Smith in his celebrated work *Wealth of Nations:* "The real price of every thing, what every thing really costs to the man who wants to acquire it, is the toil and trouble of acquiring it." He went on to be more specific: "[Goods] contain the value of a certain quantity of labour which we exchange for what is supposed at the time to contain the value of an equal quantity [of labor]" (Smith 1937[1776]:Book 1).

Western economists criticized the labor theory of value for a number of valid reasons. One obvious reason is that it is very difficult to calculate the total labor input in any commodity in an industrial production process. It is simply too complex. Even in Smith's famous pin example, there are countless labor inputs for acquiring the raw materials, acquiring the fuel, building the factory, and so forth, making it virtually impossible to disentangle the individual contributions out of the aggregate. Under these circumstances, the labor theory of value is of little analytical use.

However, many of us who study non-Western and nonmarket economic systems feel that there is still use for this concept in barter exchange, particularly in stateless societies where production is not technologically complex. In my own extensive fieldwork in societies that were largely subsistence-based in the early 1980s in the Andes, barter exchange was quite extensive for locally produced commodities. Even when cash was being used to purchase fancy goods, the question of the exchange value of locally obtained subsistence products was a serious issue. What is the value of a basket of potatoes, a bucket of fish, a pile of wool, all of which was obtained by hand a day or two earlier? What I discovered in many conversations was that people would explicitly say how many hours it cost to make or collect certain commodities. Fish and potatoes were relatively easy to calculate. The value of livestock depended on quality as well as age. Since there are only very marginal costs to increasing flocks by a few individuals, and since children were the primary herders, costs were considered relatively low. However, as Flannery, Marcus, and Reynolds (1989:100) report in their ethnoarchaeological study of indigenous people in the central Andes, there were more than a dozen terms for llamas. Each type had a separate cost in the market, and raising and keeping involved a different amount of labor. Some animals were special based upon inherent characteristics. These unique traits were not understandable with a labor theory of value but were explainable based upon supply-demand factors. However, the vast bulk of the exchange value of the livestock was indeed calculated implicitly in how much time was spent to raise and care for the animals.

The manufacture of pottery is relatively straightforward. The total number of people needed to make pots remains small, all confined to the co-operative group. Stone tools and stone preciosities are other important commodities in stateless societies. The acquisition and manufacture of these objects are easy to understand, and the total effort required to make them is easy to define. The same can be said for any number of foodstuffs, including fish, fruits, grains, and the like. Ask a farmer how much effort it takes to grow and harvest a basket of maize, and he will tell you in a very precise manner. Potters, metalworkers, and other specialists can likewise provide a fairly precise account of how much time it took to create a good. Ethnographic data suggest that there is interplay between supply and demand and the recognition of what a fair value is based upon the labor necessary to manufacture any good. The term "fair" here is critical, and I use it in an anthropological game theory sense.

Competitive Strategies in Barter Exchange

The utopian idea that people in non-Western societies did not have any kind of profit motive in exchange, an idea that permeated substantivist theory, today comes across as naive and even patronizing. It is also empirically incorrect. Quite to the contrary, most comparative behavioral and social scientists that examine this problem would maintain that actors in all kinds of exchange systems seek to maximize utility. I agree with the great Raymond Firth in his critique of Malinowski that the desire to maximize individual advantage is a human universal.[36] As anthropologists, we realize that the advantages that are maximized are not universal. Rather, different cultures value different things, and this fundamental anthropological fact must always be part of any analysis of any economy. Furthermore, since the contexts in which an exchange takes place are distinct in barter versus price-making market exchanges, the corresponding maximizing strategies are likewise different and are appropriate to the particular system in which resources circulate.

As a general rule, actors in barter markets customarily understand the exchange value of goods and services. One of the greatest barter markets in ethnographic history is that of the Kula ring. Malinowski notes, "All the trade is carried on in exactly the same way – given the article, and the communities between which it is traded, anyone would know its equivalent, rigidly prescribed by custom" (Malinowski 1921:14). Malinowski in particular was extremely fond of citing "custom" as the means by which price (equivalency) was established. But this is not a very satisfactory explanation.

Where do "customary" valuations come from? Instead of constantly shifting prices determined by individual negotiations as in price-making markets, prices in barter markets result from a broader social consensus involving buyers, sellers, producers, and political authorities. Supply and demand affects all economic systems. But in barter systems, the fundamental understanding of how much labor goes into making a product is key to setting price. This is not the time or place to develop a new labor theory of value for barter exchange. Suffice to say that for nonindustrialized production of goods in small-scale societies, people have a very accurate grasp of how much labor is required to produce most products. The customary back-and-forth between buyers, sellers, ritual experts, political authorities, and so forth generally seeks equivalency in *labor input* for these goods.

Scarcity of an otherwise obtainable item occurs when a political entity restricts access to a naturally occurring good or when commodities are created by highly specialized labor that cannot meet demand. In these instances, supply is obviously less elastic. However, in both of these instances, supply is often controlled by entities above the household. The ideal of the price-making market, in which exchange partners trade with little regard to social class, is lost where supply is directly controlled by political authorities.

A good example in archaeology and ethnohistory is obsidian. Obsidian at its source is usually vast and there is effectively an unlimited supply relative to the total possible demand. However, when a political entity or group controls the volcanic source, it can restrict access to various subgroups in a society. Households that do not have access to this source cannot barter or sell in the marketplace but instead have to work within a larger social context for procurement (production) and exchange to take place.

In barter markets, exchange partners of familiar people are maintained by elaborate social rules, another difference from price-making markets. Economic exchange is highly ritualized, much more so than in price-making markets.[37] The Kula ring of the Trobriand Islanders is one of countless examples of the ritualized nature of trade in a barter system. Individual traders have partners with whom they must trade for life. As Rolf Ziegler (1990:143) reflecting J. P. Singh Uberoi's (1971) earlier work notes, the Kula exchange is not a market type exchange, but rather it involves lifetime partnerships that are heritable. Kula traders can have many or few partners, depending upon their social status. The status of the partner shifts through time based upon the quality and quantity of the Kula items each man acquires (bracelets and necklaces). There is a big debate in the literature as to whether the Kula trade is linked to the parallel trade in domestic items. I view the two as inseparable and see the status of the Kula partners as directly linked to their success in the domestic trade, at least at the turn of the last century, when Malinowski conducted his famous ethnographic studies.[38] Status goes up or down depending upon whether one provides quality goods in sufficient quantities on trading expeditions. The quality of the Kula items reflects the value that each trader adds to the domestic exchange. The Kula ring also shows us that some traders get rich while others get poor, depending upon their status in the Kula ring. Since supply is elastic, the customary prices of commodities remain relatively stable.

The most important theoretical point here is this: *The key strategy of a trader getting rich in a barter system is not to sell more at a constantly negotiated price, as in price-making markets, but rather to increase his or her number of trade partners*. In other words, given that exchange values in barter market systems tend to remain relatively stable compared to those in price-fixing markets, where fluctuations are required for profits to be earned, *people in barter markets get rich by acquiring more partners in their factions and not by undercutting the competition with price*. Faction building is key. And faction building requires a set of strategies and skills that differ from those used in price-making markets.

The ability to acquire partners for exchange whom one can trust is a dominant theme in the ethnographic literature. For long-distance trade to be successful, one needs both related and unrelated partners across a landscape. As a general rule, if you restrict yourself to genetically related kin, you will not get rich in a barter market system. You must reach out to nonkin. May Mandelbaum Edel (1937:129) describes a fascinating institution between unrelated men of the Bachiga of East Africa. This is a pact of "ritual brotherhood" that binds two men to each other for mutual protection. According to Edel: "It is believed that this blood will swell up and kill anyone who breaks such an oath." This oath allows men to travel outside of their home territory with a degree of safety.

Evans-Pritchard describes similar blood brotherhoods among the Zande:

> Blood is often exchanged solely for commercial purposes. A man who is travelling in foreign countries will make blood-brotherhood with a native, who thereby becomes responsible for him.... When a man is travelling through a strange district of his own country he will establish a contact which will ensure his safety by finding some one of his clan, or of his mother's clan, to whom he can introduce himself as a kinsman. In a foreign country this is impossible, and here blood-brother-hood takes the place of blood relationship as the traveller's passport. (Evans-Pritchard 1933:372)

The effect of these kinds of blood brotherhoods is similar to the formation of friends among the Trobriand Islanders. It establishes a level of trust between people in different regions, provides protection, and facilitates barter exchange. Luise White (1994:359) writes that such pacts were sanctioned by very strong magic and taboo: "Any breach of reciprocity or loyalty would be punished by supernatural sanctions, sometimes enacted by super-natural agents, sometimes by the blood itself." Here we see again the importance of magic and taboo as punishment, in the evolutionary game theory sense, in keeping these kinds of dyadic co-operative relationships functioning.

From an anthropological game theory perspective, this is a social arrangement that overcomes defection issues, backed with a severe punishment for cheating. Edel noted this pact by individual choice between men from different clans. It creates a relationship that is friendly and hospitable to the one traveling, converting a potential enemy into a "cooperative trade friend." She goes on: "Whatever trade there is is carried on through pact brothers" (Edel 1937:129). One can deduce that this arrangement is costly for both parties, but as one increases the number of blood brothers, one increases one's trading opportunities and by implication increases one's prestige and access to wealth.

These kinds of formal relationships between unrelated people facilitate barter exchange. Many of the strategies of faction building that anthropologists have studied can be seen from the perspective of enhancing the size of your exchange group. People feast their partners, seek strategic alliances with other families or factions, engage in strategic intermarriage, build up their household numbers with fictive and real kin, try to accumulate deferred debts to be converted in major potlatch events, and keep their competitors at bay with force or forceful persuasion. In the Kula ring, we see this process at work:

> The exchange of gifts creates or reinforces relationships of alliance between individuals and the groups of which they are representative.... In the *kula* ring, the partnership establishes an alliance with political overtones.... It opens the channels of substantial trade and social intercourse. But a man's position in the ring is not predetermined and static. A participant starts from a position of relative capital advantage or disadvantage... from this position, if he is ambitious, he may maneuver to increase the number and scale of alliances, and hence his command over the flow of wealth and the degree to which his good will is sought, and hence his status and prestige. (Belshaw 1965:19)[39]

Malinowski further describes the complex web of reciprocity, redistribution, and labor obligation involved in Kula exchange:

> The chief gives the distribution of food, in which the others participate, and this imposes on them the duty of carrying out the expedition.... The *uvalaku* [ceremonial competitive expedition] is an enterprise partially financed by the *toli'uvalaku* [sectional chief], and therefore redounding to his credit, and bringing him honour; while the obligation imposed on others by the food distributed to them, is to carry on the expedition to a successful end. (Malinowski 1966[1922]:209)

We therefore see how leaders can use feasting and other kinds of potlatching behavior to obligate others to finance highly co-operative activities. In other words, as the social network increases, so does the number of trade "partners." The maximization of advantage is achieved by increasing the number of trading partners through noneconomic incentives, not by just earning a profit over price differentials.

Comparatively speaking, all societies have feasts of some sort. Anthropologists have long recognized the importance of feasts in labor organization. As Michael Dietler and Ingrid Herbich note:

> The use of feasts to mobilize collective labor has been a widespread and fundamental economic practice of societies around the world. In fact, variants of the practice are so strikingly omnipresent in the ethnographic and historical literature that a good case can

> be made for acknowledging it both as a virtually universal feature among agrarian societies . . . and as the nearly exclusive means of mobilizing large voluntary work projects before [capitalism]. (Dietler and Herbich 2001:240)

Feasting is one of the most common features of human sociality noted by ethnographers. In their review of nonstate societies in the Americas, Gary Feinman and Jill Neitzel (1984) found that in more than one-third (24 of 63) of all groups analyzed, feasting was a function of leadership, and about two-thirds of chiefs (41 of 63) were expected to provide ceremonies, many of which appear to have functioned in a manor similar to competitive feasting (Feinman and Neitzel 1984:Table 2.6). Almost three-quarters (46 of 63) of the chiefs were expected to provide either feasts or ceremonies. Most significantly, in more complex political contexts, where the total number of tasks assumed by chiefs was the highest, all leaders (16 of 16) had to provide ceremonies and almost all (14 of 16) reportedly hosted feasts (Feinman and Neitzel 1984:Table 2.8).

Equally compelling in this article, it is likely that many kinds of low-level competitive feasting were reported ethnographically as "ceremonies." The example of the Luiseño of southern California, a sedentary hunter-gatherer people whose chiefs were not included in Feinman and Neitzel's list as hosts of feasts, is a case in point. The chiefs of this group are described in an ethnographic account:

> Rituals were strictly governed by religious chiefs and shamans, who comprised a hierarchical power pyramid dominated by the village chief, an assistant, a council, and a secret society. . . . Most participants in rituals were paid. A guest ritual leader and his assistants – from another village or moiety – officiated. Great quantities of food and treasure were distributed at these affairs. (Bean and Shipek 1978:556)

The Luiseño people did not have a very complex political economy. They would be at the very low end of a continuum of complexity in terms of leadership authority and power. Yet even in this nonagricultural population, we can see a nascent competitive feasting, with chiefs hosting others from outside the kin group or village. In other words, Feinman and Neitzel's article indicates that virtually all of the more complex societies analyzed had feasting, and many of the less complex ones had ceremonies that can be seen as competitive feasting on a much more modest scale.

As one reads the ethnographic and historical literature, it is clear that feasts, defined broadly, constituted perhaps the most important social activity in stateless societies. Humans display an incredible capacity to create social realities that give meaning to life. There is a tendency in some quarters in the Western academy to romanticize the bygone life of the small village. These illusions aside, living in the same village with the same 200 people, and working the same fields, streams, and oceans for sustenance day after day, would be monotonous.[40] Feasts and ceremonies are not isolated events for occasional fun; they are an integral component of the social life of people in stateless societies. Numerous examples in the ethnographic literature emphasize the central importance of feasts. Even more significant, perhaps, is that the preparation for and anticipation of feasts take weeks of work beforehand and enrich the lives of the community members.

Henry Ling Roth, describing a Dyak group in Borneo, paraphrases the work of an earlier ethnographer:

> The principal festival among the Sea Dyaks is the "Gawei Burong," also called Gawei Pala, from the head which is feasted. It is given after harvest, but not every year. When a house gets a good yield of paddy, and is so inclined, the feast is organised, and to that house the neighbouring population is invited.... The preparations extend over a length of time, and cost considerable labour and trouble. Some of the arrangements are carried on with certain rites and formalities, and the impression is conveyed that something very important is about to take place. (Roth 1968[1896]:256)

A work published in 1871 by a Christian missionary, someone who at first glance would not be expected to view potlatching with a sympathetic eye, in fact notes positively that Dyak feasts brought people together for a variety of important ends. At feasts, "topics of common interest are discussed and plans formed, so that the feast assumes something of the character of a council, and affords one of the best opportunities for indulging in their intense love of 'bechara.'"[41] He goes on to say, "Sociability, friendship, love of pleasure, religious instinct, and traditional custom, are all here united" (Roth 1968[1896]:257).

Within an anthropological game theory perspective, competitive feasting and other ceremonies, imbued with religious and ritual significance, constitute a principal means of strategically rewarding people to work differently in specialized labor organizations (see Spielmann 2002). The logic is straightforward: working in a new kind of labor organization produces more surplus without affecting household subsistence activities and without requiring more time from the individuals in the community. That surplus, in turn, can be used by aspiring leaders to host ceremonies and feasts that provide goods and social occasions otherwise not available. Autonomy is preserved, because any household can pull out at will, as is evident in the ethnographic record. In short, everybody wins because with the same effort, there is a greater surplus generated and available for community members. Prestige, bigger factions, and the possibilities of becoming even more powerful reward the organizational work by the managerial leader.

Potlatching

> Only one thing enrages me, when people eat slowly and a little only of the food given by the great double chief. – Neqå'penk.em, Kwakiutl war and potlatch chief (in Boas 1966:97)

People in stateless societies differ from people in modern and certainly Western ones in the degree to which nonmaterial incentives affect their behavior. This is not because people are fundamentally different in Western versus non-Western societies. Rather, in stateless societies in particular, there is much less opportunity to store or hoard wealth. In mobile societies, this is patently self-evident. In agrarian and settled societies as well, food cannot be preserved for any great length of time, and without the use of money, it is very difficult to store any kind of perishable resource. As a result, the most effective use of surplus resources in such contexts is to give them away and obligate people to reciprocate in the future. You are in effect banking reciprocal obligations for future use.

In this light, the ethnographic record of small nonstate groups informs us that a significant amount of economic behavior in these societies is geared to the acquisition of prestige and status enhancement as much as it is focused on the acquisition of material wealth.[42] We can therefore distinguish between production and exchange for the enhancement of people's *material* well-being and that which enhances one's *social* well-being. Both have utility. Both are critical. The methods of production of social and material utility are very similar. However, the nature of exchange is quite different. Engaging in positive social and economic gift-giving behaviors to enhance one's status and prestige is here called potlatching, after the famous social institutions described by anthropologists in the northwestern part of North America among the Kwakiutl and related peoples.

We can define potlatching as a strategy of feasting conforming to culturally recognized norms of reciprocity while simultaneously indebting people and enhancing the host's status. Potlatching is a form of social competition in which goods and feasts are exchanged for political support. That support is an obligation incurred on the recipients of the goods in a potlatch. As a kind of "competitive ceremonialism" (see Cohen 2010) or "competitive feasting," potlatching is highly ritualized, with virtually all people knowing the rules of participation. Unlike barter exchange, the maximization of prestige is central, with material utility being less important, though always in the background. The main objective is political prestige; the accumulated wealth becomes the means to that end, not the other way around. Classic potlatching is also highly competitive, an occasion to shift social power between different groups. Potlatches, also known as competitive feasting or festivals, constitute a major way of promoting production and represent the means by which aspiring elites compete for prestige and influence.

As a structural form of human social organization, potlatching behaviors appear to be universal. The fundamental social mechanism of potlatching is through feasts, organized by a political authority to enhance his or her power and influence in that society. As Roth (1968[1896]:257) notes about some groups of Dyaks of Borneo (about the same time Boas first published his fieldwork about the Northwest Coast cultures), "the social character of the feast is of more practical importance than the religious, and feasting the guests occupies more attention than feeding the gods." He goes on: "The obligations of friendship are acknowledged, and hospitality even carried out to prodigality." It is an occasion, says Brooke, of "social mirth and joy," where the "dignity, the wealth, and position of the chiefs are brought prominently before the many" (Roth 1968[1896]:257). This description of Dyak feasting could be equally applied to the Kwakiutl or any other culture where the hosting of feasts is central to people's lives.

It is not surprising, therefore, that hosted people are effectively obliged to accept the invitation. As shown in the opening quote of this section for the Kwakiutl, among the Dyaks as well, refusing an invitation to a feast is considered an insult to the host: "To refuse to attend would not be regarded as any indignity done to their religion, but as a sign of ill-will to the inmates of the house" (Roth 1968[1896]:257).

Potlatching from this view is a political strategy playing off norms of reciprocity. It is a kind of elite-directed redistribution conducted with the explicit goal of obligating people in the future. By accepting the feast giver's goods and food, the participant is obligated to provide something of value in the future. In most cases in the ethnographic record, the value reciprocated is labor. Michael Dietler argues persuasively that one of the most important roles of drinking feasts in small-scale premodern societies centers on the mobilization of labor through "work-party feasts." The hosted feast reinforces reciprocal obligations and other socially prescribed exchanges, particularly of labor. Alcohol, "in effect, is a medium that allows surplus agricultural produce to be converted indirectly into labor, prestige, 'social credit,' political power, bridewealth, or durable valuables" (Dietler 1990:366–370).

As is well known in the ethnography of Big Man societies, successful feasting and ceremony serve to build up a number of reciprocal obligations. As these obligations add up, entire households can directly or indirectly be drawn into larger work units. The existence of such a process explains the quote above. It explains why Neqå′penk.em was angry at people who would not accept his gifts with alacrity – they were resisting his attempts to obligate them to him in the future.

The classic ethnographies illustrate how feasting is directly tied to chiefly authority and economic activity. Douglas Oliver's ethnography of the Solomon Islanders located near the Trobriand Islanders reads like a textbook on how leaders emerge through competitive feasting:

> Much as natives treasure their staple foods, however, they do tire of the monotony of taro.... Consequently, when feasts do take place – providing the native with rich

> puddings and succulent pork – they are appreciated all the more.... It is not to be wondered at that the Siuai honor men who organize feasts and provide food delicacies. The feast-giver need not use his own wealth to provide a banquet; he may – and frequently does – go to relatives and friends for financial aid or pigs, and this aid is not always reciprocated. But the significant factor is that the host, and not his financial backers, receives acclaim, even though guests are aware of this arrangement.
>
> A man who is ambitious for social renown generally begins by inducing his village mates to build a club-house for him, this labor he repays with a feast.... If he is still ambitious, he can attempt to extend his reputation to other communities. (Oliver 1949:14)

Oliver's description of competitive feasting continued in great detail about how chiefs could extend their power beyond the village. It is a fascinating read and deserves a long quote:

> The most effective way to do this [increase power to other villages] is to sponsor a *muminai*.... This *muminai* is a competition in which the participant is placed in a position of grave responsibility. If he succeeds he wins great renown; if he fails he is in danger of extinction.... The participant begins by modestly fêting a petty chief in a neighboring village. He invites that individual to a feast and showers him with pigs and vegetable puddings. And if the guest cannot repay within a season or two he loses prestige, and his "benefactor" looks around for bigger game. An ambitious *mumi* who is also shrewd will carefully estimate his victim's ability to repay, and will go just beyond that limit, for the repayment must be equivalent to or more bountiful than the original gift. Sometimes two closely matched rivals will compete for years, back and forth, until one of them bankrupts himself and all his relatives. After that the victor takes the measure of a more worthy rival and tries him out. There would appear to be no limit to this expansion, but actually the aggrandizer's ambitions do not extend beyond the Siuai tribal boundaries.
>
> To the spectator ethnographer, defeat in a feast competition does not seem to entail any great hardships. The victim still carries on the every-day affairs of life. He still retains his status in his own village, and continues to attend feasts in neighboring villages. He may even attach himself to his victorious rival and assist him in preparing new feasts. Metaphorically speaking, however, he is condemned to die. (Oliver 1949:14–15)

Likewise, in *Primitive Polynesian Economy*, Firth notes that feasts occur in times of economic abundance: "A chief will hold a [major] feast only if the crops have been good that season, and if he has accumulated large reserves. The first three [kinds of feasts] are regarded as obligatory but after the fakatanata [major feast] the others lie entirely at his discretion" (Firth 1965:223). He goes on to note,

> The other feasts given by a chief follow the same general form – a vast assemblage of food, a gathering of all the chiefs together, a ceremonial mutual thanksgiving, some acknowledgment of the gods, and a distribution of the food.... Each chief gives such a feast once or twice in his reign and for it he invites the other chiefs to lend their nafa, huge bowl-shaped troughs, into which taro is grated until they are heaped full. (Firth 1965:227)

Why do chiefs hold feasts in the first place if they so are costly? For Firth, after working there for so long, the answer was almost self-evident:

> A Tikopia chief [also] plays an important role in ceremonial feasts outside the series specifically associated with him as just described.... Only chiefs, their near relatives and wealthy elders can afford such a gesture [and] as the ... primary feast giver is thus usually a chief, he is able to command on general grounds a large supply of food and also reaps the major prestige of the event. (Firth 1965:229–230)

The answer, of course, is prestige: The chief's clan becomes a focus of interest and acquires prestige for a good feast. This prestige, according to Firth, is imbued with authority. That authority, even in a society in which households and villages are virtually autonomous politically and economically, can be utilized to control people's behavior. The Tikopian feast is a focal point for the assertion of that authority within the chief's lineage:

> Attendance at the ana [feast] and participation in the economic contributions does in fact help to support the Tikopia system of authority. Once elected as chief, a man is able to control the behaviour of the members of his kinship group upon some of their most vital and personal issues, such as securing adequate food, freedom of recreation, and in the last resort even their right to live. (Firth 1965:230–231)

In an article titled "Social Advancement in Guadalcanal, Solomon Islands," H. Ian Hogbin (1938) reinforces this general principle with more firsthand ethnographic observations. In this small-scale society in the Koaka district, the largest village is a mere 171 people. Hogbin describes it as having five matrilineal clans that live in about seven hamlets. These groups co-operate in war and other activities, but they are "never united under a single authority" (Hogbin 1938:289). He notes that the position of clan leader is not hereditary and that "authority depends on personal qualifications and generosity in providing feasts and otherwise distributing wealth." He goes on: "Any ambitious young man can acquire wealth if he wins the respect and approval of his relatives" (Hogbin 1938:290). He observes that the term "chief," implying some kind of authority, would not be appropriate in this case because they have no coercive power. What is clear is that status is achieved by behaving according to accepted norms and acquiring and redistributing wealth, usually in feasts.

Hogbin (1938:291–293) describes how a young man in his thirties accumulates a number of fat pigs and other things of value and then announces that the will build a big house. He invites relatives and neighbors. When it is known that there will be a big feast, the people gather to build the house. Generally unmarried men and older youths are the workers. They receive not only the feast but the obligation of the feast giver to make contributions to their future bride-prices. At some point, most or all of the villagers contribute to the construction. The project is complete when the structure is up; a large feast is given to celebrate the completion of the house. It is important to stress how central these feasts are to the community:

> This feast had been looked forward to by everybody for days and even weeks beforehand, and whenever groups of villagers met for a chat in the evenings it always formed the main topic of conversations. It was obviously a highlight of native life, a chance to have a good time. When the day arrived, the whole village was a scene of high-spirited gaiety.... Strangers came along from villages five and six miles away. (Hogbin 1938:297–298)

Hogbin (1938:298) goes on to say that the receipt of food was not the only reason that people came to the big feast. They also came to gossip and enjoy themselves. "They related to one another accounts of previous feasts." Feasting, in fact, is effectively the only means by which a man can rise to greatness. After the house is built, he expands the amount of land that is cultivated. He marries more women and continues to have feasts: "Every event of importance is celebrated with a feast." Dance feasts were particularly important in this society, going on periodically. As Hogbin (1938:301) notes, "the whole community thus owes every mwanekama [feast-giver] who holds a dance a very great debt, and his reputation is always enormously enhanced in consequence." In this society, as in so many stateless societies, status is acquired by generosity in highly structured and ritualized events. Once his status is established, the leader becomes the organizer of other communal undertakings (Hogbin 1938:302).

Competitive feasting of this nature is not restricted to the cultures of Polynesia. Perhaps the undisputed champions of such feasts in the ethnographic literature are the indigenous peoples of the Northwest Coast of North America, from which I borrow the term "potlatch." The ethnographies of Franz Boas and later scholars describe some of the most intense forms of status-seeking behavior by leaders using competitive feasting in ranked stateless societies. Boas noted that the potlatch, or more precisely the distribution of property during a potlatch, is the method of acquiring rank (Boas 1966:77). Material wealth is very important, and its possession is a cultural value, but "it is not as much the possession of wealth as the ability to give great festivals which makes wealth a desirable object to the Indian" (Boas 1966:79). Small-scale competition began with boys who were given blankets and then expected to strategically loan them to others for high interest. Older chiefs and aspiring chiefs held contests in which blankets, coppers, fish, and other valued objects were offered to rivals. Eventually, some people emerged victorious while others became incapable of repaying debts. Failure to repay a debt forced some people into a subordinate status to the holders of that debt. In the most extreme form, material wealth was actually destroyed:

> The rivalry between chiefs and clans finds its strongest expression in the destruction of property. A chief will burn blankets, a canoe, or break a copper, thus indicating his

> disregard of the amount of property destroyed and showing that his mind is stronger, his power greater, than that of his rival. If the latter is not able to destroy an equal amount of property without much delay, his name is "broken." He is vanquished by his rival and his influence with the tribe is lost, while the name of the other chief gains correspondingly in renown.
>
> ... The most expensive sort of feast is the one at which enormous quantities of fish oil ... are consumed and burnt, the so-called "grease feast." Therefore, it also raises the name of the person who can afford to give it, and the neglect to speedily return it entails a severe loss of prestige. Still more feared is the breaking of a valuable copper. (Boas 1966:93)

Boas went on describe how the potlatch escalated into fierce rivalries between competitive chiefs: "The rivalry between chiefs, when carried so far that coppers are destroyed and that grease feasts are given in order to destroy the prestige of the rival, often develop into open enmity.... While the feast is in progress the host sings a scathing song ridiculing his rival and praising his own clan, the feats of his forefathers and his own" (Boas 1966:95–96).

Ethnographic data from eastern North America also demonstrate that potlatching was an integral strategy of chiefs in the ranked agricultural societies of the Huron League. Elisabeth Tooker says that one characteristic of a successful chief was "free expenditure" of wealth (Tooker 1964:44). Feasts for friends and "leading men" of the village could include 200 to 400 people (Tooker 1964:72). She describes the great quantities of food and says that attendance was virtually compulsory, that everyone had to eat all that was given, and that other villages were often invited (Tooker 1964:72–76) – typical of competitive feasting worldwide.

Leaders can strengthen their factions and lock others into a series of obligatory reciprocities through potlatching. This increases their own higher status by providing more resources for the group as a whole. As Jeanne Arnold (1996:110) aptly paraphrases Brian Hayden (1994): "Feasts were thus a forum, a very competitive one, to give gifts, attract labor commitments, and establish debt." Leaders and the population at large benefit by regular competitive feasting and the hosting of other ceremonies that promote and maintain specialized labor organization in moderately ranked societies. As mentioned above, the surplus derived from the more productive labor and from long-distance trade can be put back into the population at large via the redistribution of politically significant goods at feasts. In turn, the population at large does not have to work more – just differently in specialized labor – and they receive goods that they could not get otherwise. Potential leaders compete with one another for factions. The successful leaders garner larger factions, give larger feasts and ceremonies, maintain larger and more complex labor organizations, and eventually eliminate their competitors.

Abraham Rosman and Paula Rubel (1971:48) note, "Among the Tlingit [of the northwest coast of North America] there is an intimate interrelationship between potlatch, funeral rites, and succession to office formerly held by the deceased." Competition occurs via dancing, eating, and rhetoric. Early ethnographers noted that speeches were highly charged, with attendees fixated on each word. At times violence would break out at perceived and real insults. In other words, potlatches were not tranquil affairs where people met to exchange and interact. A huge amount of resources – acquired from organizing labor at the household and clan level – was expended in conducting these ceremonies.

In short, barter, potlatching, and fairs can be understood in an anthropological game theory framework. We see several principles at work here. Rudimentary economies of scale, both Type 1 and Type 2, provide surplus resources. The concept of reward and reciprocity permeates the potlatch. Aspiring leaders organize the labor from these debts to create resources through specialized work groups, effectively taking advantage of the efficiencies of economies of scale. The entire potlatch is, in one regard, a massive act of costly signaling, reinforcing group cohesion and permeated with other kinds of prosocial behavior, particularly punishment for non-co-operation and taboo.

Punishment and Taboos

People who violated fishing times and rules had their goods plundered or their canoes split up. – Firth (1929:227)

Thus, if and when the group of relatives turns against a member, solidarity is demanded of them on pain of possible supernatural sanctions. – de Laguna (1952:8)

Punishment is a significant element in anthropological game theory used to maintain co-operation among individuals and groups in stateless societies. Punishment is a form of "altruistic" behavior and therefore is a difficult trait to explain from a purely evolutionary biological perspective although as I mentioned above, the concepts of kin selection and reciprocal altruism work. This is because a punisher takes on some cost that benefits the group by sanctioning antisocial behavior. In humans, the theoretical issue seems, at least to me, far less problematic. In simplest terms, a person engaged in punishment receives a payoff in two ways: first, he or she gains prestige, an intangible utility that is a prime motivator for humans, unlike any other animal. Second, there is in fact at least a marginal return to the punisher in the form of a free-rider returning to the group as an active contributor. The material reward, combined with the immaterial prestige gained, is more than enough payoff to account for such behavior by managerial leaders in small groups. This may be mathematically difficult to model because prestige is an almost impossible thing to measure.

But it is patently self-evident to any anthropologist who has spent time in small groups outside of WEIRD societies.

From an anthropological game theory perspective, the rules of proper behavior, defined by ritual, must be enforced against defectors or non-co-operators. There are two basic categories of punishment. First is the direct punishment of non-co-operators or defectors, which is crucial to keeping the group working. As countless game theory simulations have shown, and as direct ethnographic and historical data indicate, free-riding by too many individuals will set off a dynamic that is guaranteed to bring the co-operative group down. Ethnography teaches us that different groups have developed a huge array of taboos and other social norms to prevent free-riding.

The second type of punishment is directed at leaders of co-operative groups who are perceived as abusive or ineffectual. Punishment of overly aggressive leaders in stateless societies is well known in the anthropological literature, and it is a primary source of group fissioning. Christopher Boehm (1993) recounts the wide variety of social-leveling mechanisms reported on in the ethnographic literature of stateless societies. Leveling refers to the strong egalitarian ethos expressed in societies with ad hoc leadership that prevents any leader from dominating the group. This ethos reflects the tension inherent in the need for coordination of group behavior and the danger that some will try to garner coercive power.

Boehm describes adverse public opinion in small groups in particular: "For people living in small-scale moral communities, negative opinion can be psychologically troubling even when unaccompanied by other sanctions" (Boehm 1993:230). He lists a number of ethnographies describing leaders controlled by public opinion, including some well-known cases of the Tikopia, Kayapó, Ashanti, Navajo, Hottentots, Cuna, Tupinamba, Fox, Yokuts, and Mandari. A corollary of negative public opinion is criticism and ridicule. Boehm lists a number of ethnographic cases in which chiefs are put in their place if they try to order someone to do something outside of the boundaries of normative behavior. As he says, ridicule "threatens a leader's status because he cannot lead without respect" (Boehm 1993:230). This idea is found throughout the ethnographic record, from the smallest and most undifferentiated group to national governments. This is a feature of virtually any democratic political system, broadly defined. Sandra Harris (2001:451) describes how "systematic impoliteness" is a behavior rewarded by members of the British Parliament as part of the "confrontational political process." We can see here how ridicule works in small groups of all types, from the most distant village to the House of Lords.

The sanctions just listed are mild compared to the lengths to which small groups are willing to go to maintain norms of fairness. Boehm discusses "extreme" sanctions that include assassination and other forms of lethal retaliation. In a survey of 48 societies, 11 took these extreme measures. Stateless societies also ostracize and exile people who excessively break norms. It is worth reproducing his list of reasons given as pretext for punishment by people in these ethnographies:

> Of some 47 behaviors mentioned as motivating negative sanctioning, being too aggressive (13) and dominating others as leader (14) predominated, along with ineffectiveness, partiality, or unresponsiveness in a leadership role (10). Lack of generosity or monopolizing resources (5), moral transgressions (3), and meanness (2) complete the list. The great majority of these misbehaviors involve dominance or self-assertion. (Boehm 1993:231)

Ostracism is one extreme form of punishment. It effectively pushes the person out of the group, both socially and physically. In regions with very high relative population densities, it can mean being enslaved or killed. Among the Pathan hill tribes, Niloufer Mahdi (1986:295) notes that "ostracism functions simultaneously to deter behaviour that violates customary legal norms, to punish specific acts that are culturally defined as improper, and to unify the primary reference group on which individuals depend for protection and economic support."[43]

Revenge raids are common in stateless societies. In the Pathan example above, the legal norm of *badal*, or the right to avenge a wrong by killing, is a primary form of maintaining social norms. Mahdi (1986:298) makes a fascinating observation: "The man who exacts badal may not be acting consciously for the common good, but no less than a state authority, he is acting as an agent of the community in executing badal according to Pukhtunwali." In other words, this behavior can be considered a kind of irrational prosocial act that is perceived to be in the self-interest of the punisher.[44]

"Taboo" is utilized here to denote the means of enforcing these rules. Taboos are culturally constituted and are therefore quite variable across different societies. But they are found in all societies, particularly in stateless ones that lack coercive sanctions. In stateless societies, taboos are always created by whole communities and exercised by leaders at the implicit or explicit behest of the group. In the case of the Nuer, the leopard-chief uses magic or taboo to pressure rivals in a blood feud to accept payment to end the cycle of violence.

In his book on Maori economics, Firth went on at great length about how taboo was a crucial element in the economy: "The examination of a host of other beliefs and institutions shows that the significant natural resources on which the Maori was dependent for food or the raw materials of industry all had some degree of tapu attaching to them" (Firth 1929:247). Maori chiefs would taboo certain forests, protecting them from "unauthorised interference."

The same holds for the Kayans of Borneo: "The system of taboo is greatly practiced by the Kayans during the times of planting and harvesting the crops, and more especially when the paddy is being stored" (Roth 1968[1896]:420).

Meyer Fortes says that across the globe, "these emblems and insignia [of chiefs and priests] are associated with distinctive norms of conduct [that] bind those who hold an office to it by ritual sanctions." He goes on to say that "esoteric rites" composed of "taboos and other ritual observances" (Fortes 1962:67) were imposed on officeholders. In this case, taboo is used as a punishment to keep leaders from garnering too much authority and preventing abuse.

At this point we return to the theme of Chapter 1. The rich and varied ways in which stateless societies expressed religious, mythical, and ritual behaviors struck early Western travelers as odd, irrational, or exotic. Anthropological game theory demonstrates that far from being "primitive" or retrograde, these rituals expressed norms of behavior that were essential for the functioning of stateless economies. Anthropological game theory provides us the conceptual tools to see how these norms work to maintain co-operation in the absence of coercion. The behaviors and customs witnessed by our ethnographic ancestors in fact were the products of generations of trial and error in competitive environments. What these ethnographers discovered were highly rational and successful ways to structure complex economies and multigenerational co-operation.

Costly Signaling and Group Co-operation

> Sociology and anthropology [confirm] three predictions from the [costly signaling] model: (1) participation in costly rituals is associated with prosocial in-group behavior, because costly rituals transmit commitment to group beneficial beliefs/goals to participants, (2) institutions requiring costly displays are favored by cultural evolution because costly displays by members transmit higher levels of belief-commitment and thereby promote cooperation and success in inter-group or inter-institution competition, and (3) costly rituals spread culturally from less successful groups to more successful groups because costly rituals affect groups' success by increasing commitment to the group and galvanizing greater cooperation. – Henrich (n.d.:3–4)

A critical component of maintaining co-operation in small groups is known as costly signaling. As described by above (and see Henrich 2009; Smith and Bliege Bird 2005; Sosis, Kress, and Boster 2007), costly signaling increases co-operation or prosocial in-group behavior. Eric Smith (2003:407) describes it succinctly and clearly: "The basic idea is simple enough: if I wish to convince you of my sincere ongoing commitment to a common project, I can honestly signal this commitment by contributing to the common good at levels that would not be beneficial to me were I planning on defecting over the next time period." Costly signaling tells others that the persons involved are invested in co-operative behavior – reciprocity, for instance – and enhances the capacity of the group to keep and attract new members. Herb Gintis, Eric Smith, and Samuel Bowles (2001:1) describe it as "an honest signal of the member's quality as a mate, coalition partner or competitor [which] results in advantageous alliances for those signaling in this manner." Or as Eric Smith and Rebecca Bliege Bird (2005:116) describe it, "costly signaling theory proposes that expensive and often seemingly arbitrary or 'wasteful' behavioral or morphological traits are designed to convey honest information benefiting both signalers and observers."[45]

Costly signaling is a powerful concept in evolutionary game theory that explains a number of phenomena in ethnography and history. It helps explain otherwise "irrational" behavior of people who spend considerable time and resources on what on the surface appears frivolous. In reality, such activities serve to reinforce the bonds of co-operation between individuals in groups. This is particularly important when there is little or no coercion in the group.

Somewhat counterintuitively, groups that impose more requirements on their members actually seem to deal with free-riding better than those that do not. In an interesting study, Richard Sosis and Eric Bressler evaluated the nature and intensity of costly signaling among nineteenth-century US religious communities. They found that "communes that imposed costlier requirements survived longer than less demanding communes" (Sosis and Bressler 2003:211). What they termed "religiosity" correlated with increased co-operation. This is logical if one recalls the examples of the PD. Individuals with no close relationship to each other who do not share norms of co-operation are substantially more likely to defect in one-shot, zero-sum situations. However, people who do share such norms are more likely to co-operate if they think that the other people, like themselves, are bound to normative expectations of behavior, including loyalty to in-group members. There is little cost to defecting on someone who is not known, but defection on a group member carries costs from other members if the defection comes to light.

John Kantner and Kevin Vaughn (2012:69) provide an excellent summary and reformulation of the criteria for determining whether religious behaviors are examples of costly signaling or not. These criteria are: (1) the degree to which adherence to moral codes in any group varies and is not observable; (2) that devotees of any religious practice benefit from having reliable information from those who are more co-operative in the religious practice; (3) that "devotees benefit from broadcasting their adherence" such that it increases their reputation for being prosocial.

A recent study by Dimitris Xygalatas et al. (2013) in Mauritius, among people in a Hindu ritual, demonstrated that intense or extreme participation (including painful procedures) in fact promotes co-operation. They write, "These findings support long-standing anthropological conjectures about the co-operative effects of intense rituals (Durkheim, 1912). We offer the first natural demonstration that suffering predicts prosociality by capitalizing on intense, real-world stimuli that would be hard to manipulate in the laboratory" (Xygalatas et al. 2013:1604). The behaviors noted here include "body piercing with multiple needles and skewers, carrying heavy bamboo structures, and dragging carts attached by hooks to the skin for longer than four hours before climbing a mountain barefooted to reach the temple of Murugan." These are indeed extreme forms of behavior, but much what we see in the iconography in the archaeological record could be interpreted as manifestations of such phenomena.

We can expand the criteria of Kantner and Vaughn to go beyond the religion discussed by Xygalatas et al. to include any ritualized, co-operative group activity in which public displays of adherence to collective norms increase people's reputation as co-operators and keep the group functioning. The role of costly signaling is intensified in stateless societies because of the absence of money, hierarchical statuses, and the other ways to display status and wealth found in state societies.

Ruth Phillips (1978:265) describes public masking in Mende society of Sierra Leone: "Public masquerades constitute extremely important symbolic forms ... the masquerades make visible the powerful 'medicines' ... of the secret society." Both men and women have secret societies with onerous requirements for membership. The masks that people wear clearly identify the different societies among the Mende. Phillips (1978:265) makes a significant observation about the value of the masquerading rites: "Participants and audience are drawn into a common experience, which is aesthetically heightened by techniques of theater, dance, music, and the plastic arts. The resultant feeling of unity and harmony helps to overcome the threat of disunity implicit in the division of the community into separate secret factions." Here we see that performance – theater, dance, and music – form part of the ritual that leads to heightened co-operation within and between these groups.

Special Places as Costly Signalers

We have seen how people in stateless societies co-operate for common goals – co-operation between autonomous households that is structured by ritual. Throughout the ethnographic record, we see how material structures or places – men's houses, temples, various circular structures, forest clearings, special beach areas, and so forth – serve to materialize these social norms. One of the principal reasons people construct these places is to signal to the group that co-operation will pay off. Investing in these places is a means by which prosocial individuals can garner prestige and keep co-operation strong. And in a competitive context, managerial leaders can use the physical space to feast and attract followers. In short, the construction of special places is a necessary condition for the creation of successful co-operative norms in stateless societies. It is worth noting that a comparative analysis by Michael Adler and Richard Wilshusen (1990:134) found no relationship between community size and the presence of what they term "socially integrative facilities." Fully 79 percent of the tribal societies they analyzed had physically separate communal buildings. Small and medium-sized groups had evidence of these special places in the form of open plazas.[46]

A number of interesting examples in history and ethnography illustrate this principle. One of the most fascinating is the Pyramid of Ngundeng, built by a Nuer prophet of the same name in the Sudan in the late nineteenth century. E. E. Evans-Pritchard (1940:186) and Douglas Johnson (1982:125–126), using documents from the period, describe it as a "high pyramid of earth, 50 or 60 feet high, on top of which are elephants [*sic*] tusks of ivory sticking up out of the ground." These tusks formed a fence at the base of around 300 feet in circumference. Tusks were also raised on the summit of "the Mound." The pyramid was placed directly in the middle of the village, or more properly, the village grew around the Mound. Evans-Pritchard (1940:186) says that the Mound was built with ash, earth, and debris from the surrounding area. He also says that people from all around the district brought oxen for sacrifice and assisted in its building. Documents actually describe raids against the Mound itself, though the precise number raids and the identity of the perpetrators are unclear.

He likewise notes that a later prophet, called Gwek, augmented Ngundeng's pyramid. The Pyramid of Ngundeng is an excellent example of how the act of building a large construction and making it sacred is a powerful means of mobilizing and attracting people to your faction. Evans-Pritchard (1940:186) reports that the British colonial government blew up the Mound in 1928, considering it politically dangerous. This destructive act reinforces the observation that the Mound was a powerful means of bringing people together and that it served as form of resistance to colonial rule. Evans-Pritchard also mentions a similar, though smaller mound built by a neighboring group for another prophet.[47]

I find this to be a particularly fascinating example for archaeologists. We find rubble or earth-filled mounds around the world, yet there are few examples of this kind of behavior in the ethnographic or historical record. In this case, we see how a very ambitious man with a large kin-based following declared himself a prophet, built a simple mound with earth, and attracted a multitude of additional followers. The followers were acting completely voluntarily. Clearly, behind the simple act of mound building were other powerful social and political forces. By harnessing those forces and making a simple physical place – the mound – a focus of a political, religious, and social movement, a leader emerged, using this special place as a symbol of power. This strategy was adopted by a subsequent generation, which continued to use the pyramid to signal and augment its authority.

In the eastern Bororo of Brazil lives a small group in the central Matto Grosso. The group had a classic men's house that was the center of village life. As Christopher Crocker (1969:239) describes it, "the center of the transactions arising from all types of bonds is the men's house, the *baito* or *bai mana gejew* (literally, 'house in the center'), which stands in the middle of the village circle." In this very small village society with notably absent hierarchies, coordination of labor and other social activities was mediated through an institution that was physically housed in a prominent building at the center of the village.

Men's houses in fact are quite important in the ethnographic literature. More than a century ago, Hutton Webster (1908:1) surveyed the literature and noted, "The men's house is usually the largest building in a tribal settlement. It belongs in common to the villagers; it serves as a council-chamber and town hall, as a guest-house for strangers, and as the sleeping resort of the men." This observation is consistent throughout the ethnographic literature. Men's houses tend to be significantly larger and are places where ritual is focused. Data from Adler and Wilshusen strongly support the model that communal structures such as these serve to concentrate ritual activities that promote economic co-operation: "Results of this research suggest that within relatively small, politically nonstratified communities, social integrative facilities remain generalized in use. That is, the facilities commonly serve as a context for both secular and ritual activities. However, as the size of the interacting community increases, social integrative facilities of a more monumental nature tend to be constructed and used almost exclusively for ritual activities" (Adler and Wilshusen 1990:133). They note (Adler and Wilshusen 1990:137), for instance, that in a New Guinean example, the men's house is used for all internal and intervillage ritual.

Talal Asad (1972) notes that the men's house is the center of Swat Pathan political organization. As described by Fredrik Barth (1961) and reinterpreted by Asad, the physical structure in the ward or political unit in the village is central to maintaining a chief's faction. While Pathan political organization is more complex than a stateless society, the structure of co-operation being materialized in a central physical structure is the same:

> The political activities of the chief, designed to maintain or enhance his position, centre in the very important Pakhtun instruction of the men's house. In each ward there is at least one men's house, dominated by a chief; and the political, economic and recreational life of the men of the ward revolves around this common centre. Allegiance to the chief is expressed by the mere act of visiting his men's house. This allegiance is reinforced and deepened by the acceptance of hospitality from the chief. The chief is continually

giving food, and occasionally other valuables, and thus creating debts and dependence on the part of the persons who sit in his men's house. (Asad 1972:77)

The ethnographic data tell us that the Dyak of Borneo also constructed a bamboo altar during one of their agricultural cycle festivals. Roth (1968[1896]:412–413) describes how they built the altar to receive offerings from the entire group. After two days another larger one was built, and it likewise received offerings. He goes on to note that in other districts in the region, the ceremony was much more elaborate: "Imagine a lofty altar, gaily decorated, erected in the open air close to a village, and surrounded by the grand forms of our tropical palms and other fruit trees. Huge bonfires cast a ruddy glare" (Roth 1968[1896]:413). These buildings were indeed the architectural center of special places. They were central both physically in the village and socially to the community.

Among the Abelam of Papua New Guinea, Phyllis Kaberry (1941a:242; 1941b) describes the social ties between members of villages or hamlets. The hamlets are constructed around an *amei*, or ceremonial grounds. Kaberry describes this as a cleared space, about 50 × 50 feet. *Amei* are places where men converse, dances are held, yams are displayed, and ceremonies are conducted (Kaberry 1941a:243). Important herbs are planted around their sides. Kaberry notes that the domestic houses are in front of and/or parallel to the *amei*. "House-tamberan" are located around the *amei*. These are described as men's ceremonial houses that dominate the architecture of the village. They are 40 to 80 feet high, making them three times larger than domestic structures. These constructions are "peaked and painted" and are filled with ritually important objects. Wooden carvings and "curiously shaped stones" are placed around the structures. These houses are where initiation ceremonies are performed and songs are chanted. There are eight house-tamberan in this particular village.

Flannery and Marcus (2012:104) describe a similar kind of special place in the Mount Hagen region of New Guinea. They note that one of the biggest contributions by Big Men, whom they describe as having prestige but no authority, was to coordinate followers to construct a special place for congregations. This was the *moka pene*, or large, grassy plaza. This place was planted with special plants and lined by ceremonial buildings. There was a large ceremonial mound at the end of the plaza. As they aptly write, "Almost any archaeologist coming upon it would consider it a 'ceremonial complex.'" In fact, the description of this site fits extremely well a number of archaeological contexts in Formative or Neolithic contexts from around the world.

Archaeologically, the Abelam example would appear to be a village with a 0.25 ha open area surrounded by small domestic structures and larger ceremonial ones. In this example, each ceremonial structure would belong to separate kin groups in the village and would be contemporary. Such a settlement pattern is very typical of sites with the earliest ritual economies.

Patrick Kirch (2001) describes special places called *tohua* constructed by people in the Marquesas Islands in the immediate prehistoric periods. These were large, rectangular stone dance platforms where feasts and other ritualized exchange took place. Kirch describes one of these as 26 × 174 m – making it quite substantial, even megalithic, in size.

Hayden describes several historical and/or archaeological examples of feasting areas that also serve as material markers on the landscape of these kinds of special places. The Torajan of Southeast Asia erected stone stelae in a quasi-circular layout. A wooden structure was placed inside this monolithic monument. From here, meat was distributed in feasts (Hayden 2014:203, Figure 6.12). Hayden and Adams (2004) describe an archaeological example of what they interpret as ritual structures at the Keatley Creek site in the Fraser River of British Columbia. This was a winter village site constructed around 2600 to 1000 BP (Hayden and Adams 2004:84). The site was constructed by a decidedly stateless society. Two structures in particularly had evidence of feasting, including food remains and peculiar hearths. The authors interpret these structures as the location of a men's secret society, slightly away (about 200 m) from the main settlement. What is significant for us is that archaeologists were able to identify a set of structures that supported intensive feasting and possibly other activities that promoted co-operation among at least a subgroup of this society.

Hayden (2014:249, Figure 7.8) also depicts a large "stage" for feasting among the Maori, as described by missionary William Yate. This stage was a wooden scaffolding with a center beam, smaller outer beams, and a series of platforms that rose up in a cone. William Yate described it as a "pyramid, eighty feet high, and twenty feet square at the base.... These piles... when filled... present one solid mass of food: the whole is decorated with flags, and when in an elevated situation, presents a very imposing appearance" (Yate 1835:139–140). It is hard to deny that such a place would be extremely special in the eyes of the participants, and it served to attract many groups from afar.

In short, the building of special places is a necessary condition for the successful creation and maintenance of co-operative economic and social behavior. It is one of the conditions that allow *H. reciprocans* to flourish. These special places are highly variable and include ballcourts, dance grounds, kivas, men's houses, *utas*, "chiefly mounds," stone rings, and henge monuments. We could include scores more examples from the ethnographic literature. Once conditions in which the costs of co-operation are less than the costs of fissioning and defection obtain, people in stateless societies create a myriad of social organizations, backed by ritualized behavior, to enhance co-operation among their members. The construction of monumental architecture that concentrates the resources and energy of the group is a universal strategy to maintain and promote co-operation. In a social context of competition among different groups, and in the absence of some kind of overarching political authority, people will ratchet up the "magnificence" of the monumental architecture as a reaction to others. In these contexts, monumental architecture is simultaneously a means to promote co-operation and to compete with other co-operative groups in any cultural landscape. It is the core of any strategy for stateless societies to construct viable social technologies of co-operation.

6 An Anthropological Game Theory Model for the Evolution of Ritualized Economies

> Ostrom's Law: A resource arrangement that works in practice can work in theory.
>
> – Fennell (2011:9)

Lee Anne Fennel describes Elinor Ostrom's work by noting that she "challenges the conventional wisdom by examining how people interact over resources on the ground – an approach that enables her to identify recurring institutional features associated with long-term success" (Fennell 2011:9). The conventional wisdom that Fennell refers to is that of economists and theoretical biologists who develop mathematically precise and logically coherent abstract models that oftentimes have little to do with social reality. The concept of *Homo economicus* comes to mind. It is a model of human economic behavior that was elegant, bold, and precise but was profoundly wrong in most real life social circumstances, particularly for small groups.

Model building is important. A good model is particularly useful when one measures the deviations from that model. The inconsistencies and empirical misfits are precisely where we advance our knowledge. However, abstract models that do not incorporate sufficient data can be highly inaccurate. I share the frustration of Fennell of some social scientists who confront abstract models with minimal or superficial empirical content. As Brian Arthur says about many economic models, "the type of rationality assumed in economics – perfect, logical, deductive rationality – is extremely useful in generating solutions to theoretical problems. But it demands much of human behavior, much more in fact than it can usually deliver" (Arthur 1994:406).

I believe that we must assemble a rich data set and observe those recurring patterns. Proper method in comparative social sciences requires a very strong link between theoretical deduction and empirical confirmation. Economists, theoretical biologists, and economic game theorists have provided excellent models of human behavior, but I would argue along the lines above that we lack a strong empirical foundation to test and expand much of this work as it applies to premodern peoples, particularly stateless ones. As Ostrom taught us, effective economic model building *begins* with the economic practices that people do, not from a priori assumptions about humans should behave. From these empirical observations, we use theoretical concepts to help us understand those patterns, while at the same time we use the data to modify our theoretical concepts. Model building is a recursive process based upon a constant interplay of observation, creation of concepts to explain those data, new theoretically informed observations based upon those data, and the reformulation of concepts yet again.[1]

This chapter is developed in the spirit of Ostrom and Fennell's admonitions. I have assembled the ethnographic data on co-operation in largely non-Western small groups in the previous chapters. From these observations, I use contemporary theory to develop synchronic or generational scale models of how stateless societies function in the last chapter. The present chapter adds the deep diachronic component. Here, I use the empirical observations of those practices that were successful in the long term in the spirit recognized by Fennell and Arthur above. I provide a qualitative model for the evolution of ritualized economies in stateless societies over the long-term historical and archaeological record using concepts developed in the preceding chapters.

The Model

In the shift from stateless egalitarian to complex stateless societies with rank, people are not *forced* to adopt social hierarchies.

Population pressure does not force evolutionary change; rather, there is a minimal population density necessary that *allows* people to form more complex forms of co-operation. We expect an initial increase in population density, followed by the development of special places, in turn followed by a second population increase.

The strategies of co-operative groups benefit each individual in the group. These benefits cannot be obtained without co-operation.

Leadership positions in stateless societies are created by the group for group interests. Leaders are rewarded with prestige and other resources, but leaders can only persuade people to work in co-operative groups. There is no overt coercion. The community can withhold support from potential leaders who then lose their prestige and ability to organize co-operation.

Groups are fluid. People shift in and out of factions. Exogamy insures that people from different groups move across the social landscape.

For groups to be successful, they must create norms and behaviors that deal with the collective action problem of free-riding as well as create and maintain appropriate strategies to assure payoffs to individuals in that group. Effective groups both reward co-operation and punish defection.

Effective groups assure success through the ritualization of production and exchange. The strategies include feasting, potlatching, costly signaling, the creation of economies of scale and heightened exchange on one hand, and the use of taboo and other forms of punishment on the other.

Successful groups, i.e. those with the strongest norms that maintain sustained co-operation and yield high per capita payoffs, will either be imitated by others or they will absorb people into their factions. This "strategy" selection acts on the social norms and principles of co-operation that the group employs.

Archaeologically, we see how successful groups build special places in their villages or on the landscape since the early Holocene. These are places where people congregate to enhance their material and social well-being.

Group formation is cyclical. There is no inevitability in this process.

The First Great Social Transformation in Human History

The Pleistocene hunter-forager organization was the most successful social evolutionary adaptation in the history of anatomically modern humans, and it continued well into the ethnographic present in many areas of the world. Hunter-forager society was restricted to bands of no more than a few dozen people (Hill et al. 2011:1287).[2] The average size of bands they report is about 28 individuals, with a range from 5.8 to 81.0.[6] This social organization was highly adaptive, allowing our species to move into virtually every rich ecological niche in the world. It was remarkably resilient as well, surviving hundreds of generations all over the planet. In anthropological game theory terms, the great shift represents the evolution of social organization in which a small number of people consistently co-operated with each other to one in which larger numbers of both kin and nonkin co-operated across space and through time.

After the Pleistocene, in a few key places on the earth, population densities reached a critical juncture at which the cost of fissioning of hunter-gatherer-forager groups was greater than the benefits of co-operation. Around this time, a few people in a few places began to move less and started to congregate in special places that they sculpted onto the landscape. People experimented with new co-operative social organizations in places with sufficiently constant resources, either wild ones localized in a small area, or semidomesticated ones that were artificially concentrated. They did this by creating novel norms of co-operative behavior that combined sufficient payoffs to individuals in their group with norms of punishment to prevent free-riding by those same people. To modern eyes, these special places appear to be largely or wholly ritualistic or religious in nature. Katherine Spielmann (2002:202) makes this point quite well: "When populations are fairly mobile and reside in small hamlets, aggregation tends to occur in the context of communal ritual. Although these ritual loci may not be fixed on the landscape, in some cases permanent facilities are constructed."

People built megalithic monuments and great works of art while simultaneously creating social organizations never before seen in the human experience.

Archaeologically, this was manifested by the creation of special places on the landscape. These places attracted people from around the region. They congregated in these areas at specific times of the year. During these aggregations, we deduce from ethnographic and historical analogy and observe archaeologically that they worked, feasted, played, told stories, found marriage partners, gossiped, and exchanged goods. These events were conducted under a ritual calendar, usually using astronomical features such as the solstice or some major planet or the moon as a guide. In other cases, people used local timekeeping devices like biomarkers, tides, and seasonal weather.

Those who hosted the events garnered prestige and material resources. Other groups of people, being intelligent, adaptive agents, observed these norms of co-operative behaviors and the rules under which these special places functioned. They emulated those co-operative behaviors that worked well and passed these on through the generations via various kinds of cultural transmission mechanisms. As a new evolutionary dynamic set in, fueled always by sufficiently high population densities that provided the necessary conditions for sustained co-operation, some organizations disappeared while others continued to develop over periods of time.[3] This process was not unilineal. Most organizations collapsed over a generation or two, most likely due to the increasing stress of free-riding and competition but also due to exogenous factors such as climate change, resource stress, pandemic disease, or any other factor that lowered population densities below the threshold necessary for sustained co-operation. In some places, the organizations survived over millennia and grew to the point where coercive mechanisms of power were institutionalized. At this point, coercive state forms of social and political organization emerged, representing the next great transformation in human history.

The shift from a simple hunting and foraging way of life to one in which people lived in semipermanent and permanent settlements was a profound transformation in the history of our species. As understood by the theoretical framework developed in the preceding chapters, such complexity occurs when relatively large groups of related, unrelated, or distantly related people co-operate in complex ways for both their individual good and the common good.[4] Co-operation beyond the demographically small band is not easy, as we see in ethnography and as inferred from archaeology. Yet these early ancestors of ours created complex social institutions that were handed down over the generations and which permitted the sustained co-operation of the group.

It is critical to note that empirically we know that the vast bulk of humanity continued the hunter-forager way of life for quite some time. But in a few areas, people created places where they came together for various kinds of social interactions that we do not yet fully understand. Over time, as more and more people adopted these new "social technologies," complex stateless society developed.

In many instances, these special places were built by complex hunter-gatherers who did not cultivate plants or domesticate animals (Arnold 1996). This is an extremely important empirical observation – agriculture is not necessary for such complexity to develop. In the majority of other cases, however, people experimented with plants and animals, experiments that would ultimately lead to the domestication of several key species essential for the rapid growth in populations during what is called the Neolithic Demographic Transition, a process that occurred in both hemispheres.

These societies have many synonyms in the archaeological literature, including "simple chiefdoms," "intermediate societies," and "transegalitarian societies." All of these terms refer to the same general social structural principles whereby people co-operate in a much more intensive way than they did in mobile hunter-forager societies, and they do so without any kind of the state apparatus that developed later in many areas of the world.

What is even more amazing is that this same process occurred independently in many places around the globe. The cultures are different, the way in which people interacted was different, and the art, language, dress, tool kits, and social organizations were dizzyingly diverse. But underneath this diversity is a set of common phenomena that led to the rise of complex stateless societies in both the Americas and the Eastern Hemisphere. That fact that this occurred independently at least twice (in the Americas and the Old World) and probably developed independently in scores of areas around the world, indicates that under certain conditions, emergent processes gave rise to these similar social technologies. Several millennia existed between the time some of our ancestors first settled around these small, special places and the emergence of the great urban states of antiquity. This process represents a period of more than 100 generations in which people lived in complex stateless societies that nevertheless were characterized by monumental buildings, extensive trade networks, complex ritual, and all of the other the hallmarks of civil life.

This process is made all the more stark given the fact that cognitively modern humans have been on the planet for at least 50,000 years and probably many millennia more. Explaining co-operation within kin groups is relatively easy with evolutionary biological models. We have come a long way in understanding these processes among related individuals, both animal and human. But it was not until the early Holocene in the Old World or the early Formative period in the Americas that people mobilized both kin and nonkin alike to build complex social structures composed of highly co-operative agents. The question therefore goes beyond simple evolutionary biological models used for social animals. It is decidedly one of a uniquely human social phenomenon of so-called altruistic behavior – co-operating with nonkin on a sustained basis in a way that does not directly enhance inclusive fitness in the short term. We must move from a purely biological logic to one that includes both biological and economic ones. The human capacity for symbolizing and abstracting behavior is the key, as we see below.

The shift to sustained co-operation among large numbers of nonkin was nothing short of revolutionary in our biological and cultural history. And it was one of the key factors that allowed for the exponential growth in the size and complexity of human settlements in the last 10,000 years. One of the great tragedies of the twentieth century, at least for comparative social scientists, is that autonomous stateless societies had effectively ceased to exist on the planet by the 1970s. The contemporary "traditional" societies all live within larger political entities that substantially impact their economic and other behavior. As a result, we have lost many examples of viable strategies of production and exchange in stateless societies.

However, we do have available a suite of observations of stateless societies by ethnographers, historians, and other travelers, as we saw in the previous chapters. These observers recorded many aspects of these small groups, providing behavioral analogies for our models, as described in the previous chapters. These societies of course do not represent relic examples of our ancestors. Rather, they provide examples of how people in autonomous small groups could organize themselves to co-operate in the absence of states, markets, and other forms of coercion. In fact, studies of small groups of people within contemporary society also provide much of the data for our theoretical constructs. While the study of these historically documented groups cannot be used as a direct analogy to the past, as with any kind of properly done comparative social science analysis, we can at least examine how people in these contexts structured their economic activities and use these models for understanding past behavior.

A Digression on Causality in the Historical and Social Sciences

Before we get to the core discussion of the model, we need a very brief digression on some basics in epistemology. This is not really the place to begin an extensive discussion of verification in the social and historical sciences. Suffice to say that models that posit an overtly deterministic and linear causal sequence, such as in the natural sciences, are less useful in historical sciences like archaeology, paleontology, geology, and so forth. They are less useful because they do not capture empirical reality. Arguments over what comes first, population growth or sociopolitical complexity, for instance, represent textbook cases of the inadequacy of simple linear models. Humans, unlike any other living creature on the planet, reflect upon their own existential circumstances and take steps to alter those circumstances if they are able. As history and archaeology teach us, humans indeed are able to mold their physical and social environments in profoundly important and creative ways.[5]

A more precise way to put this is that as adaptive agents, people engage in behaviors that we call recursive. In mathematics, recursive formulae are ones that use the preceding term to define following terms of the sequence. This phenomenon provides for a highly varied evolutionary history. One can use the exact same algorithm, but because it begins with a different number or type of conditions, the outcomes will be different. The human capacity to symbolize or create abstractions allows us to observe others and alter our behavior. We may use the same "cultural algorithm," to stretch the metaphor perhaps a bit too much, but the outcomes will be different because we not only recognize a difference in the initial state but we actually alter to some extent those very initial and subsequent conditions. Classic adaptive behavior in humans therefore is to observe those who are doing something better and use that information to alter strategies as the need arises. People adapt to both altered environmental and social circumstances; and in a recursive way, we change those circumstances. Adaptation involves not only reacting to changing contexts as a passive agent but actually actively changing those contexts as part of the "adaptive process." This is what intelligent agents do. Such "feedback" mechanisms affect historical trajectories and modeling in ways that natural scientists simply do not face.[6]

This requires that we generally abandon deterministic logics in social science (though not always). The alternative to a deterministic logic is a probabilistic one, used in many scientific disciplines as well. Probabilistic logic is based upon the fundamental observation that in "complex causal systems . . . where multiple stochastic influences are present" (Humphreys 1986:2) simple deterministic propositions of causality are insufficient. In human societies, one of the most significant sources of stochastic variation is found in this recursive nature of human action. As Humphreys nicely points out, "the presence of multiple causal influences requires that the theories be multivariate in nature. Both the uncontrollability and our ignorance of many of those multiple influences means that the effects of those factors which are included in the analysis will affect the outcomes in a probabilistic, rather than a deterministic, fashion" (Humphreys 1986:2).

The need for probabilistic logic increases proportionally to the degree of complexity of the phenomena involved. In terrestrial mechanics, for instance, there is an absolute invariant relationship between energy applied to a mass and the trajectory of that mass. So if you shoot a perfectly round rock at a constant velocity and trajectory in an identical physical environment, it will always land at the same location. If you do it on the moon, it will go farther than on earth because of differences in gravity and friction, but the relationship between cause and effect is so simple that it can be predicted with virtually absolute mathematical certainty in a very precise linear manner.

A slightly more complex way of understanding causality in the social sciences is to posit a series of contingencies (Events a-n) that, should they co-occur, will have a high probability of resulting in Event Y. Insofar as archaeology is a comparative social and behavioral science, we find that probabilistic explanations are really the only logic we can use. Archaeology is simultaneously a natural history as well, a discipline that studies events long past. We cannot experiment and we cannot "test" a future outcome. Because of this and because we are dealing with ontologically highly complex phenomena, the logic of archaeological explanation has to be probabilistic in nature. Like paleontologists, astronomers, historical geologists, and so forth, we can only observe the vestiges of phenomena long since gone. We have developed sophisticated methodologies to observe the existing data and draw conclusions via inductive inference based upon logic and analogy to historical and living peoples. We can then create a probabilistic model based upon a series of conditionals that, if they co-occur, will result in some degree of statistically reliable possible outcomes. We then deduce implications from those inferences that are in turn tested with more systematically collected data. These new data ideally affect the subsequent inferences to be tested, which in turn create new models (not coincidentally, a highly formalized type of recursive behavior). We do not look for deterministic causes – we should be constantly refining the probabilities of many possible factors. This is how a healthy historical science works.

As a result, social scientific models with a linear sequence of events, such as population growth leads to social complexity and so on, are not viable. Rather, we must define, based upon inductive inference and deductive logic, those factors that, when they co-occurred, likely resulted in the development of social complexity. The logic is to construct a series of conditional statements (if this, then that) that increased the probability of the development of social complexity in a world filled with multiple causality and complex and occasionally stochastic processes.

A Game-Theoretic Perspective on Population Growth and Social Change

Anthropological game theory provides a probabilistic model for the emergence of complex society in the archaeological record. A key necessary condition is a landscape in which population densities are sufficiently high to promote co-operation. The benefits of co-operation increase with higher population densities. At the inflection point in which the benefits of co-operation outweigh the costs is one key appropriate condition that promotes social complexity. This is a different way to look at "population pressure" than is oftentimes seen in the canonical theoretical literature.

From an anthropological game theory perspective, population pressure does not necessarily force people to adopt new ways of living. Rather, under certain conditions, sufficient minimal densities *allow* people to adopt new ways of living. These news ways of living, of course, are co-operative social organizations that increase people's social and material well-being. From this point of view, there simply is a minimal demographic level required to reach the point where co-operation is less costly than non-co-operation. Or we could say, there is a minimal level where the benefits of co-operation outweigh the costs of non-co-operation. This very important perspective differs from many cultural ecological and processual views of the relationship between demographic levels and social change. This requires some explanation.

Most models in processual archaeology argue that there is a causal relationship between high population densities, subsistence intensification, and social complexity. In a classic article, Lawrence Keeley convincingly demonstrated the correlation between these two variables for hunter-gatherers: "The correlations between population pressure and socioeconomic complexity are found to be extremely high" (Keeley 1988:373). He goes on to describe how the mechanism of population pressure-induced cultural change occurs in classic terms:

> Population pressure ... forces hunter-gatherer groups to intensify and reorganize their economies.... Such intensification and reorganization is argued to be necessary to cope with: (1) food deficiencies or diminishing returns on subsistence labor, (2) increased competition with other social units for available resources, or (3) increased risks resulting from thinner subsistence "cushions" and/or smaller or more circumscribed exploitation territories. (Keeley 1988:374)

This is a component of the canonical model in traditional archaeological theory created in ecological and adaptationist frameworks (for reviews, see Cohen 1975; Smith 1972). This canonical approach has been used for more than two generations in archaeology and ethnography. Among one of the first to use this concept in traditional archaeology was Douglas Schwartz (1957:376) and his analysis of prehistoric change in the Grand Canyon area of the American Southwest. Since that time, this model has become entrenched in archaeological theory and logic.

This canonical model relating increased population densities to social change is that such demographic pressure forces people to make adjustments in their lifeways to accommodate the stress resulting from more people in the same area. This is usually conceived of as resource stress – lack of food and water – or an increase in competition that generates social conflict. In this view, people adapt to population growth by creating new levels of sociopolitical integration that allow them to harness the energy in the environment more efficiently. This energy efficiency is viewed as always involving a cost, since "higher" levels of sociopolitical organization involve the creation of social hierarchies. This view also assumes that coercion is present in such hierarchical relationships. From this perspective, any change away from a hunter-gatherer way of life means an increase in social inequality, more work for the bulk of the population, and a generally less satisfying life. The shift from egalitarian hunter-gatherers to societies with leaders is viewed simply as a negative that any sane person would want to avoid. Who would want to be under a chief? Who would want to give up political autonomy to an elite?

The canonical model was adopted and modified by the highly influential work of Robert Carneiro, who introduced the concept of circumscription. Here, population pressure was localized in geographically or ecologically bounded areas. One could have relatively low densities across a large region, with pockets of higher densities scattered about. These pockets become the areas in which cultural change is most likely to occur.

Carneiro was interested in the origins of state societies, defined by him as "an autonomous political unit, encompassing many communities within its territory and having a centralized government" (Carneiro 1970:733). However, his model has been applied to stateless complex societies as well. For Carneiro, war was a necessary factor in the development of complexity. Carneiro actually employs an early version of a game-theoretic logic. He argues that population pressure forces people to move from a revenge cycle to the need to acquire territory. He envisions that a community defeated in war in a circumscribed area with high population density could not flee to another area but rather had to stay in the same area. That choice came with a price, as he says. This price was "political subordination to the victor" (Carneiro 1970:735). He describes this as a process of "chiefdom" formation, not a state, implicitly expanding his model for stateless societies. This set in a new evolutionary dynamic in which the costs of accepting hierarchy were lower than the costs of "co-operating," involuntarily to be sure, with a conquering group. While this was a sophisticated modification of a simpler population-pressure-forces-adaptation concept, it still was firmly in this theoretical framework.

There was a bit of a backlash against the canonical model based on both theoretical and empirical grounds. In an article (with discussion) called "Population Pressure as a Non-Explanation," George Cowgill and Edwin Wilmsen (1975) set up the general argument against this mechanism of cultural change. These authors challenged this proposition on theoretical grounds. They first attacked the Malthusian assumption that populations would continue to expand. They then attacked the very idea that stress would naturally provoke "developmental responses" (Cowgill and Wilmsen 1975:127). Finally, they cite numerous empirical counterexamples of these assumptions.

A classic work that shifts causality and links population density with cultural practices is Esther Boserup's (1965) model for agricultural intensification. Boserup flipped the Malthusian concept of uncontrolled population growth with rigid limits in the carrying capacity of any area. She viewed carrying capacity as a more elastic constraint (Raulet 1974). As Turner and Fischer-Kowalski (2010:21963) nicely put it, "She reversed the causality, arguing that increases in population (or land) pressure trigger the development or use of technologies and management strategies to increase production commensurate with demand." Likewise, Boserup suggested that decreases in population densities would result in decreases in agricultural intensification (Turner et al. 1977:384).

The implication of Boserup's contention that a social activity such as agricultural labor intensity would fluctuate as a function of population density is quite significant. In the canonical model, there are some unilinear evolutionary hints in that it is unlikely for social hierarchies to disappear once they have established themselves. Therefore, while there may be population density fluctuations, once a threshold is reached, there is a sense that social complexity remains established in society. However, this is true for moderate change but not for radical change with social collapse. In these circumstances, any serious demographic collapse was seen to be a precipitator of sociopolitical disintegration.

In a Boserupian view, societies will vary their level of sociopolitical complexity to adjust to the existing population density and the concomitant adjustments that are required. There is actually a stronger connection between population density and social structure in the Boserupian model as opposed to the canonical one. In the latter, population pressure is a push toward greater complexity that in turn sets in a sociopolitical dynamic that favors an elite group, making it more resilient to later shifts in the cultural, demographic, and physical landscape.

Brian Hayden forcefully makes the case that population pressure is not supported by the empirical record: "A number of archaeologists have failed to find evidence of population pressure involved in domestication or the agricultural intensification process or in the development of social complexity" (Hayden 2014:117; and see Hayden 1990, 1992). He then lists more than 20 citations from around the world that support this proposition. Like Hayden, I do not accept population pressure as a lineal causal factor in most instances.

I argue that the existence of sufficiently robust population densities that increase the benefits of co-operation in any area represents the first condition that leads to sustained co-operation. This is a probabilistic model. This can be reformulated in anthropological game theory terms. One of the biggest effects of higher population density environments is an increase in the cost of free-riding and non-co-operation. This is an absolutely critical factor for the development and maintenance of special places on the landscape and for the intensification of this process over time. A fair amount of information in the ethnographic record of stateless societies demonstrates that in low-population-density contexts, people will opt for the lowest-cost option of fissioning. To repeat what we stated above, as population densities increase, the relative costs of co-operation go down while the cost of fissioning and other non-co-operative behaviors rapidly rises.

Fissioning in Stateless Societies under Demographic Growth

Group selection leads to the evolution of cooperation only if migration is sufficiently limited to sustain substantial between-group differences in the frequency of defectors.... *When the migration rate increases, levels of cooperation fall precipitously*. – Boyd et al. (2003:3531, emphasis added)

Robert Carneiro (2000:12927) notes that fissioning is "all but universal in the history of autonomous villages" an observation born out by many empirical studies. The anthropological game theory perspective adopted here is effectively Boserupian, seeing social complexity, agricultural intensification, and so forth as dependent upon population densities rather than as reactions to a Malthusian kind of population pressure and resource stress. Likewise, as seen in the above quote, some good theoretical support in evolutionary game theory links out-migration, an obvious form of fissioning, to the collapse of co-operative organizations. Population density and co-operation are positively correlated, but not in a direct linear fashion with population density being a purely independent variable.

An archaeological example here is the work of Katheryn Twiss (2008:418): "During most of the Pre-Pottery Neolithic, the extent and scale of feasting expanded as sociopolitical complexity increased. Towards the end of the period, however, populations dispersed and feasting probably declined." In this model, feasting intensity increases or declines as a response to the density of people in a region. Another way to say it is that the benefits of costly feasting are proportional to the density of population in any area.

As mentioned, in a low-population-density context, the least costly way of dealing with any kind of political, social, or even personal stress between households or lineage groups is to simply leave and establish a village in another area. Matthew Bandy (2004:322) elegantly stated this in reference to social evolutionary theory in general: "Early permanent population concentrations will frequently be unstable, with fissioning the predominant mechanism for resolving intravillage conflict." Throughout the ethnographic literature, in fact, we find that fissioning – the splitting off of households or household clusters from the larger group – is quite common. Among the Northwest groups, Federica de Laguna (1952:8) tells us, "There have also been cases of open and direct aggression within the group of kindred, but these can lead only to the splitting up of the sib and the emigration of one of the factions." Here is a classic case of simply leaving one village to avoid the costs of political dispute. Napoleon Chagnon (2012) documented numerous cases of village fissioning among the Yanomami due to intravillage fighting.

Godfrey Lienhardt (1958:114) describes this fissioning process as part of Dinka norms and customs. "It is part of Dinka political theory that when a subtribe for some reasons prospers and grows larger, it tends to draw apart politically from the tribe of which it was part and behave like a distinct tribe." In this case, it is not from stress or strife but from simple demographic growth. He goes on to note that this process happens in smaller political units as well:

> The sections of a large subtribe similarly are thought to grow politically more distant from each other as they grow larger, so that a large and prosperous section of a subtribe may break away from the other sections, with its own master of the fishing spear, its own separate wet-season camp, and eventually its own age sets. In the Dinka view, the tendency is always for their political segments, as far their agnatic genealogical segments, to grow apart from each other in the course of time and through the increase in population which they suppose time to bring. (Lienhardt 1958:114)

In this same passage, the ethnographer identifies the reason that this is possible in terms entirely consistent with anthropological game theory: "It is clear that such a theory . . . could only develop among a people who know themselves free to move away from each other; to occupy further traces of empty or weakly occupied country, and to find new pastures" (Lienhardt 1958:114).

Lienhardt goes on to describe a similar situation – one due not to prosperity but to stress. He describes how what he calls late-comers to a new place find themselves in a disadvantageous position relative to the settled peoples. He writes that when "a group of later-comers to such a camp become rich and large, its members no longer have enough space to tether their cattle on dry-standings, and resent the superior position of the first-comers" (Lienhardt 1958:115). What happens next is that an "ambitious man, perhaps a master of the fishing spear," collects some kin members or others who want to settle away. They then migrate out and create their own subtribe. In cost-benefit terms, this fissioning process is possible only in an environment in which the costs of migration are low. In the case of the Dinka, the ethnographer believes that they have a precise knowledge of this process and the consequences of migration.

Aylward Shorter (1972:151) notes, "It is true that the internal structure of the Kimbu chiefdom favoured a process of fission in which division of the chiefdom became detached as independent chiefdoms." This Tanzanian polity, discussed in a previous chapter, illustrates the process of splitting up in an environment of low population densities.

Napolean Chagnon notes that among the Yanomami a generation ago, "Kinship relatedness, patterns of marriage, and obligations based on descent seem to break down as solidarity-promoting mechanisms and villages seem to be unable to get larger than about 300 people before they fission into smaller ones" (Chagnon 2012:155). He likewise notes the danger in fissioning in relation to hostile neighbors. A village will split but live within a few meters of each other because of fear of attack: "This solution occurs when it would be hazardous for the larger group to fission and create two small, distantly located villages that would be easy prey to hostile neighbors" (Chagnon 2012:79).

Robert Carneiro (2000:12927) puts it directly: "Virtually every ethnography of an autonomous village with any time depth reports an occasion, some time in the past, when the village had fissioned." He goes on: "This fissioning . . . is all but universal in the history of autonomous villages." Carneiro's circumscription model, of course, is based upon population pressure in a confined geographical or ecological space. He likewise observes from the ethnographic record the process whereby the inability to fission will lead to greater social complexity: "The most common response we find among autonomous villages to an increase in population, if fissioning is not to occur, is the development of new social segments" (2000:12928).

The fissioning process occurs in complex hunter-gatherers as well. For instance, Randolph Widmer describes lineage fissioning among the Calusa during population growth (Widmer 1994:134–136).

High relative population densities therefore leave fewer options for leaving the group. This is one factor that significantly raises the costs of non-co-operation. Ostracism as a form of punishment does not work well in a low-population-density context but is extremely effective in landscapes with higher densities. In fact, in many documented nineteenth- and twentieth-century stateless societies, ostracism was a severe punishment. A family forced out of a group in a low-density context could establish themselves in nearby areas. In higher-density areas, this is almost impossible. Simon Powers and Laurent Lehmann (2014) work this into their model of the evolution of rank out of egalitarian societies. They describe this as dispersal cost, or "how easily individuals may leave one leader and follow another" (Powers and Lehmann 2014:6). In their model, people fission to join another group with leaders. In the model in this book, in low-density-population contexts people have the option to migrate out and maintain egalitarian norms and social organizations. But in both cases, the concept is the same: with sufficiently high densities that promote co-operation over fissioning, a selection process begins with competition for status and followers between emergent leaders of groups.

Strategy Selection in Cultural Evolution

The term "evolution" is used broadly to refer to a variety of concepts that are in fact distinct. There is no room here for a major digression on multilevel selection. It is adequate to simply note that in the history of biological and social sciences, the term evolution covers three broad kinds of evolutionary theory developed largely in the nineteenth century. These are known as Lamarckian, Spencerian, and Darwinian. These different evolutionary theories are differentiated by the nature of how species transform from one state to another.

Lamarckian refers of course to the concept of the inheritance of acquired characteristics in the tradition created by Jean-Baptiste Lamarck in the nineteenth century.[7] Strict Lamarckian evolution with its vitalistic and teleological implications is not considered viable in modern science in either biological or cultural evolution. It is moot.

Spencerian evolution centered on the notion of inherent progress and development. It conceived of evolution, in the broad sense, as a natural tendency for nature to perfect itself, from homogeneous to heterogeneous, from primitive to complex. In Spencer's concepts, all natural systems evolved toward their inherent perfect nature, like an acorn that develops into an oak tree. Everything necessary for the various stages in the entire evolutionary process exists in the acorn itself, and evolutionary process is an "unfolding" of the inherent nature of the oak tree.[8] Darwinian evolution is quite different. It is a process of descent, with modification via the process of natural selection (for an excellent discussion of these schools of thought, see Freeman 1974).[9] Of Charles Darwin's contributions, perhaps the greatest was to explain the transmutation of species via the concept of natural selection acting on random variation. I consider a social scientific theory to be a valid evolutionary one only insofar as it has a selective mechanism of some sort.

The mechanism in the evolution of co-operation is what I term strategy selection, a concept that borrows from the explanatory power of natural selection but that is altered to account for human intelligence and purposeful behavior in small groups. Humans make decisions to co-operate or not given the limited knowledge they possess and the nature of the cost-benefit landscape in which they find themselves. However, the range of available decisions under any set of challenging circumstances is not infinite. There is a limited set of viable choices available, with "optimal" being defined as the degree of success in both biological reproduction (inclusive fitness) and in creating economic resources relative to other strategies chosen and utilized by others.[10] It is not the group that is selected for; payoffs accrue to individual members of the group. Rather, it is the *strategy* of the group that provides the economic advantage that is the locus of selection.

In a competitive environment and with sufficient trial and error by adaptive actors, the concept of strategy selection ensures that ineffective choices by groups of people for subsistence organization, defense, and political organization will not survive competition from other groups. This model takes as a given that, because (1) humans are ultimately ego directed, seeking to acquire resources, prestige, and fitness, and (2) population densities are at a sufficiently robust level to foster co-operation, there will always be competition in these environments. It is important to remember that competition can take the form of in-group co-operation. In this sense, as described above, a highly competitive environment fosters a concomitant development of highly co-operative in-group norms. As we saw in previous chapters, in the absence of coercion, ritualized behavior is the key source for the construction and enforcement of these norms.

Both theoretically and empirically, similar patterns of human behavior should and do emerge in different places and times under similar circumstances. These patterns emerge not through the playing out of abstract structural principles that are beyond the knowledge of the individual; rather they emerge through a strategic economic logic of individuals in groups making decisions in a competitive environment that favors the some strategies over others. Groups that adopt the most successful strategies will dominate other groups or others will emulate the successful strategies. Groups that successfully maintain co-operation by controlling free-riding and creating more resources by working together will be at an advantage over groups that do not. Over time, groups that are more efficient in attracting people and creating and redistributing resources will grow at the expense of other groups. Total numbers in the group represent a crude but useful measure of success. Successful strategies will be imitated by others and then transmitted over the generations.

Individuals in groups that fail to adopt successful strategies will either be removed from the gene pool, be incorporated into successful groups, or out-migrate to marginal environmental niches. This cultural transmission model effectively posits that either successful co-operative groups will absorb others via coalition building or others will imitate them. In short, when several factors come together – minimal population densities, sufficient resource concentration to support semisedentism, a sufficient degree of resource constancy, and efficient economic organizations structured by ritual – complex sociological organizations develop.

This evolutionary process is not unilinear. The introduction of any variable that disturbs these core factors, such as environmental change, depopulation, or conflict, will result in the dissolution of the ritualized co-operative organization and the collapse of the co-operative group. Complex social organizations grow and dissipate over time. This is a form of chiefly cycling, a concept created for more complex contexts by Joyce Marcus (2008; also see Griffin 2011), but is also applicable to a full range of complex stateless societies.

The concept of strategy selection is similar in appearances to some multilevel or group selection models in biology, but it is reality not a group selection model in any traditional sense (e.g., Wilson and Dugatkin 1997; Wilson and Sober 1994). Rather, it is a bit more nuanced, and that nuance is extremely important. In a classic group selection model, the group is the unit of selection. In this view, individuals engage in costly prosocial or altruistic behavior to benefit the group itself.[11] MacDonald (2002[1994]:viii) provides a coherent statement of the group selectionist position in game theory terms, focusing on the free-rider issue:

> I am continually amazed at the extent to which evolutionists have been indoctrinated... against supposing that groups have any important role to play in human evolution. The problem comes about for two reasons: failure to comprehend cultural group selection, and failure to appreciate the extent to which selection between groups has shaped the human mind.... There is a critical difference between humans and other animals that renders all of the arguments against group selection moot. It is simply this: Humans are able to solve the free-rider problem by monitoring and enforcing compliance to group goals. So far as we know, animals can't do this.... Humans are able to solve the free-rider problem – the problem that organisms would tend to take advantages of group membership without paying the costs. (MacDonald 2002[1994]:viii)

In this group-centered view, altruistic behavior – costly punishments and other socially rational behaviors – keep the group functioning. This, in fact, is quite true, but the issue of altruistic behavior is key. I agree with Sanderson's (2001) critique of MacDonald to the extent that one must view the group as an aggregation of individuals. For this kind of strategy selection to work, most members of the group must benefit as individuals as well as a group. As Steven Pinker puts it, "if a person has innate traits that encourage him to contribute to the group's welfare and as a result contribute to his own welfare, group selection is unnecessary; individual selection in the context of group living is adequate" (Pinker 2012).[12]

Put another way, the group benefits insofar as the individuals in that group benefit. At this point, it becomes almost, but not quite, semantic – humans engage in short-term altruistic behavior that they also "instinctively" recognize as being in their long-term self-interest. A major debate is whether that "instinct" derives from biological natural selection and an evolved psychology or from a cultural transmission process with adaptive agents creating prosocial norms of behavior.[13] Fortunately, for our purposes, the results are the same regarding small-group behavior over deep historical time.

The Fluidity of Groups

A complicating factor illustrated by ethnography and history centers on the fluid nature of group composition in humans. Humans, unlike any other animal, are capable of switching groups and the corresponding behavioral norms unique to that group virtually overnight. In other words, animals groups are highly bounded with little fluidity whereas among humans, group identity is capable of a constant shifting. In the previous chapter, we saw how leaders in potlatching are constantly vying to attract more followers. We read in Malinowski how Kula trading partners are in theory partners for life but that in practice there is substantial fluidity as some chiefs decline or increase their social rank. The very nature of exogamy requires individuals to marry outside their group and effectively either join another one or to operate in two groups.

Group composition is constantly shifting in larger societies, where people have the ability to periodically align and realign with nonkin in mutually beneficial social relationships.

A contemporary example of how people actively switch from one group, with its set of norms, to another group with different norms is the Protestant movements in developing countries, particularly in Latin America. Conversion from Catholicism or traditional religions to Protestant ones involves a substantial shift in individual and group behavior. Such conversions are costly but also come with benefits. Goldin and Metz (1991) document profound economic changes that correlate with ideological shifts. They note, "The 'miracle' that God gave people so that they could improve economically was the enforcement of a code of behaviour already considered appropriate, but contradictory to the traditional lifestyle" (Goldin and Metz 1991:329). The code of behavior effectively constitutes a new set of rules that differed from traditional ones. These new rules served to increase the cost of non-co-operation and reward those that indeed co-operated.

Ethnographers have repeatedly documented case studies of individuals consciously opting to switch identity in order to better their economic and social positions. Anne Motley Hallum describes how one man, with the help of his wife, shifted his group identity from a traditional one to a Protestant one to improve his life: "He [the consultant] pointed proudly to his wife Hortensia and said that he drank alcohol and was bringing the family to ruin until she took him to her church and prayed for him to join the believers. Now the family was prospering in relative terms, and he was adding to their income from his occasional preaching instead of spending precious funds on alcohol" (Hallum 2003:178).

Among ethnographic cases of stateless societies, Charles Spencer (1993:42) writes about the fluidity of group membership among the Mekranoti-Kayapó in Pará state, Brazil. Citing the work of Dennis Werner (1980), Spencer writes:

> Individual villagers assess the relative merits of the various competing benjadjwyr (chiefs) and freely shift their allegiances from one faction to another, choosing to follow the leader who seems to provide best for his followers (Werner 1980:32–34). This fluidity of membership is encouraged by a lack of regular correspondence between the factional structure at any point in time. (Spencer 1993:43).

Miller and Boxberger (1994:271) report that Puget Sound chiefdoms are composed of households as the basic economic units, which combine into what they call local groups. These "local groups were fluid, and households sometimes broke off and joined other groups to improve their circumstances, ordinarily interacting with those speaking the same language." Again we see how group composition shifts as circumstances shift.

We could cite a huge number of case studies illustrating the use of the group to individual advantage and the fluid nature of that group composition. While the contemporary ethnographic vignette of Christian communities in Latin America may seem removed from the nineteenth-century Kula ring or the seventh millennium BCE in the Near East, the underlying process whereby strategic actors shift their group allegiance for strategic advantage is the same. To repeat, human group composition is not tightly bounded, particularly in small groups. Groups adopting successful strategies will attract followers and will grow disproportionally compared to other groups. Successful strategies are encoded in ritual, particularly in societies without formal mechanisms of communication. Participation in that group requires an individual to follow norms of behavior that will keep the group intact and in the final analysis will benefit all members of that group vis-à-vis other individuals and other groups. The existence of ritually mediated norms of group participation and behavior in a competitive environment is a condition for the evolution of co-operation in non-state contexts. Successful co-operative norms allow the individual to improve his or her material and social well-being while simultaneously benefiting the group.

We will now look at the archaeological evidence for the origin and development of ritualized economies among stateless societies around the world. We will look at examples from four continents and demonstrate that this process is inherent under certain demographic and social conditions. It is to these conditions and evidence that we now turn.

7 The Evolution of Ritualized Economies: The Archaeological Evidence

This chapter provides archaeological evidence that demonstrates the utility of the model developed in the previous chapter. I first describe the empirical record for the increases in population densities that we see around the globe in the critical time period in which complex societies developed after the long period of Paleolithic (Old World) or Paleoindian/early Archaic (Americas) hunter-gatherer-forager adaptations. This period is also known as the Neolithic Demographic Transition (NDT). This concept was first formulated for the Eastern Hemisphere, specifically Mesopotamia, Egypt, Anatolia, and the Levant. The NDT has now been successfully applied in the Americas, as I describe below.

Second, I look at specific examples of stateless societies in the archaeological record. A number of excellent case studies in both hemispheres allow us to define with some precision the nature of the earliest examples of this process. I describe one iconic case study each for South America, the Near East, Europe, and North America. These cases studies include the Caral and Paracas civilizations, Göbekli Tepe and later sites, henge monuments in Britain, and Watson Brake, Poverty Point, and related sites, respectively. Each case represents the fluorescence of ritualized economies in its region. I then describe the predecessors and successors of select complex stateless societies in each region illustrating the evolution of ritualized economies and co-operation in the archaeological record.

Late Pleistocene/Early Holocene Demographic Increases in the Eastern Hemisphere

Complex society most likely first emerged around 11,000 years ago in a number of areas between the Nile Valley and China. The NDT in the Old World is one of the best-documented processes in contemporary prehistoric archaeology. As Bocquet-Appel and Bar-Yosef (2008:1) note, “the signal of a relatively abrupt increase in the immature proportion of skeletons is observed in cemeteries during the foraging-farming transition. This signal is interpreted as the signature of a major demographic shift in human history, now known as the

Neolithic Demographic Transition (NDT)." Even more significant, they note that this is a worldwide phenomenon associated with the adoption of agriculture. It is found in Europe, North Africa, North America, Mesoamerica, and South America: "The two-stage NDT now appears to be a global process that is characteristic of most – if not all – early agricultural sequences worldwide" (Bocquet-Appel and Bar-Yosef 2008:1).

In the Levant, for instance, we know that the period from the Early Natufian (circa 14,000–13,000 cal BP) up to the PPNB or PPNC (circa 9000–8200 cal BP) was characterized by strong population growth on a regional level (see Guerrero et al. 2008:Table 1). As Bocquet-Appel and Bar-Yosef (2008:1) note, "the relatively rapid growth of human populations during the NDT radically transformed settlement behaviour." Ian Kuijt (2000:10–11) notes that the Neolithic in the Near East experienced "a remarkable increase in global population levels, the first appearance of large villages occupied on a year-round basis, and some of our earliest evidence for social differentiation."

The central question, therefore, is: When precisely did population densities increase relative to the construction of settled villages and the emergence of special places on the landscape? The canonical answer is that population pressure in the Late Pleistocene in the Old World pushed people to adopt agriculture, which in turn created a feedback loop that led to more population increase. Agricultural economies are implicitly associated with social complexity but it is difficult to define when agriculture appears. In some areas of the world, a suite of cultigens appear relatively quickly, whereas in other places, such as South America, cultigens are added on a very slow time scale. There is, therefore, much ambiguity about when "agriculture" begins. It is perhaps more accurate to use the term "agricultural economy" to denote the adoption of cultigens for a substantial percentage of the diet, along with settled village life, as the marker for this process.

In the model offered here, the rise in regional population densities would have provided the context for increased co-operation. Some population increase should therefore precede rise in social complexity, but we have seen from the ethnographic record that there can be a very short time between density increases and corresponding increases in co-operation, and vice versa. Of course, such chronologies at the century level in archaeology are generally not possible, and we can observe these changes only in half-millennial and occasionally shorter (as little as 100 years)[1] time frames. Therefore, to support this model, population density increases and social complexity would have to be at least covariant in archaeological time, while in theory a demographic rise should be first. The data presented for the support of the two-stage NDT are simply too imprecise to identify at the present time at which moment densities rose. However, we can say with a high degree of confidence that demographic increases at least covaried with increased complexity.

Population increases do seem to correlate with the process of domestication, including the manipulation of wild plants (arable weeds). These processes were effectively simultaneous with the first evidence of social complexity in the Near East at sites like Göbekli Tepe. The article by Riehl et al. (2013:66) reports on the incidence of cultigens and cultivated wild plants at the Neolithic site of Chogha Golan in the Zagros Mountains. They note that at the end of the Younger Dryas, circa 11,500 BP and up to around 11,000 BP in their first three archaeological horizons, there is an increase in wild barley and the presence of Triticum-type taxa. They interpret this as evidence for "probable cultivation of wild progenitor species of modern crops, including pulses such as lentil" (Riehl et al. 2013:66). It is not until about 1,000 years later that they note a "shift in the subsistence economy," with a decline in the percentages of "all taxa . . . except that of the small-seeded grasses" (Riehl et al. 2013:6). The PPNB is the period when major cultivation developed (Guererro et al. 2008:59). Taken at face value, these data would support the model in this book.

Bellwood and Oxenham note that during the Neolithic, population density increased rapidly and would have promoted migration: "The transition from hunting and gathering to farming was not always rapid in millennial terms, but the two-way feedback between food production and population density in these regions eventually snowballed into considerable population growth and a need for some members of each population, perhaps the younger ones, to seek new farm land outside their home range" (Bellwood and Oxenham 2008:17).

While these authors argue that population growth resulted from the adoption of agriculture and was a push to complexity, their observation can also be used to support the model in this argument. The migration of Neolithic farmers out of their home territories was costly, and thus the obvious benefits of staying and co-operating in larger numbers would have gone up relative to migration.

Zheng et al. (2012) published a very intriguing analysis of mtDNA derived from modern populations in Africa, Europe, and America. They then examined the lineages and calculated periods of demographic expansion. They concluded that 11 of 15 lineages in Africa, and all lineages in America and Europe, witnessed major demographic expansion before agriculture and the Neolithic period. They suggest that the rapid increase and amelioration of the climate after the Last Glacial Maximum, provided the context for rapid demographic growth. Needless to say, these data and conclusions conform quite well to the model presented here. The efficacy of these methods, particularly the underlying assumptions of rates of change, will of course have to be tested with future research.

Mark Collard et al. (2010) analyzed carbon dates as a proxy for population levels in Britain. They discovered a sharp increase in population levels between 6000 and 5610 cal BP.[2] This correlates to the first evidence of agriculture in the region. After this, there was a drop up to 4400 cal BP, with a subsequent

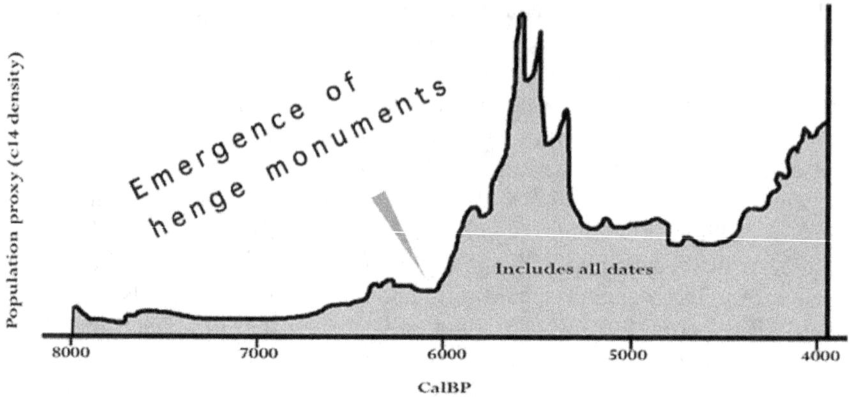

Figure 7.1. Population density proxy for prehistoric Britain. Adapted from Collard et al. (2010).

rise after that time. The curve is quite impressive (Figure 7.1). It shows a very slow but steady increase up to 6500 cal BP, then a 500-year period of elevated populations, followed by a dramatic upward swing to four to five times higher than the levels at 6500. The drop after 5610 is precipitous, falling to levels that are still three times higher than that at 6500. In other words, the increase in population densities correlates with the advent of agriculture and the first evidence of special places and complex stateless societies in Britain. The slow and steady rise clearly precedes the development of the henge monuments at around 6000 cal BP. The levels at 5500 and later are all substantially higher than the pre–6500 cal BP period. In other words, these data very clearly support the notion that complexity in stateless societies begins coincident or after a significant rise in population density in Britain.

Another aDNA study comes to similar conclusions for Africa: "The correspondence between the timing of genetic evidence for a sharp increase in population size at 4,600 ya in our Holocene sample of sub-Saharan Africans and the archaeological evidence for origins of agriculture in western Africa is quite close" (Gignoux et al. 2011). Yet the same study does not find demographic increases with agriculture in Europe: "The absence of a distinct Holocene-period expansion suggests that the adoption of agriculture by indigenous hunter-gatherer Europeans was slow at best and did not contribute substantially to the initial geographic and demographic expansion of agriculture in Europe" (Gignoux et al. 2011). The population growth, according to these data, did in fact increase around 6500–6000 BP, a time well after cultigens were introduced into Europe. Not coincidentally, this is the period at or somewhat before the rise of the henge monuments that first appeared in the region (Henge monuments are the first indictors of special places in the region as described below). In Europe at least, the initial population density rise covaries with social complexity, the development of special places, and not necessarily with agriculture. The relationships are indeed complex and require much more finely tuned chronologies to test in the future.

In short, there is a general correlation between population densities and the rise of social complexity in the areas of the Old World that are best studied – the Near East and Europe. The pattern is this: there is first a population density increase, followed by increased social complexity which is then followed by a very rapid population increase. There is little question that once agricultural economies are established, population growth rapidly increases. However, the data support the idea that initial social complexity first requires a population density increase to minimum levels.

Scholars who hold different theories about these processes will of course argue that one of these factors precedes the other. Traditional theorists of population pressure models will insist that all of the major population increases were first, while those who see the adoption of agriculture to be the cause of fertility increases will argue the opposite. The brute truth is that, at the present time, our data are only unambiguous enough to say that there is a general correlation between population density increases, agricultural economies, and the initial development of social complexity. This pattern is sufficient to satisfy the criterion that a minimal level of population density is required for the evolution of co-operation via ritualized economies in the model developed here.

Mid-Holocene Demographic Increases in the Americas

Peros et al. provide an excellent overview of the broad population trends in North America from the Paleoindian period up to the demographic collapse that occurred with European contact. They summarize their findings: "Population was low and relatively constant during the Paleo-Indian period, and began to increase during the middle Archaic, roughly 6000 cal yr BP. The population of the continent then underwent a rapid increase beginning ~2000 cal yr BP, roughly coincident with the transition between the Early and Middle Woodland Periods in eastern North America" (Peros et al. 2009:8).

The data seen in Figure 7.2 strongly support the model in this book. The first increase in the rate of growth began at circa 7000 BP (a millennium earlier than the authors state but which is evident in their data), a period prior to the establishment of agricultural economies. The second increase, between 3000 and 2000 BP (a millennium later than cited by the authors but confirmed by their graph), precedes or corresponds to the full adoption of settled village life based upon agricultural economies in a number of areas in North America.

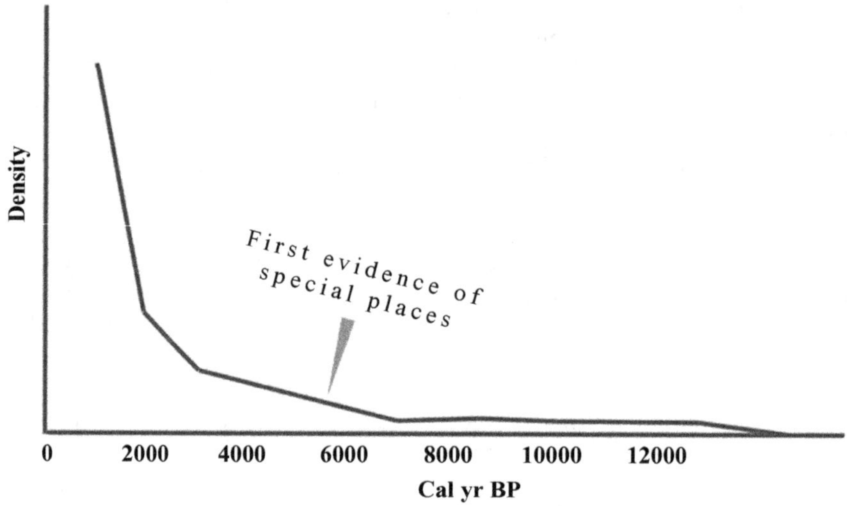

Figure 7.2. Population density trend for prehistoric North America. Adapted from Peros et al. (2009).

In the Old World, the "two-stage NDT" refers to a pattern in which the first stage is characterized by an increase in populations, followed centuries later by a decrease. Matthew Bandy (2005) wrote a seminal article applying the two-stage NDT concept to three areas in the Americas. He found in the Titicaca Basin, the Basin of Mexico, and the Valley of Oaxaca that, "in all three cases, an initial very high rate of population growth is followed by a drop in the growth rate that occurs 600–800 years after what in the Old World would be called the Neolithic transition" (Bandy 2005:S113). Therefore, the pattern of a significant rate of population increase at or before the development of social complexity holds in the Americas in the areas studied as well. What is critical for our model is that the higher population densities correlate to increasing social complexity. As in the Old World, there is a general correlation between population densities, plant domestication, settled village lifestyles, and the rise of social complexity in the New World.

The nature of plant and animal domestication in the Americas differs from the Near East/Levant. The pattern of plant domestication in the Andes, for instance, is one of early adoption of certain crops and then a slow and selective use from region to region. Domesticated tubers have been found as early as 8000–6000 BCE. Maize is found domesticated in the northern Andes as early as 6000 BCE, and it spread to the south, reaching the south central Andes by at least 2000 BCE. Camelids were domesticated by at least 4000 BCE. Deborah Pearsall describes this slow process:

> First, in the Early Holocene there were frequent and dispersed plant domestications, some of which were advantageous and left a trace in the record. Second, early domesticates spread, sometimes widely, through social interactions among forager/horticulturalists. Cultivation was small-scale, in well-watered settings. Third, increasingly productive crops fueled population growth, which led to the spread of societies into new habitats that were dependent on agriculture, and creation of built environments for farming. (Pearsall 2008:119)

We again see a general correlation between the adoption of agricultural economies and population density increases. But as we just said, the evolutionary pathway of plant domestication and cultural complexity in South America is different than in areas studied in the Old World. As Robert Tykot and John Staller (2002:666) note, the social history of domestication in at least Andean South America "contrasts with that of the Near East, where village settlements appeared first, followed by agriculture and, even later, ceramics." For reasons yet poorly understood, Andean peoples were much more selective in the timing and kinds of domesticated plants and animals they incorporated into their economies. Our data are limited, but what we can say is that there is a general covariation between the *use* of domesticated plants and animals on a fairly widespread scale, population increase, and increased sedentism. Any "clean" lineal models of causality using these three variables simply do not work well in the Andes. One possibility is that maritime resources were so vast. This stable and predictable resource base mimics agriculture from an economic perspective. Marine resources provided a rich source of food outside of plant agriculture, permitting the development of sedentary societies along the oceans near rivers (see Moseley 1975).

David Wilson's (1983:255) survey of the Santa Valley in Peru provides an illustration of this complex pattern. Figure 7.3 shows his reconstruction of the population size from approximately 2100 BCE up to the present. Even though cultigens were available in the region before 2000 BCE, we do not see any major growth spike until circa 100 BCE, almost a millennium *after* the introduction of irrigation agriculture. Prior to 1000 BCE, population levels slowly but steadily rose. We see in this valley the introduction of ritualized landscapes and economies about one to two centuries *after* the first-century BCE growth spike (Wilson 1988).

The vast western Andean desert stretches along the coast for more than 2,500 km.[3] Several dozen rivers punctuate this dry region. For thousands of years these rivers have been used to irrigate the dry lands. Valleys on the Andean coast represent classic instances of circumscription. In a way, these are "linear oases" that cut through unproductive deserts. Sprawling, virtually lifeless desert landscapes separate these oases, creating substantial barriers to intervalley communication. These valleys are naturally spaced dozens of kilometers apart, providing extremely rich zones separated by an otherwise barren

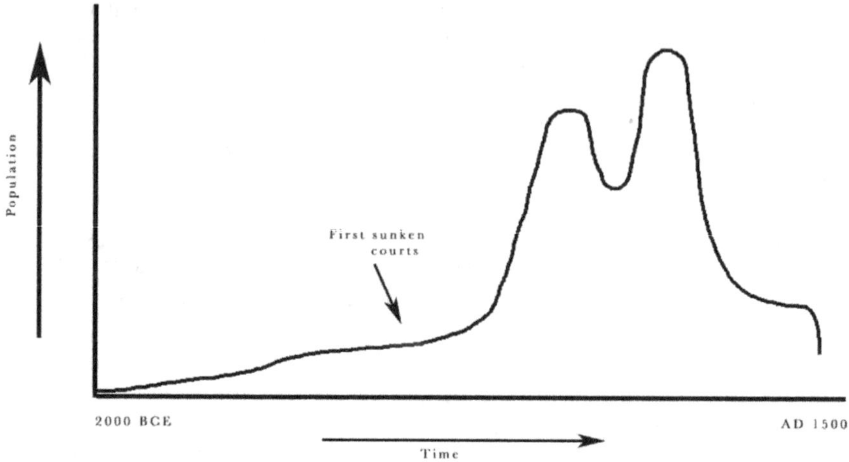

Figure 7.3. Population trend for prehistoric Santa Valley, Peru. Adapted from Wilson (1983).

landscape. In particular, the coastal areas where the rivers discharge are very rich places that combine highly productive riverine environments with equally productive marine resources. These data from the Santa Valley support the model in this book that population density increases provide the context for the evolution of complex co-operative social organizations in a circumscribed environment.

Our work in the Titicaca Basin of southern Peru and northwestern Bolivia provides confirmation of a similar pattern (Stanish 2003, 2006; Griffin and Stanish 2007). Cultigens were available well before settled village life, but we see a slow and steady population growth from 2000 BCE up to the Late Intermediate period 3,500 years later. Evidence of ritualized economies exist as early as 1400 BCE (Cohen 2010; Hastorf 2005; and see below), coming after a period of establishment of settled villages and steady demographic increases beginning at least at 1800 BCE and probably earlier (Aldenderfer 1989). In the Titicaca region, there was a several-centuries-long period between settled village life, population increases, and the first evidence of ritually significant constructions, particularly sunken courts (Cohen 2010; Stanish 2003).

Intensive survey and excavation work on the prehistory of the Hawaiian Islands provides some very precise data. Due to some excellent chronological controls, as well as the availability of historic accounts, we have a good demographic profile from the fourth century CE up to the present. Tom Dye (1994:3) reconstructed the demographic trend, as seen in Figure 7.4. His data show that the first ratchet in population size began a generation or so before

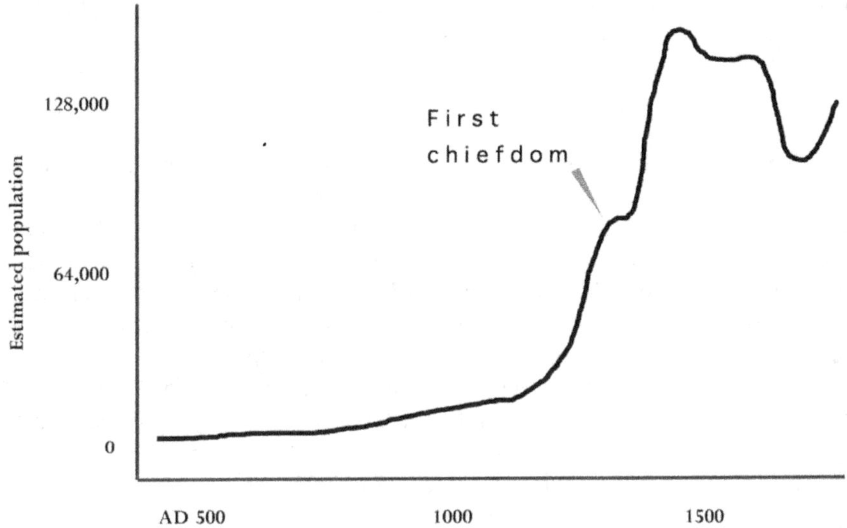

Figure 7.4. Population trend for prehistoric and historic-period Hawaii. Adapted from Dye (1994).

1227 CE. This was preceded by several centuries of initial occupation and very slow but steady population increases. As reconstructed by Patrick Kirch (1994:528), the chiefdom or early complex polity formation process took about 200 years, beginning around 1338. Here is a clear indication that population densities preceded significant co-operation into larger political units. These data support the model here in that there was a steady rise in population prior to the emergence of complex settlements, a necessary condition that favorably altered the costs and benefits of co-operation.

According to Michael Kolb (1997:275), the period from 1000 to 1400 CE was characterized by "low-density dispersed settlement" composed of small household and small-scale agriculture. Small polities, optimally located in fertile areas, characterized the political organization at this time. In other words, it was in this context of low-level political complexity that the population began to climb, preceding the following consolidation and unification periods, circa 1400 to 1650 CE. Again, we see additional evidence that population densities increased prior to substantial political complexity.

In sum, we find that in areas with good demographic data, there is a pattern of slow, steady growth followed by a ratcheting up of population levels. The data can be interpreted to support either a canonical model of population pressure or one adopted here in which a minimum population density that allows for increased co-operation is reached. However, in some instances where chronological controls are excellent, we find good support for an initial population increase that precedes the development of complexity.

The Earliest Special Places

The nature of the first special places, evidence for the first complex stateless societies, is effectively an empirical question. Before we survey the literature, it is worth preparing our minds with a brief "thought experiment" about the hypothetical kinds of settlements that we might find that were built by the world's earliest complex societies. As people settled, and as demographic conditions for widespread co-operation among people in settled areas developed, there are in theory a number of potential scenarios. This experiment assumes only two kinds of architecture and space – ritual and domestic. It does not assume that these categories overlap; nor does it assume that people substantially mix the domestic and ritual in the same area. In other words, we proceed with only two categories of space in this hypothetical experiment. Given this, there are logically only a small number of potential scenarios.

One possibility is that people could simply enhance their domestic structures as a correlate to the development of more social complexity. As society had access to more resources with sedentary or semisedentary lifestyles, we can hypothesize the development of bigger and better houses, clustered in villages for defense and/or greater economic and social interaction. If households were the primary unit of production, both social and material, then we can hypothesize that individual domestic structures would be become more elaborate without any significant differentiation between them in the more complex villages. We would expect to see differences between villages – between richer and poorer – but not within villages. The hypothetical settlement would have larger and better-made houses, perhaps built out of more durable materials. We would see an increase in art and an elaboration of architecture in the domestic unit. A model of larger and finer domestic structures, clustered in larger and wealthier settlements, therefore represents one possible scenario. Ritual would not be a factor until the development of more complex, hierarchical states.

Another possibility is the elaboration of individual ritual structures alongside domestic ones. In this scenario, related to the first, extended families would include ritual space as part of enhanced architectural complexes in and around private dwellings. Ritualization would increase but would be managed by the larger domestic units. Here we would see a gradual increase in domestic space quality alongside the development of ever more elaborate ritual space adjacent to or incorporated in these domestic contexts.

A third possibility would be the elaboration of ritual space within villages while domestic space remained relatively unchanged. The entire village, or perhaps subsections of the village, would build a single, large, and elaborate structure in the middle of a permanently occupied village. There may or may not be enhancements to the domestic areas, but the ritual focus would be communal and controlled by a single group resident in a village. As with the above hypothetical models, there would be no differentiation between domestic structures in the villages, but a few villages would be more elaborate. The ritual areas would be the focus of the entire village community.

A fourth possibility in this thought experiment would be the construction of some kind of ritual space and place in the landscape, outside of any village, that was shared between villages and other settlements. We would see a regional dynamic, in which each regional cluster of settlements constructed their own place in the landscape.

A final possibility would be some kind of hybrid configuration, with ritual space created within and outside of villages, paralleled or not by some kind of enhancement of the domestic space.

We can now turn to the archaeological record to see how closely or not these hypothetical models conform to empirical reality. What we in fact find in all of our case studies is that the third and fourth of our hypothetical scenarios are most common, albeit with some modifications. We now turn to these empirical data.

Early Complexity in Andean South America

I start with Andean South America first because it is my area of specialty, and, along with many areas in North America, it is one of the most intensively studied regions of the world using explicitly scientific archaeological research designs. In particular, we have a long tradition of regional research, including very large surveys and extensive test excavations in regions of primary state formation. These kinds of data are ideal for testing and illustrating the ideas in this book.

The first humans entered South America around 15,000 years ago, almost certainly via the corridors available from northeast Asia into what is now Alaska and British Columbia. From these modest migrations, people quickly moved across the landscape, taking advantage of the vast and untapped resources of two continents. Populations enjoyed modest growth throughout the millennia during the Paleoindian period. In the central Andes, we call the subsequent time period, circa 8500–1500 BCE, the Archaic (Aldenderfer 1989). The Archaic can be conveniently divided into early, middle, late, and terminal, with dates varying between regions and among investigators, depending upon the criteria used to construct the chronologies (Figures 7.5 and 7.6).

The Early Villages of the North Peruvian Coast

At the present time, the earliest documented monumental construction in the Andes is found at the site of Sechín Bajo in the Casma Valley (Fuchs et al.

Late Horizon	**AD 1476-1532**
Late Intermediate Period	**AD 1100-1476**
Middle Horizon	**AD 600-1100**
Early Intermediate Period	**100 BCE-AD 600**
Early Horizon	**800-100 BCE**
Initial Period	**1800-800 BCE**
Late Archaic	**3200-1800 BCE**
Middle Archaic	**6000-3200 BCE**
Early Archaic	**12000-6000 BCE**

Figure 7.5. Chronological chart for the Andes.

2012:Table 1). The archaeological team recovered several early dates at the base of one of the principal monumental constructions. One date of a hearth came in at 4487–4332 BCE. This hearth (not midden) was in a sealed context, and the subsequent construction phase yielded dates with a maximum range (2 sigma) between 3510 and 2915. This construction or series of constructions is associated with a sunken court and most likely was built as early as 3400 BCE over an as yet undetermined occupation that was perhaps 1,000 years earlier. According to Henning Bischof, "Around 3400 BCE (calibrated), the First Building of Sechín Bajo was built with a particular type of elongated and truncated hand-formed pyramidal adobes along with other larger [bolsiforme] ones" (Bischof 2012:15).[4]

Another possible very early monumental site in the Andes was excavated by Jerry Moore in far northern Peru (Moore 2007a). El Porvenir consists of about 3 ha of mounds and platforms and was first occupied around the early fifth millennium BCE, along with other sites in the region. A lower floor in one of the mounds has a ^{14}C date of 4710–4220 BCE (Moore 2007b:11). One

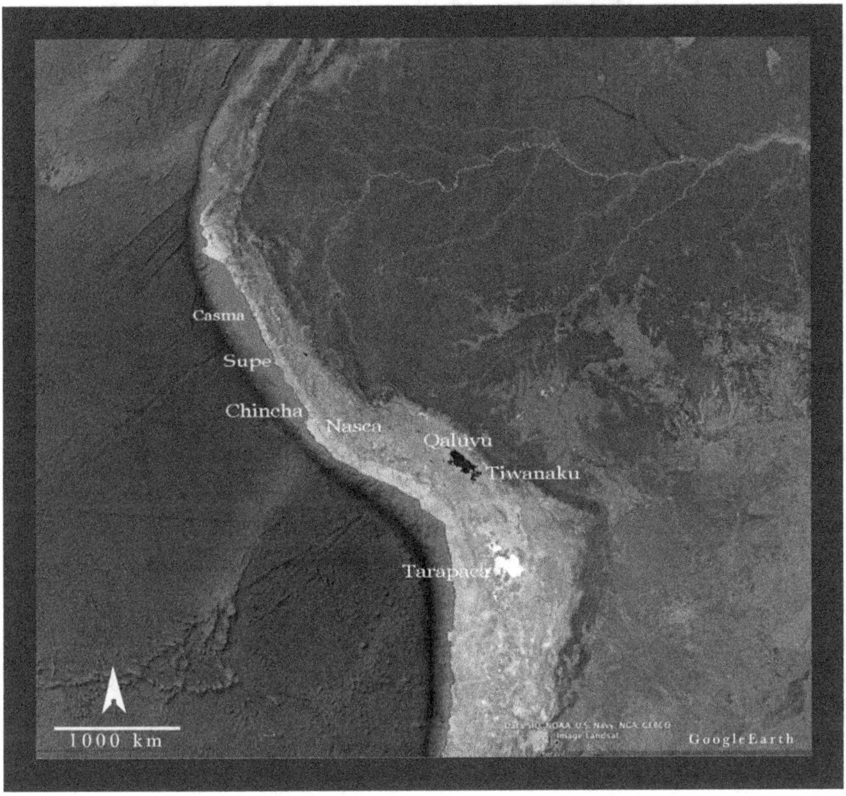

Figure 7.6. Map of the Andes showing locations of sites in text.

of the earliest structures in this context is a 5-m-diameter circular building. This is quite large for the Terminal Archaic period of the region. Since there is a pattern of house mounds built around an open plaza much later (1300–400 BCE) in the same area, it is possible that the larger circular structure could indicate a very early process in which a communal/ceremonial structure was built in a cluster of smaller structures. Unfortunately, we do not have broad enough exposure of the fifth- and fourth-millennium-BCE levels to definitively state that there is some kind of nondomestic structure of any consequence in the site. The circumstantial evidence, on the other hand, is compelling.

The village of Real Alto in Ecuador, located in the Chanduy Valley, was first occupied around 4800–4000 BCE. Around 3900, the site contained about 12 to 15 simple structures of pole and thatch (Quilter 2013:82; Slovak 2003). Jeffrey Quilter notes that it was arranged in a circle or possibly an open-ended circle, creating a U shape. This site was a small hamlet but one arranged on what may be a conscious plan to enclose space. According to the investigators, by 2500 the site held up to 100 houses arranged in a rectangle, all surrounding a central plaza. Most significantly, four mounds in this plaza were "for communal use" (Quilter 2013:83–84). Two of the mounds have evidence of feasting and burials, suggesting a strategy of creating a space for hosting events and anchoring social groups in that space.

The site of Aspero, located on the northern edge of the Supe River adjacent to the Pacific Ocean, was contemporary with at least part of the major Caral settlement described below. The earliest phases of monumental construction at Aspero began around 2800–2000 BCE (Feldman 1980, 1985; Moseley 1992:117).[5] One large monument is the Huaca de los Idolos, a flat-topped pyramid 1,500 square meters in size used for ritual display (Feldman 1985; Moseley 1992:115). Along with this pyramid, Aspero has 12–15 ha of domestic midden areas and 17 other pyramids between 1 and 4 m high. Excavations at the site reveal a pattern of continually rebuilt constructions by a resident population, a pattern found at many sites throughout the coastal valleys at this time. The architecture of Aspero, El Paraíso, and others illustrates the principle of openness, public performance, and visibility. The pyramids are built low on the beach, where they can be viewed from a wide area. Importantly, they are built low, below the flanking hills. The pyramids could have been built higher up on the ridge, but the clear architectural intent was to place these buildings in a highly visible area.[6]

In sum, we find the first modest constructions of small mounded special places beginning in the late fourth and certainly in the mid-third millennium BCE in the northern Andean coast. The center of this early process was concentrated between the valleys of Casma and Supe.

The Site of Caral in Third Millennium BCE Peru

The justifiably famous site of Caral in the Supe Valley of north coastal Peru in the central Andes (Figures 7.6 and 7.7) is the earliest major center in the Americas that represents the full expression of a special place constructed on the landscape by a complex stateless society. It is the first manifestation of a ritualized economy on a large scale, although, as we will see below, there are indications of complexity in earlier settlements, with evidence of nondomestic architecture designed to host and attract people. These earlier sites represent the cultural context in which Caral emerged and give us insight into the nature of the process whereby ritualized economies evolve.

Reconnaissance and excavations by Haas et al. (2004) in the Pativilca and Fortaleza Valleys just north of Supe where Caral is located discovered 16 sites with monumental architecture dating to the Late Archaic. In their report on

Figure 7.7. Location of Caral.

the almost 100 carbon dates on at least 12 well-dated sites (Haas et al. 2004: Table 1), we see that they are almost identical to Caral in the date range in the Late Archaic. Some dates are earlier, such as the site of Punta y Suela, which has three dates from approximately 6200 to 5400 BCE. Since earlier middens are found under some of these mounded platform constructions, we do not know if these mounds are in fact as early as the early seventh millennium BCE or if they are artifacts of construction practices that incorporated very old midden into platform fill. Haas et al. (2004:1020) argue that the period "between 3200 and 2500 cal. BC, with 27 dates, marks the clear appearance of large-scale communal construction and population aggregation" in this region. What is clear is that there are settlements with monumental architecture in the same age range as Caral, but it is not nearly as complex. It is also likely that a few small monumental sites in this valley were in fact loci of ritualized monumental construction very early on, making them bona fide precursors to Caral. It is also quite likely that many of the monumental sites found in Supe, above and below Caral, are also examples of early special places that predate Caral. It appears that the "Caral phenomenon" has its origins in small, single-mound platform sites in the region dating back at least to the middle or early fourth millennium BCE.

Caral itself consists of a series of pyramids and other structures spread across a pampa in the upper valley in what is best understood as a ritualized landscape. The dating of the main monuments of the settlement is a bit controversial due to the various problems associated with dating fill-and-reuse mounds. The earliest monumental architecture most likely appears in the last centuries of the second millennium BCE, circa 2600. Additional structures were subsequently built, and existing ones were modified over the centuries. The range of dates for the construction of monumental architecture on the site, therefore, is around 2600 to 1700 BCE (Shady and Kleihege 2008; Shady et al. 2001).

The celebrated archaeologist Ruth Shady is the lead researcher and indomitable force in unlocking the scientific mysteries of this great prehistoric monument. She has published extensively on the site cluster and has described many ritual aspects of the architecture. Shady and her team have been working in the Supe Valley for decades, publishing a huge corpus of data (Shady 2000; Shady and Leyva 2003; Shady et al. 2001; Shady et al. 2009), including both systematic survey and excavations. The bulk of the following discussion derives from her work and that of her associates.[7]

Caral is the largest of at least two dozen sites located in the upper valley area (Figures 7.8 and 7.9). The vast majority of these sites are most likely

Figure 7.8. Aerial image of Caral in the Supe Valley.

Figure 7.9. Photograph of Caral.

contemporary, at least in part, with Caral or are very close in time. This conclusion is based upon the surface characteristics of the sites as well as some carbon dates run on samples from test excavations. According to Shady,[8] virtually all the sites up- and downriver from Caral have monumental architecture. Some of these are quite humble, constituting small sites with one nondomestic architectural construction. Others are a bit larger, but Caral itself is substantially larger than any other site at its height. Although we are not certain of the precise dates of these sites, it is a very good assumption that they are chronologically close to each other in the late Middle and Late Archaic periods.

At this time, virtually none of these earlier, small sites have been extensively excavated and reported. Based upon analogies from other areas, however, it is highly likely that a number of small sites with monumental architecture are earlier than Caral. From the perspective of labor investment and complexity, these humble sites look very much like the kinds of constructions we see in the ethnographic record. In other words, small sites like Watson Brake, in which individual groups created special places to congregate, were the focus of a local community, perhaps no more than one or two days away.

The structures at the site are indeed impressive. Most are not completely artificial platforms but rather are sculpted hills. The builders at Caral used a time-honored technique known as shicra bag construction. Shicra bags are made with nets of vegetal-fiber ropes that contain rocks. These bags were filled and then used as the base of the construction of the monumental architecture. They ranged from about 14 to 48 kg, with some weighing substantially more (Shady et al. 2009:38). The shicra bags are curious in that they are not necessarily crucial to build up the fill for a platform. One can carry rocks in a manta or cloth shawl like many people still do today in construction in rural areas. Likewise, a simple bunch of reeds tied together makes a very easy means for two or more people to carry a number of rocks. There is also substantial additional work to gather the vegetal fiber to weave and fill the bags.

The shicra bags, therefore, did not just help in the construction but seem to have had some other kind of social function. What the bags provide is a way to account for the amount of labor contributed by any individual or group. In other words, the shicra bags most likely allowed a rough accounting of the contribution of any group of people. They could be gathered in one spot and then transported to the construction site. Instead of tearing open the bags for reuse after dumping the stones, the entire shicra was left in place. Another consequence of this is that the precise contribution of each person or group was permanently recorded, insofar as they could remember the bags they had brought. An interesting project would be to see if there are any subtle differences between the shicra bags in each construction.

We see this in other features of Andean life. Michael Moseley proposed that marks on adobe bricks in the much later Moche state represent a way to account for tax contributions of bricks to build a large structure. I also witnessed a community building effort on the Island of the Sun in Bolivia many years ago. In this instance, the community members spent many hours figuring out how to precisely assign the work assignment – leveling a hillside for a building – to five groups before the work began. Once they decided what was fair, the work went quickly, with men, women, and children all contributing. It was vitally important to establish the precise amount of earth to be moved by each group, reinforcing the notion that co-operative labor in a noncoercive political environment requires adherence to principles of fairness.[9] I would therefore suggest that the shicra bag construction technique not only facilitated the building of platforms but did it in a way that ensured fairness.

A key architectural feature of Caral is the sunken patio.[10] These semisubterranean structures are usually associated with mounds with flat summits and entrances made of stone staircases. They range in size from a few meters to up to 18 m in diameter. We see at Caral and contemporary coastal sites the early use of the circular sunken patio associated with platform mounds and other enclosures, such as rectangular structures. As Sheila Pozorski and Thomas Pozorski note (2008:612), it is most likely that the earliest use of this sunken patio architectural tradition was in the Supe and surrounding valleys during the Late Archaic, with a major expression at Caral.

Sunken patios are also often paired with staired pyramids and plazas. The patio and staired pyramid combination, along with some kind of open space, creates a very specific kind of "theatrical" experience for the participant. It is likely that the courts were relatively open, though there might have been some partial roofing on the sides.

David Carballo et al. (2014) note the theatrical importance of plazas and high-status mounded structures. In the case of La Laguna, a Formative-period settlement near modern Mexico City, they highlight the significance of architecture in co-operative behavior in what they refer to as "suprahousehold" projects. Consistent with the model in this book, they note that events carried out in these architectural areas are "appropriate for the embedding of labour within ritual networks" (Carballo et al. 2014:142):

> This high-status residence was constructed on an artificially flattened promontory overlooking a formal ceremonial centre with the two plaza groups that are the focus of this study: the central plaza is a formally bounded, rectangular plaza situated between the largest temple and the ball-court; the eastern plaza is an open area lacking formal boundaries located in the space between the three primary temples. Multiple lines of evidence indicate that the eastern plaza served suprahousehold functions of food storage, preparation and consumption associated with community festivals and rituals. (Carballo et al. 2014:144)

We see parallels in the architecture of Late Archaic Peru. As one approaches the pyramid, one pathway leads downs into the patio space and one emerges on the other side, usually to a stepped pyramid. One ascends the steps in a prescribed way. Evidence indicates that there was some kind of structure at the top. At the contemporary site of Aspero, for instance, excavators found structures that contained a variety of goods at the top of the main pyramid. The architecture and the common pattern of patios, plazas, and platform mounds is best interpreted as structuring ritual movement around the core of the site area. While there is obviously more than just labor mobilization involved, the promotion of co-operation among a group of individuals engaged in common labor tasks stands as a likely outcome of such festivals.

At Caral, several platform mounds align in a manner that creates a central rectangular plaza area or promenade of sorts. Each platform has architecture that invites a kind of procession or structured movement. In other words, the basic architectural plan of Caral is one in which a number of individual platforms create a space that is ideal for ceremony, potlatching, feasting, and large public ritual displays.

Only a few structures at Caral have evidence for permanent occupation. In fact, this work provides little evidence for intensive, long-term use, such as deep middens, storage structures, and cemeteries. It is not insignificant that poorly made perishable structures have been found between the pyramids at the site, additional evidence that the site was occupied for short periods.

We agree with Shady (2000) and Flannery and Marcus (2012:243) that each temple complex was most likely associated with a multifamily residence and that each complex was built and maintained by a separate corporate group of some sort. Excavations by Shady and her team (Shady and Leyva 2003) likewise discovered numerous ritual objects at Caral, including flutes, figurines, *Spondylus*, human sacrifices, stelae, carved bones, and plants with no domestic use (e.g., coca) and from different ecological zones, including the tropical forest.

A number of geoglyphs in the vicinity of Caral are located in a pampa away from agricultural fields. Likewise, there is a large geoglyph at the nearby and contemporary site of Chupacigarro in the same valley (Shady and Kleihege 2008:144). These stone lines and cleared areas complement the architecture of the site to create a ritual landscape, an ideal setting for processions and other orchestrated movements (and see Moore 1996a). We agree with Anthony Aveni (1990:112; 2000a), Michael Moseley (1992), Aurelio Rodríguez (1997), and others who argue that these kinds of lines, such as found most famously in Nasca, were intended to be walked on. Helaine Silverman and Donald Proulx (2002:192) argue that much later Nasca lines most likely orchestrated processions, "dancing single-file across the Pampa and the valley hillsides and then physically executing the geoglyphs at the end of that particular ceremonial episode" (and see, discussing the lines as ritual pathways, Silverman 1990).

There is intriguing evidence that the architects of Caral used at least the June solstice sunset to mark time in some fashion. As seen in Figure 7.10 all the principal pyramids at the site appear to align with the solstice. Since many of the monuments have been reconstructed a bit and others are covered or damaged, the measurements are necessarily less precise than if we had the original walls exposed on the surface. However, using the relatively clear air images from Google Earth, we can see that the azimuth of the solstice – 294° – seems to be an underlying principle in the placement of the most important buildings.

The Piramide Mayor, in the northern part of the site, is oriented on its long axis to the solstice sunset. Likewise, at complex marked "A" on Figure 7.10, while very difficult to measure due to the imprecise ground conditions, the approximate alignments are 297° ± 3° on the longer axis and 28° ± 3° on the shorter one. The "perpendicular angle" for the June solstice is 24°. That is, rectangular structures with a long axis of approximately 24° indicate that the short axis would align with the June 21 sunset at 294°. The inherent error in these measurement would therefore easily cover the solstice and suggests strongly that the complex is based upon a solstice-oriented plan.

The two pyramids of "C" and "D" present a fascinating pattern. These are slightly trapezoidal, and the average or center of their angles almost perfectly

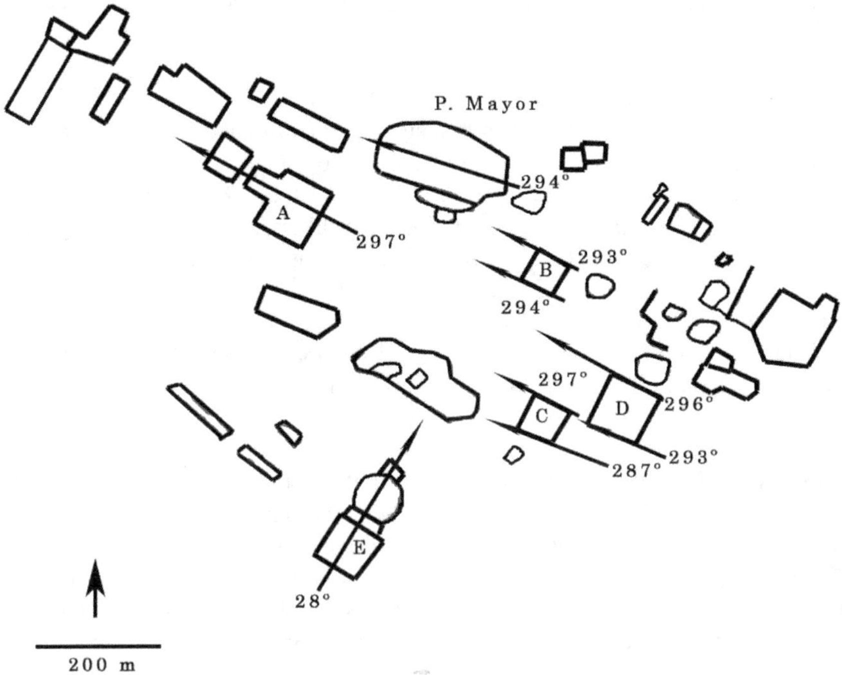

Figure 7.10. Solstice alignments of principal buildings at Caral.

marks the solstice. That is, facing the west, the C Pyramid has approximate angles of 287° and 297°. (Again, these structures are reconstructed and are slightly disturbed on the surface, so the measurements are approximate.) The center alignment of the pyramid therefore is approximately 292°, again very close to the solstice sunset line. Pyramid D likewise has angles of 296° and 292°, which of course from the perspective from the ground to the east means that the two angles perfectly bracket the setting sun. This phenomenon has been discovered in the Chincha Valley associated with the Paracas culture as is described in detail below. Finally, Pyramid marked "E" has a long axis angle of approximately 28°, again meaning that the short axis precisely faces the setting sun during the winter (June) solstice.

Beyond these individual structures, the site layout itself follows an axis that aligns with the June solstice. It is difficult to say if this is intentional, but at the very least it is highly coincidental to the point that it is unlikely to be random. It is significant that the river segment adjacent to the site also happens to naturally "align" with the solstice. If the architects at the site simply followed the river segment, then it is possible to explain the pyramid orientations as a simple adaptation to the local topography.

Like Caral, a number of Late Archaic sites in the Peruvian North Coast show evidence of site planning around astronomical phenomena. The solstices once again are prominent in structuring the sites, and most likely the social events that occurred in these special places. As J. Malville (2014:169) notes, "the earliest large-scale structures in Peru that suggest intentional astronomical orientations are found during the Late Archaic in the Supe Valley of Norte Chico and in the Casma Valley dating to 2500–3500 BCE." This includes Sechín Bajo, although that particular orientation most likely dates to the later Initial period. Robert Benfer (2012) has likewise documented numerous early sites that have solstice orientations in some form. In short, the solstices, both the June sunset and December sunrise, clearly were used by Late Archaic and later peoples to align their monumental structures in many cases. As mentioned in previous chapters, I view these devices as means to mark events of regional significance.

The best explanation for the Caral site is that it represents a settlement with distinct platform mounds controlled by distinct, albeit related groups. We are not clear if the settlement was permanent or periodical. Caral was, among other things, the place of ritualized periodic barter fairs, perhaps some of the first in the Andes, where people met from coast and highland to exchange goods, information, marriage partners, and the like in the Late Archaic. There is substantial archaeological evidence for exchange at the site. Furthermore, it is important to remember that Caral was not the only site in the region with these features. Sites such as Miraya, Lurihuasi, and Pando (Shady et al. 2009) are also candidates to have hosted regional barter fairs. It is likewise important to contrast these *chaupi yunga* settlements with contemporary ones on the coast such as Aspero. In these coastal sites, there is clear evidence for extensive and permanent settlement, alongside the ritual architecture. This strongly suggests a model in which permanent coastal populations constructed sites such as Caral in the upper valleys as special places where people congregated in prescribed times, regulated almost certainly by the solstice and possibly other astronomical phenomena.

The first substantial settlement in the Andes built by complex stateless societies was in effect a place to bring disparate peoples together from around the region. The settlement of Caral is more than a place to live. It is a very special place, imbued with ritual significance. It is a theatrical place on the landscape designed in part to host large numbers of people on a periodic basis. This interpretation is backed by architectural evidence and excavation data. Significantly, there is little to no evidence of coercion at the site; far from it, Caral represents the culmination of several centuries of small-scale efforts to build special monuments to periodically attract people in small villages up and down the Supe Valley and beyond.

Successors of Caral

The coast was not the only region with early monumental sites contemporary with Caral. During the late preceramic, a widespread building and ritual tradition developed among a number of highland communities as well as the coastal areas (Stanish 2001:46). These traditions developed slightly after the apogee of Caral and continued until the Initial period, around 1800 BCE. Most prominent is the one called the Kotosh Religious Tradition (Burger and Burger 1980). Richard Burger (1995:47) describes two artificial mounds and a series of superimposed temples at the type site of Kotosh, located near Huánuco (Onuki 2012). The highest mound was 14 m high and had a three-tiered platform with numerous chambers built into the base. There are at least 11 chambers and possibly up to 100 chambers at Kotosh itself (Burger 1995:48). A prominent feature of this architectural tradition is small buildings, usually plastered and decorated with sunken or semisubterranean fire pits. One of the most spectacular of these preceramic structures is the Temple of the Crossed Hands at Kotosh. The Kotosh Religious Tradition is found in many other sites in the highlands (Quilter 1991:42).

The Initial period in Peru is traditionally considered to be the time when a widespread ceremonial architectural form – U-shaped building – developed. This tradition stretches from the northern Andes into as far south as the Titicaca Basin (Stanish 2001). The ideal layout of the U-shaped architectural tradition was a high, flat-topped pyramid mound flanked by two projecting linear structures to form a large U. Perhaps the largest settlement of this time period, El Paraíso, is located 2 km from the coast in the Chillón Valley of the Central Andes and is part of this tradition. According to Jeffrey Quilter (1985:294) and Michael Moseley (1992:119), the major construction at the site was in progress by 2000 BCE, and it continued to be occupied for two to four centuries. The 100,000 tons of stone masonry construction are found in at least seven mounds that form a giant U shape over 58 ha (Quilter 1985:279; Quilter and Stocker 1983). The site has a huge, 7-ha plaza located between the arms of the U. Many structures at El Paraíso were elaborately decorated. In particular, one structure was painted red and had a bright-orange burned floor with evidence of fire rituals. Moseley (1992:120) notes that artifacts include red pigment grinders, bird feathers, unfired figurines, and fruit tree branches, all suggestive of elaborate and repetitive rituals. Earlier, we believed there was little evidence for permanent habitation at the site. However, later work has indicated that it indeed had a resident population.

The site of Huaca La Florida, located 11 km inland in the Rímac Valley, is one of the oldest of the classic U-shaped structures so far studied (von Hagen and Morris 1998:51). The main pyramid is 17 m high, and the two projecting structures rise from the base to a height of 4 m for approximately 500 m of their length. Construction at the site began in the eighteenth century BCE. Burger estimates that the site required 6.7 million person-days of labor. He notes that it is not even the largest of the U-shaped sites on the coast. The poorly known site of San Jacinto in the Chancay Valley is four times as large, with a 30-ha plaza and 2,000,000 m^3 of fill (Burger 1995:61).

While centered on the central coast of Peru, this U-shaped architectural tradition has been noted as far south as the Lake Titicaca Basin (Stanish and Steadman 1994:13) and as far north as Piura (Guffroy 1989). It represents an architectural style that encourages visibility of performance by a large group of people. (See Moore [1996a, 1996b] for a nuanced discussion of these principles.) The flanking arms as well as the main pyramid provide excellent elevated platforms from which to view activities inside the U.

Another architectural tradition is known as Cupisnique and is characterized by low platform pyramids, large stairways, and rectangular sunken plazas. Colonnades and elaborate painted sculptures distinguish this architecture (Burger 1995:92). The architectural complex known as Huaca de los Reyes at the site of Caballo Muerto is emblematic of this late Initial-period style. Ware-feline motifs executed as adobe friezes adorn this pyramid (Conklin 1996). Again, all these features suggest elaborate and repeated rituals conducted in the open.

The U-shaped architectural tradition represents an elaboration of the earlier ones started at Caral and related sites. The U shape became the standard form to express the principles of the ritualized economy – performance, exchange, meeting places, barter – throughout the coast. In other areas, such as the Titicaca Basin, the square or rectangular sunken patio was adopted in the mid-second millennium BCE. This tradition flourished throughout the south central Andean highlands from circa 1400 BCE, reaching its apogee in the powerful Tiwanaku state in the first millennium CE.

Geoglyphs as Part of the Ritualized Landscape in the Andes

The ancient Andean peoples created a very sophisticated artistic tradition based on geoglyphs (Ruggles 2014). Geoglyphs are large (generally more than 5 m) figures created on a landscape. Nasca-like lines are also found up and down the Pacific watershed, from north-central Chile into northern Peru. Luis Briones (2006) calculates that Chile alone has at least 5,000 geoglyphs, most of which are concentrated in the north of the country. It is not a coincidence that we find lines associated with Caral in the Supe Valley. We also find linear geoglyphs in valleys such as Casma where early ritualized economies emerged.

The area from Nasca to Chincha contains an unusually high density of preserved ancient geoglyphs for both climatic and cultural reasons. The climate in these desert areas – lacking heavy precipitation, subsequent runoff, and much wind erosion – preserves the most fragile of ground drawings. The cultures of the region took advantage of this stable, open, and barren environment, intensively using the geoglyph technology for centuries and creating a palimpsest of geometrical designs and naturalistic figures. While geoglyphs are broadly recognized as a coastal phenomenon, it is likely that they were also created using different techniques in other environmental regions; they unfortunately do not survive harsher climates. The most famous geoglyphs are representations of animals and anthropomorphic figures. In the aggregate, however, the geoglyphs are overwhelmingly geometric in form, and in this geometric category, the vast majority are lines and trapezoids.

Geoglyphs in fact are found throughout the ancient world, from Arabia to Mongolia, from northern Europe to Australia and California. They may be created with piled earth, such as the famous Hopewell Mounds of the American Midwest. They are often carved out of a landscape using an extractive technique: topsoil is removed to expose a different-colored surface. They can be built with rocks in what is called an additive technique. Or all of these techniques can be used together to create hybrid forms.

There are many explanations for the Andean geoglyphs. María Reiche (1949, 1993), the iconic scholar of Nasca, argued that the lines were astronomical aids. She was unable to convincingly demonstrate this relationship with field data. Johan Reinhard (1986) argues that lines marked sacred mountains, using ethnographic and historical data on Andean religion and belief as a model. In a similar vein, Gary Urton (1990) and Anthony Aveni (1990) suggest that the lines were similar to the Inca *ceque* system of ritual procession. The majority of researchers conclude that the geoglyphs were central in ceremonies involving dances, pilgrimages, ritual runs, and processions (Rodríguez 1997:12; Ruggles and Saunders 2012; Silverman 1990:453; Silverman and Proulx 2002:192). Fieldwork by scores of archaeologists has turned up evidence that both supports and at least partially refutes all of these theories in particular circumstances (Aveni 2000a; Johnson, Proulx, and Mabee 2002; Lambers and Sauerbier 2006; Reinhard 1986; Ruggles and Saunders 2012; Unkel et al. 2007).

The central Andes is home to some of the most elaborate and famous geoglyphs in the world. Here, ancient peoples created a wide variety of geoglyph types. The UNESCO World Heritage lines of Nasca are well known. Nasca lines are both geometric – straight lines, rectangles, and trapezoids – as well as figural forms such as humans, animals, and insects. Some authors (particularly Benfer 2011) believe that many Initial-period mounds along the coast were built in figural forms effectively presaging the Nasca lines by many centuries.

Geoglyphs, in fact, represent one of the most important means by which complex stateless societies created ritualized landscapes, and they should be considered a very important "social" technology. In this sense, they are like any other technology in the premodern world that can be used in a variety of ways. The use of geoglyphs, in fact, is extremely common in stateless societies. There is no one "function" of geoglyphs. Rather, like any other technology, say pottery, they can be used for a variety of reasons. Just as pottery can be used to cook, to bury people, to decorate, and as gifts, geoglyphs have a wide variety of potential uses. Curiously, geoglyph use virtually disappears with the development of states, a theme we will touch on later.

Evidence of Raiding in the Initial Period of the North: Cerro Sechín

The site of Cerro Sechín is located in the Casma Valley. (Cerro Sechín is a different site than Sechín Bajo or Sechín Alto, though they are in the same valley) (Figure 7.11). Cerro Sechín is famous for its carvings of captured people, presumably male warriors about to be sacrificed, as well as trophy heads and other evidence of violence. Cerro Sechín, therefore, holds the oldest unequivocal iconographic evidence of organized violence in the Andes.

Jerry Moore (1996b:48–50) provides an outstanding summary of the architecture of the site, which is only briefly summarized here. As Moore says, the chronology of the site is complex and highly contentious. The initial buildings at Cerro Sechín were placed on a stepped platform with three levels (Samaniego et al. 1985:173; Tello 1956). I agree with Samaniego et al. (1985) that the

Figure 7.11. Archival photograph of the Casma Valley circa 1956.

aggregate evidence suggests that the temple and associated art date to before 1300 BCE. These scholars provide a series of ^{14}C dates from an area in front of the main temple, two of which are considerably old, around 5000 BCE. I agree with Moore (1996b:48) that the probable date of the temple is before 3740 +/− 40 BP, as reported in Samaniego et al. (1985:Table 1). The context of this date is a "temple," according to the excavators. This is an important date because it calibrates to 2285–2029 BCE.[11]

The bulk of the architecture of Cerro Sechín is visible from the front, the sides, and the hill above. There are some possible covered structures, but these are possibly later additions, as are the stone carvings (Burger 1995:79–80). The original construction was, according to Burger (1995:79), a terraced platform that measured about 34 m on a side. There was a "summit complex" that is difficult to define (Samaniego et al. 1985:167). Likewise, there was a circular sunken court in front of the platform, possibly quite early in date. While it is difficult to determine the precise nature of the early architecture, it appears that the site was originally characterized by an earthen platform at the base of a hill, where most or all of the activities could be viewed.

Some of the most outstanding features of Cerro Sechín are the numerous carvings in stone on the outer wall of the pyramid or "temple" (Figures 7.12–7.16). These early Initial-period carvings depict macabre scenes of conflict, including decapitations, trophy heads, and body parts, plus warriors and victims

Figure 7.12. Site of Cerro Sechín, Casma Valley.

Figure 7.13. Wall depicting captives or trophies at Cerro Sechín.

Figure 7.14. Detail from stele panel at the site of Cerro Sechín depicting trophy heads.

Figure 7.15. Detail from stele panel at the site of Cerro Sechín depicting a processional figure.

in various states of subjugation. The vast majority of the depictions represent victims, while less than 5 percent of the images show chiefs or leaders dressed with hats, staffs, headdresses, and so forth. In this regard, it is fascinating to read the description of indigenous patterns of raiding among the Tagalog peoples in the Philippines as described by Laura Junker (1999:345): "head taking, human sacrifice, and the use of enemy body parts as war trophies were tied to status displays in postwar feasts and elite mortuary ritual."

Elizabeth Arkush and Tiffany Tung nicely characterize the nature of organized violence as depicted in Cerro Sechín:

> Although Cerro Sechín's gruesome iconography has sparked much archaeological speculation, it says little about the prevalence or severity of warfare. The carvings do, however, present the first clear link between dismembered body parts and violence, and they suggest that an ideology of domination through violence was promulgated to Casma's residents. There was not much threat of outside attack, for there is no defensive

Figure 7.16. Detail from stele panel at the site of Cerro Sechín depicting an elite personage.

> settlement near the river confluence or lower down the valley; indeed, it is hard to imagine that adjacent valleys could have posed a threat at this time. Those at Cerro Sechín may have been aggressors, enacting or threatening violence upon others in the region. But Casma's warlike iconography does not coincide well with independent settlement pattern evidence for warfare – the first of many such disjunctures. (Arkush and Tung 2013:322)

Arkush and Tung, correctly in my view, point out that this iconography is not evidence for widespread war. Rather, it is clear evidence for classic kinds of raiding by the people at Cerro Sechín, backed by intense costly signaling in the dress and architecture. I believe, therefore, that the calibrated date of 2285–2029 BCE, cited above, for the latest period in which the Cerro Sechín temple was built, is extremely important.[12] It in effect is the minimally earliest time period for the existence of organized raiding in the northern Peruvian coast. The scenes at Cerro Sechín indicate neither large-scale warfare nor some purely ritual activities. Rather, as Samaniego and colleagues (1985) have also noted, the stone carvings probably represent raiding behaviors.

The significance of Cerro Sechín is that we see a shift from Caral in the Late Archaic circa 2600–2300 BCE, where evidence for raiding is minimal, to the early Initial period, where raiding is important enough to be the central, dominating theme of the art and architecture of a major settlement. Our latest data therefore indicate that within a few centuries of the building of the pyramids at

Caral, widespread and formal raiding became a major activity for the people in this region. Of course, it is possible that less formal raiding occurred prior to that depicted at Cerro Sechín.

Ritualized Economies in the Far South of the Central Andes: Tarapacá

The archaeological evidence tells us that the origins of ritualized economies based upon the creation of special places to foster barter exchange and other activities appear in the intermediate ecological zone referred to as the *chaupi yungas* in the North Coast of Peru as early as the Late Archaic, circa 3000–1800 BCE. This same process occurred at the geographical extreme of Caral in the south central Andes much later in time. Here we find a similar, albeit more modest, empirical pattern in the northern Atacama Desert in the Tarapacá Valley (Figures 7.6 and 7.17).

The Tarapacá Valley in northern Chile is located about 1,300 km south of Caral. We find the first settlements built by complex stateless societies in this region almost two millennia later than Caral. This is an area of dry deserts and rivers, with very low water volumes compared to the drainages in the north.

Figure 7.17. View of the Tarapacá pampa from the east.

Figure 7.18. Large structure found at the site of Pircas.

The principal settlements in the Tarapacá Valley are known as Caserones and Pircas and date to the Formative period of this region, circa 400 BCE–300 CE (Nuñéz 1984).[13] The *chaupi yungas* of this region are found about 1,300 m above sea level in an area optimal to mediating exchange between the highlands and the coast. While the time frame and geography are distinct from the Supe and Casma Valleys, the patterns of how the first complex societies developed are similar to Caral millennia before.

Pircas is located opposite Caserones to the north on the pampa (Figures 7.18 and 7.19). It was situated in an absolutely dry and lifeless desert. It has no water, vegetation, or other natural resources that could sustain human settlement. The village of Pircas, therefore, is without question a temporary settlement that housed people for very brief periods, perhaps a few weeks at most. Water is available from the river below, several kilometers away. But there are no other features that could support any permanent occupation.

Pircas structures are characterized by large, round buildings reaching up to 10 m in diameter (Figure 7.18). Colleen Zori (2011:733) describes these as "circular and linear rock features thought to have been temporary camps occupied by groups moving seasonally through the valley between the highlands and the coast." The pampa in which they are located has lines similar to the Nasca–Ica–Palpa areas. As seen in Figure 7.17, a very long pair of converging lines cuts across the desert toward the town of Huara. While many of the "lines" are in

Figure 7.19. Site of Caserones opposite the valley from Pircas.

fact modern trails or late prehistoric or modern agricultural features, a number of these geoglyphs are virtually identical to the lines found in the more famous valleys to the north.[14] Geoglyphs throughout the zone around Pircas include circles, animal figures, and humans (Figures 7.20 and 7.21).

A generation ago, Lautaro Nuñéz (1984) excavated some of the large structures at Pircas. He obtained two dates in one structure: 1880 ± 60 (20 BCE–318 CE, 95.4 percent) in the lowest level and 1450 ± 90 (410–770 CE, 95.4 percent) in a higher one at the same structure. A human tissue from another structure at the site complex had an early date of 2420 ± 80(776–388 BCE, 95.4 percent). These data indicate Formative-period use of the Pirca structures.

Nuñéz's work demonstrated that the structures contained a wide variety of remains, including maize, cotton, camelid coprolites, large baskets, tropical feathers, turbans, and rugs, in multiple occupation levels. As seen in his article (Nuñéz 1984:Figure 8), there are wooden ritual objects and possible drug paraphernalia. He also found basalt cores, proving that there was lithic manufacture at the site as well (Nuñéz 1984:Figure 9). Again, these rich archaeological deposits are found in sites with no immediate access to water, agricultural resources, or grazing areas. This area would not have been easily settled for any length of time. The archaeological data, both excavation and settlement, strongly suggest temporary camps containing a rich variety of exotic and high-value goods.

Figure 7.20. Geoglyph above the Tarapacá Valley along a prehistoric trail.

Figure 7.21. A massive geoglyph, known as El Gigante, in the Tarapacá pampa near the site of Pircas.

Caserones is located on the edge of a pampa, with the end of the river valley less than one km away. It is a large, walled village site that represents the largest pre-Inca settlement in the region. The defensive wall was constructed sometime after the first occupations of the village. This is demonstrated by the fact that the wall cuts through other remains and was constructed with midden. There is therefore little question that there was some kind of conflict for at least part of the time in which Caserones was occupied.

Caserones likewise has a Formative component that makes it partially contemporary with the Pircas settlement. In other words, for more than one millennia, from about 400 BCE up to around 600 CE, a large permanent settlement with classic features of sedentism – large middens, access to substantial agricultural fields, substantial architecture, and the like – was located in the pampa. At the same time, large but scattered temporary structures were located in the pampa to the south, across the river from Caserones. These structures in Pircas sites are interpreted here as the occasional residences of itinerant traders in caravans, who periodically set up camp across from Caserones to trade. The entire landscape of the Pircas settlement area was crisscrossed with geoglyphs.

In short, it is clear that the intermediate zone between coast and highlands – the *chaupi yungas* pampas in the Tarapacá drainage – was the location of the first special places in the area, hosted in a ritualized landscape that attracted visitors from the altiplano, sierra, and coast. Barter trade was almost certainly a major part of this system. Colleen Zori and Erika Brandt (2012:403) aptly argue, "Rock art data suggest that trade was increasingly embedded in ritually sanctioned events involving groups from different ecological zones."[15] At Caral, albeit on a different scale, each of the large circular structures was associated with a corporate group, used periodically during feasts and fairs. In this model, we see the Pircas settlements as the locus of the periodic regional fairs, while the Caserones site would have been the permanent village, where goods could be collected throughout the year to be traded in the periodic barter fairs.

This interpretation is supported by data from the Guatacondo area about 125 km to the south of Tarapacá. Here Gray Graffam, Mario Rivera, and Alvaro Carevič (1996:103) discovered unambiguous evidence of copper production and exchange in the Formative. There is therefore evidence for significant copper trade in this key time period in the Atacama in the first centuries BCE at settled villages like Guatacondo. Dates with good context for copper smelting waste cluster between 200 BCE and 80 CE, dates consistent with the Pircas/Caserones sites.

What we therefore see at the other end of the central Andes is a surprisingly similar pattern in the first complex societies in the region.

The Apogee of the Ritualized Economy: The Paracas People in the South Central Coast of Peru

The earliest late Middle Archaic sites in the Supe Valley and sites like Sechín Bajo in Casma represent the very beginnings of ritualized economies in the central Andes. These early experiments, characterized by single platform mounds and sunken plazas, culminated in sites like Caral a millennium later. Caral represents a cultural process whereby dozens of distinct groups created a large special place in which many people aggregated and built ceremonial mounds for substantial periods of time. This process is seen later on in places like the Atacama, where population densities did not reach levels sufficient to support complex stateless societies until hundreds of generations later.

This pattern of creating special places continued throughout the central Andean region. The end of this process is in effect marked by the emergence of state-level complexity. This is a process, not an event, and it is a pattern that emerges and recedes through a long period. We see, for instance, in the Early Horizon (900–200 BCE), the development of huge, permanent settlements with resident populations, including Chavín and Pukara.

I now turn to an example of the end of this process, a case study of a ritualized economy in its fullest expression. We look at Paracas, a culture that developed around 800 BCE and continued until circa 100 BCE on the Peruvian south central coast. The Paracas people settled valleys from Chincha in the north to the Nasca in the south. The environment where the Paracas people created their civilization is some of the driest land in the world. Yet, cut through by rivers from the mountains, the valleys provided the ecological setting for huge populations that converted this dry desert into some of the richest agricultural land on the planet (Figure 7.22).

The Paracas type site is on the eponymous peninsula that juts into the Pacific Ocean to the west. Some scholars suggested that Chincha was, in fact, the home of the Paracas phenomenon (Lumbreras 2001; Moseley 1992). Our (Stanish et al. 2014) research in the region in fact supports this hypothesis. Chincha is almost certainly the valley where the Paracas phenomenon developed (Figure 7.6). The two rivers in the Chincha Valley provide one of the largest volumes of water in any drainage on the Pacific coast. Irrigation canals were dug by at least 800 BCE and probably earlier. Most of the earliest peoples in the valley likely settled the rich river bottoms, where they constructed patches of irrigated land. This is a pattern seen throughout the world in the Neolithic and Formative periods in the Old World and the Americas. Early villages settled near abundant wild resources (complex hunter-gatherers) or where agriculture could provide equally rich concentrated resources. Most settlement patterns in these contexts show a roughly equidistant distribution of sites that optimize the productive land between villages of similar sizes. Over time, some settlements

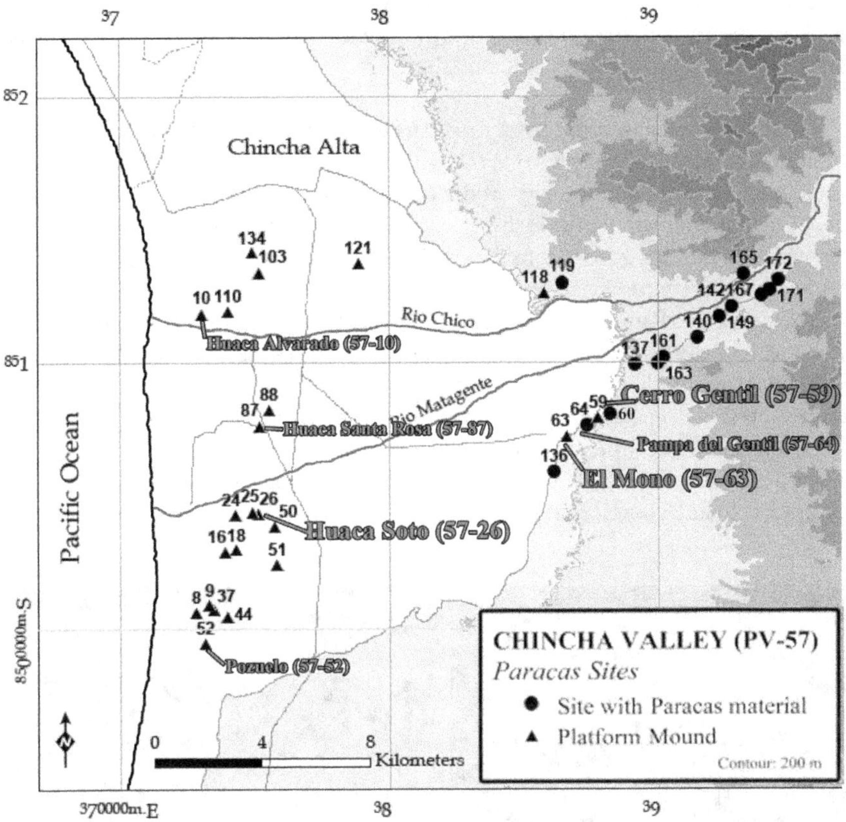

Figure 7.22. Map of the Chincha Valley and surrounding areas. Redrawn from Canziani (1992).

grew larger. Site sizes and other features differentiated into more complex patterns of social interaction.

In the Andes, the corresponding irrigation systems grew as well, similar to other areas of the world. By the beginning of the Early Horizon, circa 900 BCE, it is likely that people had started building valley-wide irrigation systems, settlement data that strongly suggest some kind of regional co-operation between otherwise autonomous social units. This required people to move their settlements to the hillsides above the valley bottom and out of the irrigated land. This pattern continued up to the early modern period throughout the Peruvian coast.

There is little evidence at the present time that we can use to reconstruct the earliest periods in the Chincha area. It is almost certain that these sites are either buried under later ones or have been covered by elaborate irrigation systems and agricultural terraces. There are undoubtedly older cultures in the region that go back to the Archaic period of the fourth millennium BCE. However, there has been little work on these phases, and there are virtually no reported sites in the Chincha area.

The later Paracas-period settlement system is much better known. Paracas, in the Chincha Valley, represents the fullest expression of a complex stateless society in an ecologically and product-rich area of the world. There was not only an abundant supply of freshwater, but the people had access to rich marine resources and other goods from the coast. The carrying capacity of the Chincha region with irrigation systems is enormous.

The art of the Paracas is justifiably famous (Figure 7.23). This early culture was particularly expert in textiles and pottery, and objects from looted tombs have been acquired by museums around the world. As we will see, iconographic representations on Paracas art help us understand norms in this society that helped maintain co-operation at a sustained level.

In the case of Paracas, we see a society sufficiently complex that indicators of social coercion can be detected in the archaeological record. The

Figure 7.23. An example of Paracas art.

transition to state societies is not a discrete event but a process in which coercive mechanisms take generations to coalesce and institutionalize. This transition is uneven, and we see in areas of state development a fluid process in which coercive social structures develop and collapse as kin groups resist loss of autonomy. But unlike well-documented later states like Moche, Wari, and Tiwanaku, Paracas still retains the fundamental features of a complex but nonstate political entity. It represents the end process of nonstate mechanisms of social, economic, and political co-operation – the need to persuade people to participate in any particular coordinated group activity.

The Paracas-period occupation in Chincha consists of dozens of pyramids and numerous smaller sites (Canziani 1992; Lumbreras 2001; Wallace 1971). The pyramids or platform mounds range in size from 20 to 150 m at their base and 6 to 14 m high. There are about nine pyramids near the coast and number of smaller ones in the high pampa above the river bottom. This latter group of pyramids are located about 17 km from the coast, putting them firmly in the *chaupi yungas* zone of this region.

Survey in the upper Chincha Valley discovered a number of linear geoglyphs on the pampas above the town of Carmen (Figure 7.24). A most striking aspect of the distribution of the geoglyphs is the degree to which the lines visually converge on known Paracas-period platform mounds (Stanish et al. 2014). As seen by the hypothetical projections (Figure 7.25), there are four clusters of

Figure 7.24. Linear geoglyphs in the pampa above Chincha.

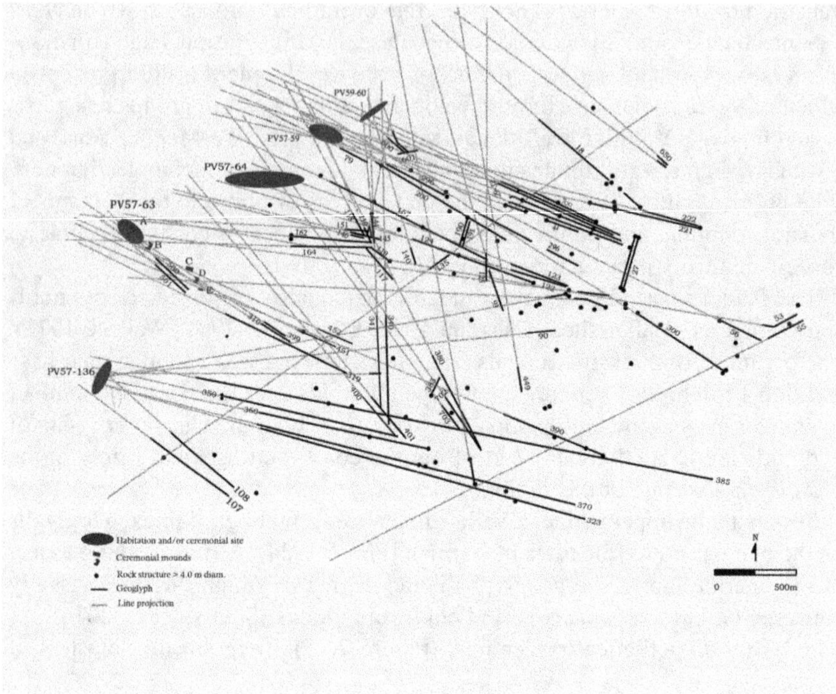

Figure 7.25. Linear geoglyphs and their relationship to settlements in the upper Chincha Valley. From Stanish et al. (2014).

linear geoglyphs around several Paracas sites in the upper pampas. All four clusters have a pair of lines or sets of paired lines that visually "frame" the major site from a distance. The technique was to build a pair of radiating lines in the form of a V, which at ground level from the apex of the lines, appear to be two straight lines due to perspective projection distortion. This technique was used to mark significant features, specifically pyramids and the setting sun during June solstices. Three of four clusters contain long lines that mark the June solstice sunset, and several of these lines also frame known Paracas platform mounds and sites. Additional line segments frame known Paracas sites in the mid-valley cluster but do not appear to be solstice markers.[16]

Archaeologists also found a number of stone structures adjacent to the principal geoglyph throughout the desert pampa landscape (Figures 7.26–7.28). All line clusters have large (approx. 4–12 m in diameter) circular structures that allowed for enhanced viewing of the sites from a distance, particularly those that frame Paracas platform mounds. Some structures are clearly associated with lines, while others are not. Many of the circular structures are virtually identical to double-ringed *chulpas*, or aboveground tombs known from

Figure 7.26. Stone mound in the Chincha pampa with the platform mound of Cerro del Gentil in the background.

the southern Peruvian highlands. Other structures are similar to what has been discovered and excavated in Palpa by Unkel et al. (2012) and are referred to as *refugios* by Helaine Silverman (2002a:83) in Nasca and Ingenio. Unkel et al. (2012) describe the results of excavations in four structures associated with geoglyphs, all of which dated to early and middle Nasca times. These structures had stone platforms and wooden posts, along with Nasca-period pottery.

Figure 7.27. Split mound type in the Chincha pampa.

Figure 7.28. Circular mound type in the Chincha pampa.

Lines converge at five principal sites in the pampa of Chincha.[17] One of these is known as Cerro del Gentil (Canziani 2009:149; Tantaleán et al. 2014). It is the largest site in the upper valley region, though smaller than the contemporary pyramid mounds in the lower valley. Located on the end of a long ridge, the site consists of two platform mounds, one about 35 × 80 m and the other roughly 20 × 30 m. The larger pyramid measures about 25 m at its base and is roughly 10 m on a side at the top (Figure 7.29). A sunken court in front of the pyramid (to the west) measures approximately 12 × 12 m. Below this court are two to three levels of terraces that completed the formal architectural complex. Canziani (2009) argues that Cerro del Gentil was built in a traditional Paracas plan, with a strict east–west orientation, and he refers to it as a smaller representation of the larger Paracas settlements in the valley below.

Figure 7.29. Platform mound site of Cerro del Gentil.

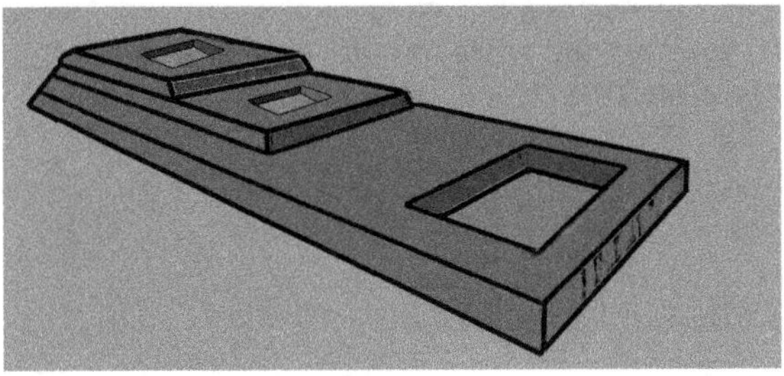

Figure 7.30. Schematic view of Paracas platform mounds in Chincha. Adapted from Canziani (2009).

Excavations by Henry Tantaleán and his team in one of these pyramids yielded a trove of artifacts deposited in a sunken patio. The site of Cerro del Gentil (PV57–59) is one of five settlements in the pampa onto which lines converged during the Paracas period.[18] It is a large platform mound with three levels (Figure 7.30). The base level measures 50 × 120 m at its maximum and conforms to a classic Paracas architectural pattern in the valley.

There is a sunken patio in each level. The patio is part of the same tradition that developed one and a half millennia before in the Late Archaic at sites like Caral. In the Paracas culture, these patios reach their fullest expression. The patio at Cerro del Gentil is roughly square. It was plastered and painted in at least black-and-white designs.

Tantaleán's team excavated the middle patio and discovered a plastered court with an incredibly rich array of offerings, including textiles, pottery, mummies, foodstuffs, engraved gourds, wooden objects, cane objects, and miscellaneous pieces. All the identifiable iconography on the art objects was Paracas in style. Carbon dates on two objects were consistent with a Paracas date. We found little evidence of domestic habitation directly on or near the site, though there was an apparent village a few hundred meters below in the adjacent *quebrada*.

We conclude that the site of Cerro del Gentil was a ritual location for the deposit of high-value objects. What is fascinating is that virtually every object is unique. Even the technique of making a basket was different in each case, with people using different warps and wefts and different leaf colors. No pottery offering or textile is similar to any other. It is likely that the offerings were made over some period of time, at least a generation or so. It is quite likely that objects were periodically removed and replaced. In short, this court on this platform mound was used in some kind of feast or potlatch ceremony. We do not know what the other courts contain, but it is likely that these were used in a similar manner.

We analyzed the structures off the top of the pyramid and along the sides and flat areas around the base. We did not find any habitation structures that dated to the Paracas period. Nor did we find much residential debris on the side terraces. We did find massive quantities of marine shell waste. It appears that people consumed or prepared the meat of these mussels on or near the pyramid. In short, we have evidence of a ceremonial mound above the valley, with little evidence of habitation.[19]

The Site Cluster of Chococota or El Mono

The mound group of Chococata is located in the Chincha Valley in the upper pampas. It is composed of five mounds. It is located in a similar topographical setting as Cerro del Gentil. We discovered at least 14 lines converging into the site area. Three of these lines also mark the solstice. Two mounds with U-shaped architecture (Monos B and D on Figure 7.31) precisely face the winter, or June, Southern Hemisphere solstice on the short axis. The largest platform mound structure of the site, Mono A, aligns along its long axis with the solstice as well.

Figure 7.31. Site area of Chococota, or El Mono.

Marcus (2008) notes that stateless societies use rituals with institutions such as men's societies, kivas, and ancestor cults. She notes that these rituals "at a minimum ... would include dancing, singing, chanting, and the wearing of costumes, the latter often representing ancestors or supernatural beings" (Marcus 2008:256). These behaviors have been documented in a number of cultures, as Marcus notes, including in the Near East in the Neolithic and in Formative Mexico (Marcus and Flannery 1996, 2004). To this we can add the Paracas peoples. Paracas art is replete with figures dancing, playing flutes, and in apparent procession. It depicts complex ceremonies and people elaborately dressed in masks, tunics, and headwear. There is little question that ceremonial mounds such as Cerro del Gentil were places where such activities occurred.

The Large, Contemporary Pyramid Complexes on the Coast

The work of several previous researchers has identified at least nine very large Paracas pyramids on the coastal plain of Chincha (Figure 7.22). These sites are an order of magnitude larger than those such as Cerro del Gentil in the upper valley. One of these sites is called Huaca Soto and was described in great detail by Canziani (1992:91). He noted that it was 200 m long by 70 m wide, reaching 15 m high in the west side. He also observed that Huaca Soto is associated with two other pyramids, all oriented along an east–west axis. Like its smaller counterpart at Cerro del Gentil, Soto has three platforms that ascend to the west, effectively facing the ocean that is a mere 3.75 km away. Canziani notes that each of these stepped platforms has a sunken court similar to Cerro del Gentil. He also observed that the pyramid was not built in a single episode. Rather, there is clear evidence of remodeling and expanding the pyramid complexes over time. This is demonstrated by the different construction episodes exposed from road and house construction on and near the edge of the site.

Huaca Soto was recently excavated by Benjamin Nigra and Kelita Pérez (Figure 7.32). Their excavations confirmed the observations of Canziani that the pyramid was built over many different epochs, with a major expansion beginning as early as the early Paracas period, circa 800 BCE. The artifact patterns discovered at this very large structure were very different than those found in Cerro del Gentil. There was very little material deposited in the court area. At Cerro del Gentil, we did not find a single *cuy* (guinea pig) bone; Nigra and Pérez found very large numbers in Soto. The Soto court was "clean," with very few artifacts, and was a bit restricted, both visually and physically. In contrast, the Cerro del Gentil was much more open and accessible.

There was little evidence of domestic habitation around Cerro del Gentil. In contrast, there was significant evidence of habitation structures at the base of the Soto pyramid up to the base of the second pyramid structure in the Soto complex. We excavated three test pits between these two structures and found

Figure 7.32. Site of Huaca Soto, Chincha Valley.

midden and other garbage in this area. The artifacts included pottery in Paracas style. In other words, the Soto pyramids are much larger, are located in prime agricultural land near the ocean and have substantial residential structures adjacent to the platforms. They are by any definition a dense and complex form of settlement that emerged in this nonstate context. These large settlements maintained the smaller platform mound complexes in the pampa.

A Paracas Ritual Landscape in Chincha

The archaeological data in Chincha indicate that Paracas peoples engineered a carefully structured landscape to demarcate areas and times for intense social activities focused on sites like Cerro de Gentil. We are quite certain that many of the geoglyphs were intended to be walked on and thus were used to direct processions or other kinds of orchestrated movements over the landscape. It is likely, as Lambers and Sauerbier (2006:249) put it for the Palpa Valley to the south, that "a continuous and manifold chain of activities took place on geoglyph sites." Many archaeologists and geographers agree that these sites were used for processions, pilgrimages, and/or other kinds of orchestrated journeys. This is reinforced by Paracas art iconography, which consistently demonstrates lines of people in elaborate dress.

We can view the Paracas geoglyph and settlement complex in Chincha as designed to be a grand "theater" for periodic market fairs and other potlatching events during Paracas times. In this model, the people in the large settlements in the valley bottom constructed elaborate, ritually mediated landscapes in the pampas above to attract followers and visitors in grand potlatches. The elaborate dress noted on the iconography strongly suggests that costly signaling was used extensively in these feasts and fairs.

The location of this theatrical landscape is known as the *chaupi yungas*, a place where cold and warm climates merge between the productive highlands and coast. It is a barren place, untouched in 2,200 years. This dry and desolate pampa has no economically significant natural resources. There are no minerals, no water, no way to irrigate without modern pumps, and no natural vegetation. It was, therefore, the perfect place to construct a ceremonial space for periodic fairs, processionals, and pilgrimages in ancient times – it is upland from the rich and populous coastal strip irrigated by highland waters, and it is directly below the fertile highlands. Caravan routes still descend to the coast, allowing people, animals, goods, and ideas to travel back and forth. It is in an optimal central place for travel in both north–south and east–west directions (Griffin 2013). The pampa is a neutral place, owned by no one but used by all in the past. In short, it is an economically useless but ritually priceless place that was central to the society of these ancient Peruvian people (Figure 7.33). Below, we describe a ritual procession from the highlands to the pampa Carmen. The ovals represent mounds, the polygon represents a major site, and the white lines trace the path from the low hill in the east down to the site.

We have reconstructed the chain of activities in Chincha during Late Paracas times. Our work discovered hundreds of linear geoglyphs or lines etched in this desert plateau (Figures 7.25). We likewise discovered a range of ancient

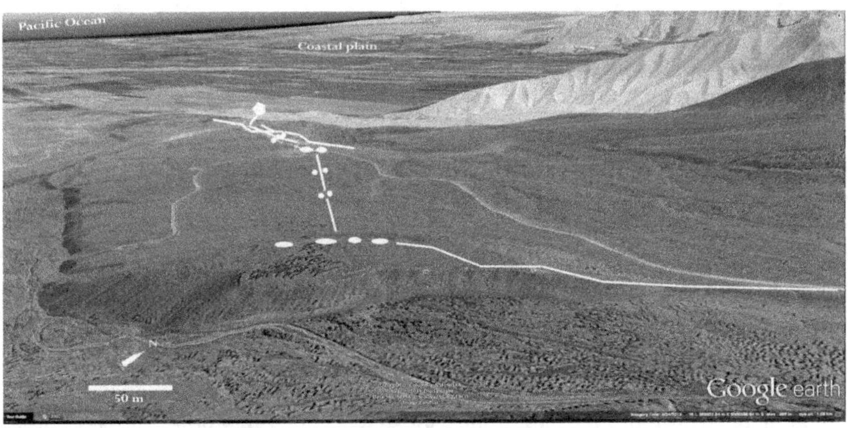

Figure 7.33. Procession route from the highlands to settlements in the Chincha pampa.

Figure 7.34. Rectangular box structure in the Chincha pampa.

mounds built between and among the lines. These mounds have many shapes. As mentioned earlier, most are simply round piles of carefully constructed fieldstones ranging from 1 to 6 m in diameter (Figure 7.34). Others have clear shapes, such as the "split eye" (Figure 7.28), the "rectangular box" (Figure 7.34), the "doughnut," the "arc," and the "pairs."

A number of the mounds and lines are clearly associated with the winter (June 21) solstice. U-shaped and rectangular structures faced the setting solstice sun in antiquity. Linear geoglyphs convened on the solstice sunset as well. The rectangular box structure, as seen in Figure 7.34, has lines that clearly hit the solstice. There is little question that the marking of at least the winter solstice was part of this theater.

One of the processional routes begins in the far eastern section of the Gentil pampa landscape. At 350 m above sea level, on a low hill, ancient peoples constructed a series of platforms. The largest platform is rectangular and visually bounds the winter solstice sunset on both sides for a period of a few days. That is, if you stand directly behind the rectangle, the lines on either end run to both sides of the setting sun during the solstice. Another pair of mounds frames a long line, seen only from the hill above. The line is too indistinct to be seen on the ground. Therefore we can deduce that the movement from the base of the hill toward the large monuments would have been directed by people from the hill. The line leads to a larger pair of mounds. The short axis of this pair (the perpendicular sight line to the west) aligns with the largest set of paired mounds in the entire pampa, about 100 m away. One descends from the small pair to the larger. These large mounds are doughnut shaped; they had holes in the middle that held some kind of pole, flag, or other vertical object. At this point, the two large mounds align, again perfectly, to two lower but wider mounds to the west.

An "arc" mound between these two circular mounds aligns to the solstice. The two large mounds, in turn, are aligned 280°. As one turns slightly to the right, one encounters a pair of long linear geoglyphs, about 15 m wide, that open up and frame the site of pampa de Gentil. The lines are clearly directing the participant toward this site. Along the way are numerous mounds with different alignments and architecture. As one nears the site, the number of mounds proliferates. This high density of additional mounds could indicate a series of meeting places, campgrounds, or staging areas for different social activities.

The integrated geoglyph and settlement clusters in the upper Chincha Valley should be viewed in this sociopolitical and economic context. We see that each cluster was integrated politically with a larger settlement in the lower valley and served as a place of exchange between the highlands and the coast. This built landscape was most likely one component of a set of strategies used by groups within this competitive environment to attract and entice people. Periodic feasts also served to reinforce norms of co-operation. If we can allow some speculation, we can imagine how significant these events were in the lives of the people of the time. As we saw in Chapter 5, the anticipation and intensified social activities that surround large feasts are some of the most important components of people's lives and identities. These feasts and fairs were crucial elements in those lives and were the means by which complex economic activities connecting people from across the landscape were organized.

We conclude that the "Paracas peoples engineered a carefully structured, ritualized landscape to demarcate areas and times for key ritual and social activities" (Stanish et al. 2014:1). We showed how the solstice was an important marker of time in this ritual landscape, similar to other contemporary sites on the Peruvian coast, most particularly the UNESCO heritage site of Chankillo (Ghezzi and Ruggles 2007, 2011).

This analysis leads us to propose that people in the lower valley settlements demarcated places and times for interregional interaction between highlanders and coastal populations and that these places were exemplified by sites like Cerro del Gentil. These ritualized movements and astronomically defined schedules encouraged and promoted this interaction. The ritualized landscape publically concentrated the mounds and settlements as focal points for social gatherings, but it was also a product of these gatherings. The sixteenth-century Quechua term *jach'a allitataña*, defined as "a fair or solemn market," best captures the essence of this ancient area. The act of creating geoglyphs within the broader ritualized landscape and effectively creating ritual pathways organized in time by at least the winter solstice was a key component of individual participation in such events. It certainly was a significant component of co-operative strategies within communities as they competed with other settlements using potlatching-style feasting to garner more people to their factions and to create a complex regional political system.

Early Complexity in the Near East

Göbekli Tepe in Anatolia

What we find in at least the greater Near East, the area where the first complex sites presently known developed, is that the third and fourth of our hypothetical scenarios are most common.[20] As in the Americas, many of these early sites conforming to these scenarios were built by complex hunter-gatherers and not settled villagers. Many of the earliest structures were clearly ritual in nature and were located in places that were not optimal for the acquisition of resources. Perhaps the most famous of these is the site of Göbekli Tepe in Anatolia.

The earliest evidence, in fact, for complex society among stateless societies on the planet, at the present time, comes from Anatolia. Göbekli Tepe was discovered in a survey in the 1960s but was not recognized as an important settlement at the time. In the 1990s, archaeologist Klaus Schmidt "rediscovered" the site, realizing that it was much earlier and more important than previously thought. Excavations indicate that the site began around 9500 BCE and continued for a couple of millennia.

Göbekli Tepe is located on a hill, described as a "natural gathering spot" by science writer Andrew Curry (2008:278) (Figures 7.35 and 7.36). It covers a substantial 9 ha, with at least 20 oval complexes containing stone stele more than 2 m high (Schmidt 2003:5). At any one point, it is likely that no more than

Figure 7.35. Artist's hypothetical reconstruction of the landscape of sites like Göbekli Tepe. Illustration by Katy Killackey.

Figure 7.36. Artistic rendering of the hill of Göbekli Tepe.

a handful of temples were actually functioning. The earliest phases at Göbekli Tepe consist of circular structures on a hill, with each structure containing two monumental pillars, or stelae (Figure 7.37). Walls and more pillars surround these circular structures. These pillars are shaped like T's and are between 3 and 6 m in height and up to 10 tonnes in weight. Many of these stelae are carved with elaborate images of animals. In the aggregate, the stone pillars and walls create a massive and impressive monumental landscape.

There is a quarry near the site that is the source of the stone monoliths (Banning 2011:624). We can see good evidence of a sophisticated system to preform the blocks before they were transferred to the hill where they were presumably carved before being set into the ground. This would be a Type 2 kind of economy of scale that would have required a minimum number of people to mine and carry blocks to the hilltop. E. B. Banning (2011) calculates that the larger pillars would have required around a dozen people to quarry and move to position with an absolute minimum of about six or seven workers. We likewise see cupules ground into the limestone on the site area. These appear to be mortars to grind either seeds or perhaps rocks to extract pigment. Mortars with larger diameters carved into the limestone bedrock were found at the site as well. There are also portable mortars found in and around the structures. The presence of these features indicates that people were here for more than

Figure 7.37. Artistic rendering of one of the circular structures at Göbekli Tepe.

transient periods of time and that activities beyond ritual were conducted in the site area.

The environment at the time that the site was occupied was rich, with an abundance of wild plants, animals, and migratory fowl in the region, though the hilltop location of Göbekli Tepe was not a likely hunting ground. This early period, known as the Pre-Pottery Neolithic A (PPNA), was a time before significant domesticates and widespread agriculture, making the inhabitants of the Göbekli Tepe area classic complex hunter-gatherers. The main excavator of Göbekli Tepe believes that it is "simply" a temple structure without any other function. But other scientists see evidence of people living there at least part of the year. Analysis of animal bone from the site indicates that gazelle made up more than half of the consumed animals present. Excavators also found wild boar, wild sheep, red deer, vultures, cranes, ducks, and geese (Curry 2008). All these animals were wild game; none were domesticated, and almost all were apparently eaten.

There is evidence of significant production of stone tools at Göbekli Tepe (Schmidt 2000). Whether the site was used only part of the year or merely as a transit place, there is little question that people at least manufactured tools and consumed substantial quantities of animals. Schmidt himself dismissed any notion that the site was occupied as a village. However, there is significant evidence that Göbekli Tepe was not just a "religious" center, but that it included many prosaic activities.

Most important, as described by Oliver Dietrich et al. (2012), are clear signs of feasting at the site, including the consumption of alcohol. This fits in with theoretical arguments made by Marion Benz and Alexander Gramsch (2006) and others that feasting is an integral part of the shift to Neolithic lifeways in the region. The current data suggest that Göbekli Tepe was a center of a ritualized economy, including at least stone manufacture, grinding, exchange, and periodic feasting. Simultaneous feasting and intensified economic production were two of the primary activities of this site.

We can envision the periodic gatherings at the site of people from around the region. The site was located on a hill, with possibly a few trails to the top. The Göbekli Tepe site is not unique. It is reported that there are dozens more sites like this in the region and we expect to find many more.[21] Bahattin Çelik (2011) reports that the site of Karahan Tepe, located about 40 km to the southeast of Göbekli Tepe, has similar carved pillars, carved cupules, and other features like the more famous site. It is highly likely that more of these sites will be discovered and described in the near future.

Figure 7.35 shows a hypothetical reconstruction of these kinds of early sites in the region characterized by carved stelae and located on low, nonfortified hills. We can imagine temporary tents made of animal skin at the base of the hill, with each social group clustered with their own. People would bring exotic materials collected from their seasonal rounds. Feasts and communal production of stone tools, ornaments, and other commodities would occur in and between the structures at the top of the hill. Forms of highly ritualized exchange would take place in the structures. The nature of the architecture and its placement on the landscape ensures that people cannot cheat the group – norms of behavior demand that all exchange take place at predetermined times in predetermined places, and any free-riding would be immediately obvious.

Based upon existing data and the theoretical discussion in the preceding chapters, I suggest that the temples were most likely constructed at various times by distinct social groups, most likely kin groups. The placement on a large hill that was not rich in resources ensures that it was a kind of "neutral ground," outside of any group's core territory. The placement on a hill allows for a certain kind of theatricality in that people stayed at the base in temporary camps would ascend the hill in prescribed ways and at prescribed times.

The site was open and visible, an architectural plan that reinforced norms of fairness in any kind of ritualized exchange (Stanish and Haley 2005). It is likely that only a handful of temples functioned at any one period, each built and maintained by separate groups. During special times of the year, people would congregate at Göbekli Tepe and engage in intense social, economic, and ceremonial interactions, including feasting and exchange. People would bring stone materials from distant places, where they would collectively manufacture lithic objects for reciprocal exchange with other groups. This same strategy was used with other object types that do not survive in the archaeological record very well. Göbekli Tepe is an outstanding example of a meeting place for various groups throughout the landscape. It is highly ritualized, visible, and dominant in the landscape. And it is one of at least a dozen more in the area that has been identified. Built by people who were at least seminomadic hunter-foragers, Göbekli Tepe represents one site in a probable regional system of special places built and used episodically throughout the millennia.

Other Sites in the Region

At about the same time that Göbekli Tepe was founded, the site of WF 16 developed as a major center in the Wadi Fainan region in Jordan (Mithen and Finlayson 2000). The bulk of the buildings at the site are, according to the excavators, typical of other semisubterranean and semicircular structures in the PPNA of the Levant (Finlayson et al. 2011:8183). One structure was noticeably different. By 9000 BCE, the people using WF 16 had built an impressive, 19 × 22 m adobe structure (Structure O75) with a plaster floor and decorated benches. Other features in this structure include a pit lined with plaster, broken stone bowls, mortars, parallel gullies, raised platforms, decorated walls, evidence of replastering events, evidence of feasting, and benches (Finlayson et al. 2011:8183; Mithen et al. 2011). This structure has the hallmarks of a ritual and feasting area and is distinguished from other structures on the settlement. Finlayson et al. describe other PPNA sites in the region with structures similarly differentiated from the other structures on their sites. They interpret these data as indicative that "ritualized communal activity may have been important in assembling the work force needed for harvesting" (Finlayson et al. 2011:8187).

Tatiana Kornienko (2009) nicely describes a number of public monumental buildings in northern Mesopotamia at several sites, including Hallan Çemi, Jerf el-Ahmar, Tell 'Abr 3, Dja'de el-Mughara, Nevali Çori, Çayönü Tepesi, and Göbekli Tepe. Most of these are a bit later than Göbekli Tepe but they are in the same artistic traditions. She describes how each of these sites has a set of special buildings that share common characteristics: they occupy a special location in the settlement; these structures are continually rebuilt or renewed over generations; they are larger than other structures in the community; they contain massive stones as benches or stelae; they all have evidence of artistic adornments such as pillars, pilasters, or other sculpted objects; they do not have evidence of domestic use; and there is evidence of ritual ceremonies at these buildings (Kornienko 2009:85–88, 96). She also describes Hallan Çemi as having a 15-m-diameter open area that was used from the PPNA through several periods. This open area was in the center of the settlement and is a textbook case of a public area where co-operation-inducing ritualized behaviors could be carried out. This is similar to the site of Çayönü Tepesi, which has an open area of about 1000 m^2. This interpretation is supported by Çambel's (as quoted in Kornienko 2009:89) observation that "the square [at Çayönü Tepesi] could have been used for the ritual carving of animal carcasses during collective feasts, since there were many animal bones and fragments of specialized tools found here." In both cases, we see an open area for public ceremony and feasting.

Feasting, Trading, and Raiding

We have already mentioned the good evidence for feasting at Göbekli Tepe. Hayden (1996) argues that there is evidence for competitive feasting in the Natufian a millennium or two before sites like Göbekli Tepe were founded. Ofer Bar-Yosef (1998:163) describes a 9-m-diameter semicircular structure at the Natufian site of Ain Mallaha in the Levant. This structure was at least twice the size of other buildings on the site, though it was not made with any special materials. Bar-Yosef interprets the structure as possible evidence for communal activities.

Kornienko describes very compelling evidence of feasting at the early Neolithic site of Hallan Çemi. She writes,

> Large numbers of animal bones and river pebbles that had been cracked in fire were found on the square's surface. Among the remains were large, well-preserved parts of animal carcasses, and three horned sheep skulls were found lying in a row. This discovery, along with the large number of bones and evidence of the continuous use of the central square throughout several generations, are indications of the symbolic rituals once performed here. Archaeologists suggest that the area was the community's general meeting place and also the location of ritual feasts. (Kornienko 2009:83–85)

Katheryn Twiss marshals data for extensive feasting in the southern Levantine Pre-Pottery Neolithic. She likewise notes that it is correlated to population increases consistent with the model proposed here – that there is a minimum level of population density necessary to support the development of the ritualized economy: "Feasting increased as the transition to agriculture progressed, populations grew, and the foundations of socioeconomic complexity were laid" (Twiss 2008:438).

We have good evidence of specialized craft industries and trade in the early complex societies in the Levant: "Stone beads become numerous and diverse only in the Neolithic.... In the Pre-Pottery Neolithic A (PPNA) stone body ornaments begin to appear in abundance. By the Middle Pre-Pottery Neolithic B (PPNB) most sites contain them and they become more numerous and diverse in the PPNB and Early Late Neolithic (ELN)" (Wright and Garrard 2003:267). Kornienko (2009:83) describes the presence of imported copper ore and obsidian only around the special ritual buildings at Hallan Çemi. This suggests that the distribution of this rare import to manufacture other objects was associated with the special monumental architecture.[22] At the PPNA site of Dhra' in the southern Dead Sea in Jordan (circa 9500 BCE), Kuijt (2001:107) describes a short-lived site that covers about 1 ha with structures. He notes that the lithic assemblage includes a high number of "projectile points, awls and borers, wood-working tools, and ground and pecked stone pestles and bowls." Excavations at Jericho reveal that large quantities of imported obsidian and greenstone were brought in from Anatolia. These data have been interpreted as "testimony to the wealth of the site and its role as a distribution centre" for prestige goods (Naveh 2003:85; and see Bar-Yosef 1986). At Hallan Çemi, the presence of exotic obsidian and copper indicates active regional trade (Kornienko 2009:83). The PPNA occupation of Jericho has obsidian from Anatolia and greenstone beads (Naveh 2003:85). Wright and Garrard (2003:279) conclude, "Thus, relative to the PPNB, the ELN sites indicate intensification in craft production, with an increased element of (site) specialization."

The question of the existence of raiding in the Levant, Anatolia, and the Near East in general in this time period is highly debated. I am fairly confident that raiding occurred in the seventh millennium BCE in the region, based upon bioarchaeological evidence (e.g., Clare et al. 2008; Glencross and Boz 2013). Glencross and Boz (2013:103), for instance, note that the "current synthesis of bioarchaeological data suggests small-scale fighting and feuding through the Neolithic . . . [and that] the evidence for violence and conflict currently remains sparse and only suggests conflict on a small scale."

The famous wall at Jericho, dating to the PPNA, has been variously interpreted as a ritual feature, as a practical feature for water control (Bar-Yosef 1986), and/or as defensive. It is 3.6 m high and about 1.8 m at its base. It surrounded most of the settlement.[23] There is also a tower abutting the inside of the wall, about 8 m high and 9 m wide at the base (Bar-Yosef 1986:157). On the surface, this looks like a very effective wall to keep animals in and people out. The width of the wall in particular is ideally suited for defense. The wall is a typical kind of construction that would be very effective in a context of low-level, largely nonlethal raiding. However, many scholars dispute a defensive interpretation of the wall (Bar-Yosef 1986), and I agree with Keeley, Fontana, and Quick (2007:84) that "Jericho's PPNA/B ditch, wall, and tower logically and probably had a defensive function but until more of its early Neolithic enceinte is uncovered, a completely nondefensive purpose for these structures remains plausible."

There is little evidence for raiding or organized conflict at the site of Göbekli Tepe, though such evidence emerges later on. It remains to be seen if indeed there was no raiding in this earliest period or if we have not conducted sufficient research.

The Archaic in North America

Throughout the Americas, there are many examples of the first complex societies that developed in any particular region. In North America, some of the earliest special places are found in Louisiana and date to around 3400–3000 BCE at mound sites like Watson Brake, described below (Saunders et al. 1997).[24] These include earthen mounds, causeways, and shell mounds. Cheryl Claassen (2010:197) offers a somewhat speculative and controversial argument that shell mound sites in the Ohio River valley were in fact Archaic-period special places. She makes a persuasive argument that at least the bluff-top shell mounds were areas of feasting and ceremony, suggesting that the presence of dog burials, caches of artifacts, apparently sacrificed bodies, and other evidence indicates activities other than subsistence procurement. These sites were constructed between 4500 and 2000 BCE (Claassen 2010:197), making them roughly contemporary with the Watson Brake phenomenon for at least 400 years. Michael Russo (1994) likewise identifies several dozen possible Archaic mound sites in the lower Mississippi Valley.

The most elaborate early special place can be found in the lower Mississippi Valley not far from Watson Brake at the site of Poverty Point. Built more than a millennium later than Watson Brake or the Ohio Valley shell mounds, Poverty Point is an icon of early complexity in nonagricultural societies. This site is, in my opinion, as important as Göbekli Tepe is in the Old World, a complex, planned site built and maintained by hunter-foragers.

Poverty Point (1700–1300 BCE)

The site of Poverty Point in Louisiana (Figure 7.38) is one of the oldest of the large settlement complexes in the Americas. The bulk of the site was most likely built from approximately the seventeenth to the fourteenth century BCE (Connolly 2006; Gibson 2000; Kidder et al. 2008:9). It consists of one very large platform mound, called Mound A or the Bird Mound, several smaller ones and a series of concentric rings in a core area of approximately 3 km^2. The Bird Mound was constructed very quickly. In the words of Ken Sassaman (2008:6), interpreting the results of T. R. Kidder's work, it was constructed "virtually instantaneously."

Kidder et al. describe Mound A (Figure 7.39) as

> the largest earthwork at the site and the second-largest earthen mound in eastern North America. Situated at the western edge of the ridge system, the mound is roughly T-shaped when viewed from above.... The western half of the mound consists of an

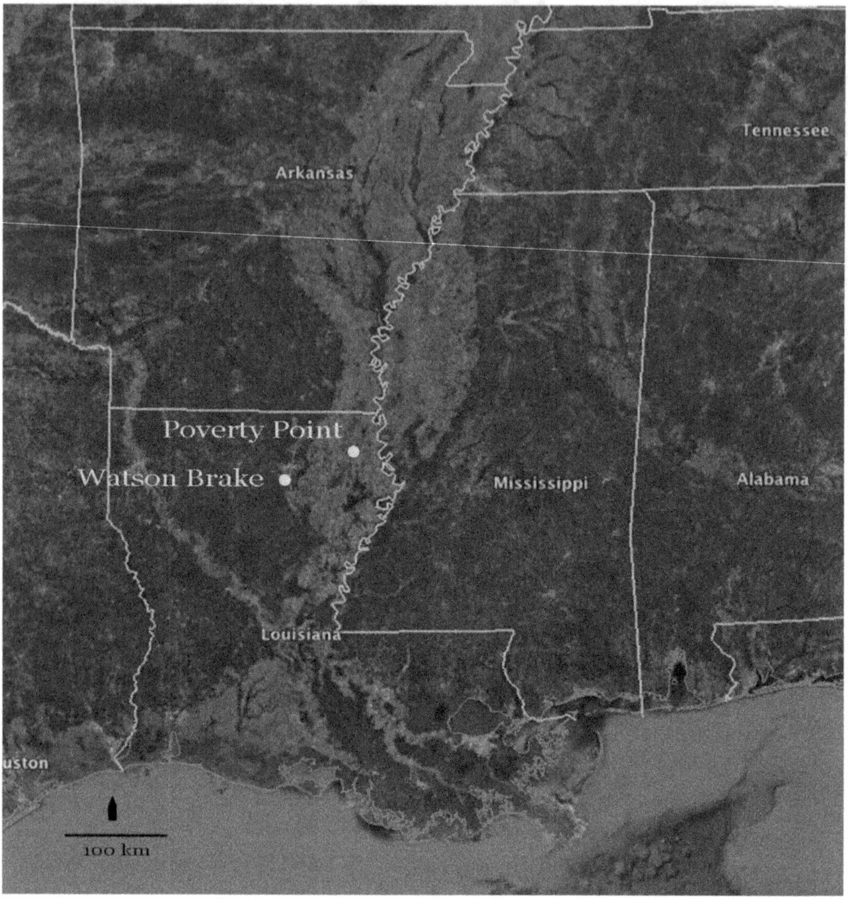

Figure 7.38. Location of Poverty Point.

elongated cone rising 22 m above the land surface. Attached to the cone is a flat platform by a ramp-like feature. The mound stretches nearly 210 m north to south and approximately the same dimensions east to west. The total volume of the mound is ~238,000 m^3. (Kidder et al. 2008:10)

The site also includes a series of six concentric earthworks that form a semicircle adjacent to a riverbank. These linear mounded structures are 1 to 3 m high and 20 to 40 m wide (Kidder et al. 2008:9). The earthworks cover an area over 1 km in diameter (Figure 7.40). Circular structures inside the plaza area are defined by the semicircles of earthworks. Research indicates that these ringed mounds held domestic structures made of pole and thatch.

Figure 7.39. Bird Mound, or Mound A, at Poverty Point.

Figure 7.40. View of Bird Mound at Poverty Point (background) from the first concentric ridge (looking to the mound in the background).

There is substantial evidence of specialized lithic artifact manufacture and probably exchange at the site. Webb (1968:312) tells us, "Over 2300 plummets have been found through excavations and from the efforts of private collectors. Roughly eighty percent of these plummets are made from hematite." In fact, magnetite and hematite plummets are common in the Late Archaic period in this region (Lipo et al. 2012:84). Webb furthermore describes this stone industry:

> The lapidary industry at Poverty Point is characterized by a great variety of objects, excellent workmanship, new techniques, and the use of such exotic materials as red jasper, red and green talc, banded slate, quartz, quartzites, hematite, magnetite, limonite, galena, feldspar, amethyst, fluorite in yellow, green, and purple shades, and many unidentified colored or translucent stones. Together with copper, the sources of materials cover much of the central United States. (Webb 1977:48)[25]

There is substantial evidence that raw materials for at least stone objects were imported from a huge area. Kenneth Sassaman (2005:352) indicates that a number of materials from a 1,000-km radius were found at the site, including novaculite, hematite, magnetite, quartz crystal, soapstone, northern gray chert, galena, and copper. A large cache of soapstone was found at the site. These materials were used to manufacture beautiful objects in the shape of animals, projectile points, and the like (Ortmann 2007). A site museum exhibit includes other nonlocal raw materials, including schist, clinker, petrified wood, talc, shale, Catahoula sandstone, greenstone, fluorite, cannel coal, marine shell, slate, Ozark white chert, and red jasper.

Poverty Point has abundant quantities of baked clay objects. These include clay balls, figurines, and a number of decorated objects. The latter category includes cubes and round objects with a variety of motifs. The objects are highly varied in style and color. Figurines were manufactured of clays with many colors, such as red, white, and dark reds. The figurines also represent a wide range of traditions.

Excavations and surface collections at Poverty Point have discovered a very rich collection of carved objects, such as the famous zoomorphic beads (Webb 1971). These were made largely out of imported red jasper. The inhabitants of Poverty Point also manufactured magnetite and hematite plummets. Atlatl weights were common and were also made out of a variety of materials. According to the site museum exhibit, there were more than 8,000 spear points at the site, with evidence of lithic manufacture at the site itself and evidence of the import of raw materials throughout the region. I agree with Jon Gibson, who sees the site as the center of a vast trade network that integrated communities from the Mississippi Valley region.

At the contemporary Slate site in Humphreys County, Mississippi, located around 100 km to the east, James Lauro and Geoffrey Lehmann report that:

The Poverty Point component is the result of special activities focused upon the production of lapidary items principally from slate, but including other materials such as quartz crystals, quartz, bauxite, quartz diorite, and jasper. The lapidary items include disc, tubular and barrel-shaped beads, effigies, and miniatures. Other artifacts diagnostic of the Poverty Point culture include micro-tools and cores, a consistent variety of projectile points and other chipped stone implements, boat stones, pendants, perforated gorgets, pitted stones, and plummets. (Lauro and Lehmann 1982:iv)

Astronomical Alignments at Poverty Point

There is some controversy (as there is with most sites in the Americas) as to whether there are astronomical alignments at Poverty Point. Clear intentional breaks between the ring earthworks create radiating pathways from the center of the semicircle to approximately 185°, 240°, 295°, and 335° (Figure 7.41), with an uncertainty of at least +/-5°. The second and third pathway lines align with the December and June solstices at circa 1500 BCE respectively. The location of the setting sun during these times has not appreciably changed in the last 4,000 years. Furthermore, the sight line bifurcating Mound A, from the east, is 263°. This 263° azimuth marks the sunset during the vernal equinox from this spot in antiquity.[26] It likewise hits the approximate center of the architectural complex as defined by the semicircular earthworks. As Kenneth Brecher and William Haag write:

For the latitude of Poverty Point, the summer and winter solstice sunset azimuths are 241° and 299°, respectively, in good agreement with the orientations of the southwest and northwest avenues. Such a solstitial alignment, while not surprising, seems hard

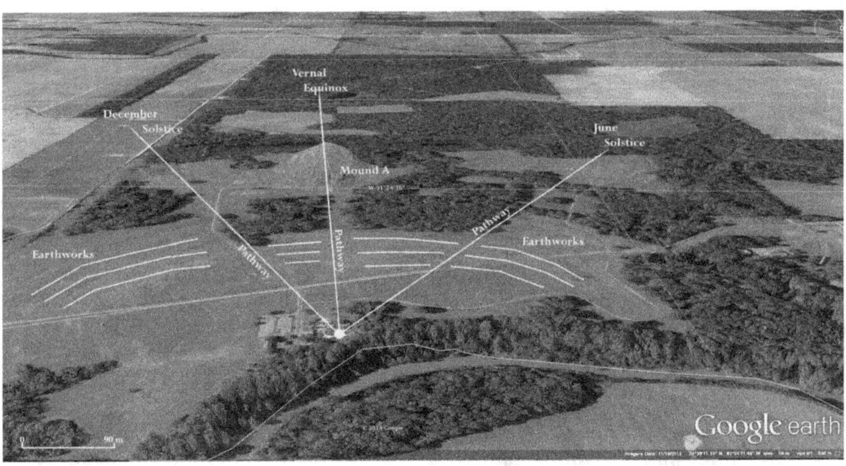

Figure 7.41. Astronomical alignments at Poverty Point.

to doubt in the Poverty Point earthwork. Whether or not intentionally included in the design of the earthwork by its builders is, of course, not known. (Brecher and Haag 1983:162)

Anthony Aveni (2003:173), arguably the dean of serious archaeoastronomy in North America, concurs on the existence of alignments at the site. Yet regional specialists seem surprisingly resistant to what looks to me to be an obvious, intentional construction to mark two of some of the most striking celestial features in any sky. Brecher and Haag, quoted in the above paragraph, hedge their interpretation rhetorically, asking if the placement was intentional. Virtually none of the general overviews of Poverty Point, as well as the otherwise outstanding label copy in the site museum, mention these alignments. I have yet to find any mention of the vernal equinox mentioned in the literature, though I am sure that many people have noticed this alignment.

Robert Purrington (1983) explicitly argues that the pathways are not aligned with solar phenomena. He calculated six important alignments at the site: 185° along one pathway to the Lower Jackson Mound; 236° through another pathway to the hypothetical December solstice; 260° to Mound A; 290° through a third pathway to the hypothetical June solstice; 305° along the pathway to Mound B; and 336° along the pathway toward Motley Mound. He properly notes that there is an uncertainty of ±3°.

Of course, today, we have off the shelf software and Google Earth to help us out to levels of precision unimaginable a few decades ago, making our work far easier than it was just a decade ago. We can see that the line bifurcating Mound A in a westerly direction is in fact 263°, within the range that Purrington calculated over three decades ago. The rest of the alignments are subject to bias as to where one locates the "center" of the architectural complex and how accurately one can measure the precise line through the mound gaps that created the pathways. Furthermore, where one marks the sunset on the horizon injects a bit more uncertainty, up to around 1°. These uncertainties allow an investigator to tweak the data in a variety of ways, depending upon the preferred interpretation. In the case of Poverty Point, the range of possibilities is actually limited to a few degrees in either direction in the pathway location. Where one locates the center is difficult. I chose the site of the modern museum adjacent to the river cut. This is where the three sight lines of the solstices and equinox would meet, but it excludes the northwest–southeast trending line that connects the two other earthwork breaks.

Within this limited range of possibilities, there are quite clear alignments at Poverty Point. What is obvious to me is that two pathways and one sight line to the major pyramid begin in the center of the concentric rings and mark the June and December solstices and the vernal equinox, respectively. These astronomical alignments have been used by other indigenous cultures to structure activities within their social worlds. The idea that this could be just chance seems highly unlikely, particularly given the fact that Mound A was built out of geometric proportion relative to the rest of the site. There is little doubt that the people at Poverty Point built pathways to the solstices and that they created a pyramid that may have had many functions. One of those functions was to mark the vernal equinox. The pyramid is wide at the top, with what would have been a flat top and rapidly descending sides. The precise azimuths on the flank edges of the upper part of the pyramid are 259° and 266°. This would mark a period of about 15 days 3,500 years ago, with the equinox in the middle of this period. As I describe for the South American case studies I strongly believe that these solar alignments were designed to mark periods of ceremony and feast days at these special places.

I do reject the idea that Poverty Point is a large and precise calendar to mark days or even hours. The alignments were not intended to be a precise timepiece. Rather, solstice marking around the world in stateless societies is a very obvious and almost universal means of knowing within a day or two the time of great festive events. By "universal," I simply mean that the solstice azimuths do not vary much at all over very large regions. Therefore people throughout a vast region, such as the entire North American continent, would experience the same azimuth position of the solstices.

Part of the ritualized calendar that structured the events at the site included the obvious sunsets of the solstices as well as the three-month period between the winter and summer solstice, the equinox. Excavation data tell us that Mound A was almost certainly contemporary with at least some of the ring earthworks. In this regard, the irregular, off-centered placement of the large platform Mound A is fully understandable by the desire of the architects to mark the equinox between the two solstices at the expense of geometrical perfection.

More will be said below about the role of solstice and equinox markers in premodern, stateless societies. Suffice to say that Poverty Point has a fairly sophisticated set of architectural markers that define many things that we do not fully understand (such as the two pathways to the northwest and southeast). Some of these pathways, in contrast, are clearly understandable as markers of major solar phenomena, which was part of the theatrical allure of this settlement for people over a vast region.

Poverty Point as a Regional Special Place

Archaeological data synthesized by Kidder et al. (2008:11) indicate that "Poverty Point witnessed significant population aggregations for brief periods of time and the duration of much of the construction probably was limited to a period of less than one hundred years." There is abundant evidence for the importation of objects from a vast area in a context of nonagricultural, hunter-foraging society. There is evidence for temporary residence at a settlement with sight lines to the solstices and vernal equinox. Domestic structures were built in highly regular rings in front of a major pyramid. The site of Poverty Point is a massive built landscape designed to attract people from many distances for precise periods of time defined largely by solstice indicators. It is the quintessence of a special place built by a stateless, complex nonagricultural society to structure economic and social life.

I find the words of Jon Gibson and Mark Melancon to be at least half of the story of Poverty Point:

> We contend that feast-sponsoring ritual built Poverty Point, not directly but indirectly. How? By bringing together, physically and spiritually, family and friend, neighbor and acquaintance, ally and potential adversary, and filling them with powerful rallying casuistry, joyous friendship, and uplifting unity – the core sentiments that define a people and its allies socially and politically. (Gibson and Melancon 2004:190)

The other half of the story is economic. Poverty Point was also a place where people imported, produced, and bartered goods. The intense social interaction of the festivals at Poverty Point was also an incentive to keep working in cooperative groups far from the site. Poverty Point was a major special place where people congregated for a number of reasons to enhance both their material and social well-being.

We now discuss earlier sites in the Mississippi Valley and in the American Southwest to place Poverty Point culture in perspective.

The Precursors and Contemporary Settlements of Poverty Point

Watson Brake (3400–3000 BCE)

Poverty Point is a comparatively huge site, built by nonagriculturalists as a special place where people congregated for what appear to be long periods of time. Yet Poverty Point is far from the earliest special place in the region. The justly famous site of Watson Brake, located around 75 km to the west of Poverty Point in the Quachita region, is substantially older. It was first occupied around 4000 BCE, with construction of a circular mound complex beginning around 3500 (Saunders et al. 1997, 2005). The formal mounds were built on existing midden, suggesting that there was a modest site at the area prior to the construction of the mounds. Saunders et al. (1997:645) provide extensive carbon dates that indicate an initial occupation just after 4000 BCE, with mound building starting around 500 years later. The mound building process ended another 500 years later, around 3000 BCE.

Watson Brake (Figure 7.42) does not look like a typical semipermanent settlement made by a hunter-forager society. It looks like a planned complex of

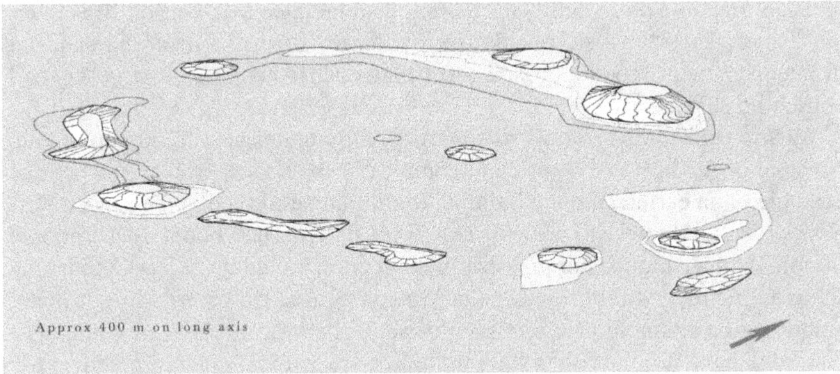

Figure 7.42. Reconstruction of Watson Brake.

11 earthen mounds with abutting ridges that cover an area of approximately 8 ha. Elevated walkways or causeways connect the site. The shape is ovoid, providing an open space within the ring of mounds. The ring is about 300 m wide. Watson Brake is of course substantially smaller than Poverty Point and was abandoned well before the latter was founded. According to Saunders et al. (2005), there is evidence of site use in all seasons, but it is still not certain if the site was permanently settled or if it was just periodically used in all seasons.

Saunders (2012:36) reports that the site area of Watson Brake was occupied for possibly centuries prior to the construction of the mounds. This is significant. It indicates that people had been using this area and presumably other areas nearby for many generations and that at a certain moment Watson Brake was the area where elaborate mounds were constructed.

It is quite significant that Watson Brake has an abundance of microdrills used for bead production. According to Johnson (2000), blades and drills were a hallmark of the lithic assemblage, indicating a site focused on specialized production of these tools using local materials. There is no evidence for an extensive exchange of distant raw materials or goods. Saunders et al. (2005:651–652) likewise report substantial numbers of flaked lithics, beads, and drills. They also found some bone tools, bone beads, and a number of fired earthen objects. Saunders et al. (2005) also report a rich range of fauna and botanical remains, indicating intensive local exploitation.

Watson Brake represents the beginning of the tradition of building special places in the Mississippian culture area. John Clark (2004) and Sassaman and Heckenberger (2004) argue that Watson Brake is largely ceremonial. Saunders et al. (2005:665), in contrast, argue for more "daily, secular events" taking place at the site. I would interpret this as an early special place, based upon the layout and presence of a formal plaza area and the fact that mounds without any practical function were built. Our current data indicate that Watson Brake was most likely an area where people lived at least part of the time and engaged in intensive bead production and most likely other activities that go beyond household subsistence.

Watson Brake it is certainly not unique. As Saunders et al. (2005:662) note, "in spite of the impressive size and complexity of Watson Brake, it is but one example of an earthen mound building tradition that thrived in the Lower Mississippi Valley between ca. 6000 and 5000 B. P." They count 14 additional Middle Archaic mounds of this size in the region, including at least 3 that may share an architectural design seen at Watson Brake. However, while it is not unique, is an example of a rare type of site. The vast majority of contemporary sites have no more than one mound or are without surface architecture. However the sites of Frenchman's Bend Mounds and Hedgepeth Mounds have five and six mounds, respectively. They are contemporary at least in part with Watson's Brake (see Saunders 2012).

Current data indicate a hiatus of mound building in the lower Mississippi River valley between 5000 and 3700 BP (Saunders et al. 2005:663). If this pattern holds, it suggests some interesting cyclical patterns that we have only begun to fully understand.

The American Southeast

The St. Johns River valley in northeast Florida is home to a series of Archaic shell mounds, originally interpreted as simply large refuse heaps of hunter-foragers (Randall 2008:13). Randall's summary of the data from these sites indicates that people first used these sites around 7300 cal BP in the preceramic Archaic. These early occupations were not special places but rather were used for more prosaic uses. Later occupations include a ceramic-period one at 4700–3600 cal BP, followed by agriculturalists at 3600–500 cal BP (Randall 2008:13).

There are thousands of shell mounds in the southeast, and they have been particularly well studied in the coastal regions of Florida. Bruce Piatek (1994) noted a while ago that the Tomoka mound complex in coastal Florida did not fit the traditional understanding of the Archaic period in North America. Michael Russo (1994) likewise noted how these complex sites did not conform to notions of cultural complexity linked to agriculture. The site of Tomoka, for instance, has a number (at least nine) of conical mounds built during the Late Archaic. The site is long and narrow, stretching over 1.5 km by about 100 m wide. The largest mound is 3 m high and more than 30 m in diameter (Piatek 1994:110).

Sassaman and Randall (2012:58) note two basic complex shell mound forms. One type is composed of crescent-shaped ridges, while the second type takes a

U form. The U-shaped mounds appear to be later. Most significantly, they have investigated numerous mounds in northeast Florida and conclude that many show different episodes of construction and use. They argue, "There is a regularity and rhythm to these events that suggests some manner of cyclical practice" (Sassaman and Randall 2012:61). These observations indicate these shell mounds were not just prosaic accumulations of midden remains but were intentionally created monuments used for periodic ceremonies. These are classic special places designed to promote the congregation of people from around the area.

Randall (2008:13) observes that these Archaic "shell mounds were truly monumental in scale and significance." He goes on to say, "Reconstruction of the site has established that 7,000 years ago communities created and reproduced linear settlements characterized by multiple domestic middens, some possibly contemporaneous" (2008:13). He describes some of these as "three shell ridges in a U-shape, each ridge rising upwards of 8 m above the floodplain and extending over 300 m inland or along the waterfront. [Wyman] had observed similar facades at scores of isolated linear or crescent-shaped shell ridges, typically 150 m long and 6 m high, throughout the middle and upper reaches of the river" (Randall 2008:13).

Lawrence Aten observes that the mounds, in this case at the Harris Creek site in the St. Johns River area, appear to have been used for repeated ceremonies for the community. This interpretation is consistent with a function of the mounds as feasting and ritual areas, the kinds of rituals that are organized by leaders to keep people co-operating:

> Mt. Taylor mound was indeed initiated for the purpose of building a raised platform for communal activities. This plan may or may not have initially envisioned mortuary use, but once started the internment aspect of communal activity was kept in the "background," i.e., on the sides and rear of the mound, not on its summit. While it is possible the summit platform was used at times for domestic purposes (possibly only for individuals or families of special status), reshaping the front slope in the interval between Layer 7 and Layer 8 reinforces the impression that this mound was not simply an enlarging heap of Archaic village refuse, but rather one that was maintained by the community for certain purposes. (Aten 1999:178)

It is not surprising that Quinn et al. discovered that there was a brisk trade of nonlocal foods from the Harris Creek assemblage. Again, this suggests a site where people congregated to meet and to feast, with the complex architecture the means by which this interaction was facilitated. In fact, Rebecca Saunders (2004) argues that there is good evidence of repeated feasting at the U-shaped sites.

Another noteworthy mound is a ceramic-period one at Silver Glen Run, also located in northern coastal Florida. This huge U-shaped mound is more or less

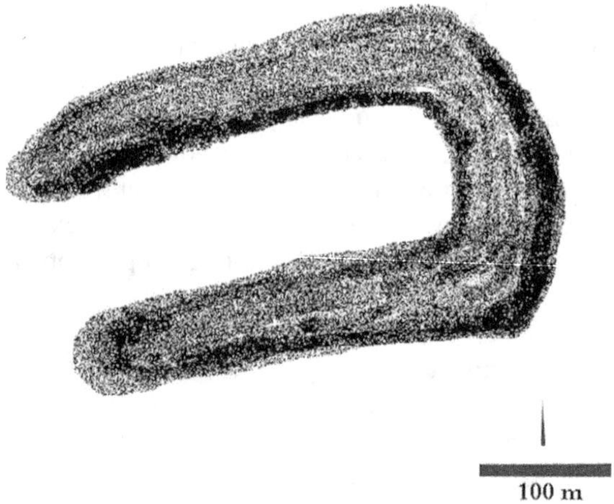

Figure 7.43. Reconstruction of Silver Glen Mound. From Randall (2008:Figure 3).

contemporary with Poverty Point, more than 1,000 km to the west (Figure 7.43). The site was built over earlier cemeteries. Randall notes that only four sites in the area had fancy, high-quality pottery and that these four were most likely built in U shapes as well. They were also spaced about 20 to 30 km apart. He states, "These Orange-period monuments recall the spatial organization of Orange-period coastal shell rings, but on a much grander scale" (Randall 2008:16). In other words, there were numerous smaller domestic shell midden sites, along with four obvious "special places" spaced about a day's walk apart along the river and built in a similar architectural style.

Randall also notes a brisk trade and exchange of objects and technologies in this region in this time. Randall and Sassaman's concept of a "new world order" is particularly cogent in light of our discussion concerning the emergence of complex, co-operative organizations in stateless societies. Randall sees a number of complex economic and social relationships between distant settlements:

> The alliances that were maintained through bannerstone exchange likely facilitated the introduction of either potters or pottery production from coastal communities beginning 4,600 years ago at the onset of the Orange period (Sassaman 2004). The appearance of Orange fiber-tempered pottery was not simply an addition to traditional subsistence technologies, but instead represents a "new world order" in the organisation of regional ritual and domestic practice (Randall and Sassaman 2007). New patterns of settlement and monumental construction, in which sacred and secular places were spatially segregated, emerged from the coalescence of once-separated coastal and interior populations. (Randall 2008:15)

Michael Russo (2008) likewise reports that Archaic-period nonagricultural peoples constructed U-shaped shell mounds on the coasts of Georgia, South Carolina, and Florida. The U-shaped structures are particularly noteworthy as places for congregations. He describes the Archaic site of Tick Island (8VO24):[27] "A sand mound built by preceramic Archaic peoples was placed on top of an existing preceramic Archaic shell midden and mantled with preceramic Archaic shell midden" (Russo 1994:94). The site dates to around 2700 BP and is interpreted by Russo as a definitive ceremonial mound. Saunders and Russo (2011:38) conclude, "By 4500 B.P. in all three regions of Florida, estuarine subsistence bases supported population nucleation and the creation of monumental architecture." Saunders and Russo (2011) describe a number of complex, Archaic-period monumental mound sites in the Florida coastal region. In other words, we have in the Archaic period of the southeast Atlantic region complex hunter-foragers living in settled regions in the very rich estuary areas that created a number of special places that I interpret to be ceremonial sites to host significant social events.[28] These social events were part of the increased co-operation between and within various groups at these centers. By 3500 BP, this process had come to an end, as the regional system was disrupted and monumental construction ceased.

Raiding, Trading, and Feasting

There is little doubt that the Archaic period in North America was a time in which a number of peoples throughout a vast area constructed special places on the landscape. These places were marked by monumental constructions and represented places where people from a large region converged. There is also substantial evidence of raiding in the Archaic of the American Midwest and Southeast. David Dye characterizes the Middle and Late Archaic as a period of "blood feuds": "By 5000 BC, conflicts appear to have intensified with the development of revenge-based blood feuding in resource-rich areas" (2013:138). He notes a pattern of trophy taking in feuding episodes in the Late Archaic in the mid-South and Midwest. Ken Sassaman (1995:187) notes the elevated frequencies of bone fractures and points embedded in skeletons during the Middle Archaic of the Southeast. Robert Mensforth (2001) notes strong evidence for human trophy capture in the Archaic period of the Ohio and Kentucky regions. The list of empirical cases from this observation is quite large.

The evidence for a correlation between feasting and extensive regional exchange is likewise very strong. Vernon Knight (2001) explicitly links the construction and use of platform mounds with feasting in eastern North America in the Woodland period, circa 100 BCE–700 CE. The platform mound (Bird Mound) at Poverty Point can be profitably interpreted in the same way. This is similar to what my colleague Haley and I have suggested for platform mounds in the coast of Peru. The flat surfaces make ideal places for feasting and other ceremonies, and these can be conducted in a way that all participants can observe.

Anderson et al. (2007:457) sum it up: "The Middle Archaic period in the Southeast... from 8000 to 5000 ^{14}C yr BP... is a time of appreciable culture change. During this interval monumental construction began in a number of areas, long-distance exchange networks emerged, evidence for warfare appeared, and experimentation with agriculture was initiated."

Henge, Barrow, and Circle Monuments in Northwestern Europe

Henge monuments are defined as settlements with circular or oval areas enclosed by a bank and internal ditch. These circular or ovoid banks and ditches have at least one, and often two entrances (Atkinson et al. 1951; Renfrew 1983:157). Henges with two entrances are often opposed to each other. A bank, according to the Oxford English Dictionary entry of "henge" is "a portion of the surface of the ground raised or thrown up into a ridge or shelf; a lengthened mound with steeply sloping sides." Stone circles, of course, are megalithic monuments composed of upright stones placed in a circle on a landscape. Barrows are the elongated mounds that were used largely for burials.

The earliest evidence of complex society is therefore found in the numerous Neolithic barrows, henges, causewayed enclosures, and stone circle sites throughout Western Europe. The first monuments date to the early Neolithic, around 4000 BCE. This is roughly coincident with the introduction of agriculture in the area and is a defining feature of the Neolithic lifeway. Alasdair Whittle et al. (2011:690) summarize a huge amount of data and conclude that latter Neolithic causeway enclosure building started around 3700 BCE. These monuments are found throughout the British Isles, from the far southwest through central England and most of Scotland, as well as throughout Ireland. They are also found in Portugal, Spain, Brittany, and north to Scandinavia. Hundreds of such monuments have been documented, and it is most likely that hundreds more have been destroyed over the centuries.

Colin Renfrew (1983:152) noted that the shift from the earlier simple barrows to more elaborate henge monuments in Western Europe around the middle of the fourth millennium correlated to the rise of social complexity. Referring to the excavations at Neolithic Hambledon Hill, he points to evidence that sites such as these were "the meeting places of scattered communities from the surrounding areas" (Renfrew 1983:154) prior to the development of the more elaborate monuments. This interpretation is supported by recent data from Roger Mercer and Frances Healy (2014). They interpret the massive data that they and their colleagues collected about the Neolithic occupation of the site:

> It is argued above that people came to the hill for short periods of time, sometimes in the early summer or autumn, sometimes up to a few hundred at a time, sometimes in smaller numbers. They came from distances of up to 45 km away to the south, and perhaps as much as 80 km away to the north-west. They brought with them already disarticulated human remains, wheat, barley, hazelnuts, red deer antler and pots from the area where they lived, a few pots and axeheads exchanged from farther to the south-west, and livestock on the hoof, especially cattle.... While on the hill, people carried on the everyday activities of life, but also did everything that a scattered population would do when gathered together. There is tangible evidence of feasting.... Some gatherings equally tangibly included construction of new earthworks and maintenance of existing ones; a few were occasions of attack; and carefully placed deposits were made in both pits and ditches, some of them reopened for the purpose. (Mercer and Healy 2008:761–762)

They go on and note that other sites were similar: "The evidence from Hambledon is similar to that from other causewayed enclosures, not only in outline, but in the minute details of ditch recutting and of the assembly of deposits. This bespeaks beliefs and traditions shared over a large tract of territory" (Mercer and Healy 2008:763).

Renfrew notes that the earlier barrow sites were not merely tombs, but were in the words of Andrew Fleming, they were "tombs for the living." His observations presage the arguments in this book that the monuments were not just burial places but were "public centers" (Renfrew 1983:159). Such centers served as meeting places on the landscape for politically autonomous settlements, "perhaps the locus for an entire range of religious rites relating the community as a whole to its ancestors as all as to its more recent dead" (Renfrew 1983:159). In work at a Neolithic barrow at the site of Quanterness in the Orkney Islands, Renfrew demonstrated how one large barrow held the remains of at least 157 individuals (Renfrew, Chesterman, and Aitken 1979). Taking into account unexcavated areas, Renfrew estimated that the tomb held around 400 individuals (Renfrew 1983:160). Like the ones to the south, this chambered tomb held disarticulated bones that had been collected from other areas. Again, this emphasizes the social nature of the tombs: they were a place to collect and honor the dead, a place to socialize and reinforce social ties, and I would argue a place where the norms of co-operation were repeatedly reinforced and sustained by economic labor specialization that provided for increased per capita production.

These barrow and henge monuments are not rare, though the former were much more common than the latter. Aubrey Burl cites a figure of approximately 1,300 megalithic ring monuments in Britain, Ireland, and Brittany alone (Burl 2000:48). Colin Richards (1996:320) observes that henge sites were located "on the floor of natural bowls or valleys, and in close proximity to water." There are exceptions, of course, but he points out that the henge monuments are located in areas with a set of common characteristics. He therefore argues, along with

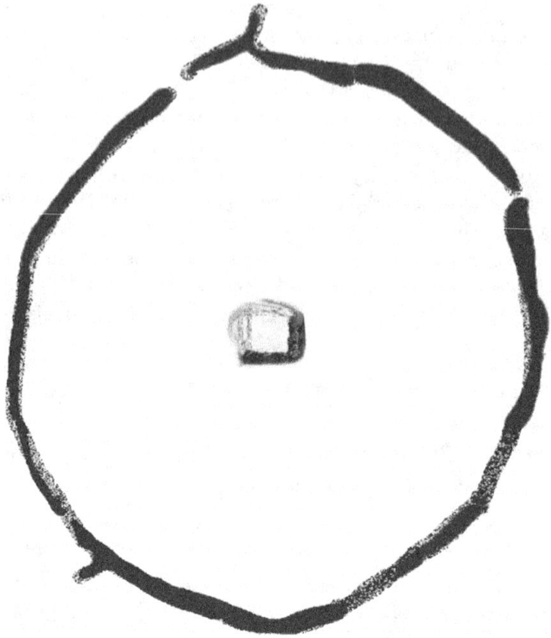

Figure 7.44. Ground plan of Dubbleby Howe.

Ian Hodder (1982), that common principles structure the architecture of the houses, graves, and henge monuments (Richards 1996:320).

Alex Gibson (2014) describes in some detail the henge site of Duggleby Howe in Yorkshire. The chronological control at this site is quite excellent, and it provides an insight into the construction sequence of a classic henge monument. This site has a large mound, about 38 m in diameter at the base, with a 14-m-diameter flat top and standing a bit over 6 m high. It is encircled by an almost 2.5-m-deep ditch which is about 5.25 to 7.75 m wide (Figure 7.44). There were burials at the site as early as the thirty-sixth century BCE. About 500 years later, the mound was constructed (around the twenty-ninth century cal BCE). The ditch seems to be about 400 years later, and, along with the enlarging of the mound, it transformed the site into a major monument (Gibson 2014:10). As Gibson writes, "In the second half of the third millennium the site was transformed from a visually striking but comparatively modest barrow into a monumental round barrow much as we see [on the surface] today," being "surrounded by a large causewayed quarry ditch" (Gibson 2014:12–13).

Several observations about Duggleby Howe stand out. First, the site was probably one of many such barrows[29] in the region. For some reason, this area was chosen by the inhabitants to be enlarged, and this occurred roughly around the time of other sites, such as Stonehenge, Durrington Walls, and Avebury. As Gibson (2014:12) notes, it was at this time that Silbury Hill was also rebuilt into one of the larger Neolithic mounds in the British Isles. One can suggest therefore that there was a general process in which a few settlements were successful in attracting larger number of people. This is reflected in a regional pattern of larger but fewer major mound complexes through time.

Gibson (2014:12) also discovered that the enclosure area itself at Duggleby Howe was "devoid of visible archaeological activity" from the times either before or after construction of the ditch. Gibson observed that there was no continued reworking of the site area, as we see at Stonehenge. Rather, he discovered distinct periods of use and disuse, followed by a renewed interest in the site. People consciously chose to mark this modest barrow with a large quarry ditch and turn it into a ritually significant place of congregation and exchange. Huge earthworks were constructed around other existing monuments in the region around this time as well.

Spielmann (2008) nicely describes the ritual significance of the enclosures in early Neolithic Britain. She notes that researchers have found at least 50 such segmented, earthen enclosures and describes a "general template" in which a series of ringed embankments "became the loci of ritual deposits," including cattle bone, flint, polished axes, flint-knapping debris, pottery, feasting remains, and many offerings in pits (Spielmann 2008:40). She cites work by Richard Bradley and Mark Edmonds, who note that the enclosures have very high concentrations of axes and pottery, oftentimes found as offerings in external pits.

The famous site of Stonehenge in Britain is an icon of European archaeology. It is part of this henge tradition, with the core-site construction of a ditch beginning around 3100 and lasting to around 2300 BCE. The ditch was originally about 100 m in diameter, growing over time into the site we see today. The site as seen on the surface today dates to the middle and later centuries of the third millennium BCE (Parker-Pearson et al. 2007). It has monumental antecedents as far back as the beginning of the third millennium BCE. The nearby site of Durrington Walls, larger than Stonehenge, is most likely contemporary with the apogee of the Stonehenge monument in the mid- to late third millennium (Parker-Pearson et al. 2007:626). It is located about two km to the northeast.

Researchers have painstakingly worked out the theatrical aspects of the "Stonehenge Bowl" over the years. This includes the existence of 16 long barrows, at least two curseses (parallel embankments with adjacent ditches that form what looks like a race course), avenues, wood and stone henges, and causeways (Parker-Pearson et al. 2006:237). Parker-Pearson et al.'s observations about the hill on which the site of Durrington Walls is found are very relevant to this discussion: "Whether this hill was a place of gathering prior to ceremonial processions is not known, but it is a distinctive landmark visible from many miles away. For the thousands of people involved in building and celebrating at Stonehenge and Durrington Walls, their temporary campsites or settlements over the winter months could have sprawled across the wider landscape around these monuments" (Parker-Pearson et al. 2006:237). It is fascinating that these avenues were resurfaced with flint cobbles, indicating long-term use and reuse. Clearly, Stonehenge cannot be viewed in isolation from its surrounding landscape. It was central to a vast Neolithic complex.

The various solstice alignments in Stonehenge and other henge monuments are well attested in the literature. Other more elaborate astronomical schemes have been suggested with less success. Certainly, however, solstice marking permeates the layout of many henge monuments, not just the Stonehenge area. Parker-Pearson et al. (2006:239) note that both sunrises and sunsets for both the summer and winter solstices have been documented for the area. Following our observations in other sites around the world, the people who used Stonehenge and surrounding sites timed at least some of their processions with solstices.

The most recent research in the vicinity of Stonehenge promises to fundamentally alter our understanding of this region. Work by David Jacques and his colleagues (Jacques and Phillips 2014) suggests a much deeper occupation of the area back to the Mesolithic, circa the ninth millennium. At a site (Blick Mead) about a 20-minute walk from the main site of Stonehenge, this team discovered good evidence for feasting:

> Worked flint, intensely fired burnt flint and the remains of aurochs, wild boar and red deer indicate that hunting, butchery, cooking and food consumption took place close to a base at the spring. The cooking of salmon/trout and toad points to intensive use of food resources in the area, and hints at lengthy stays here. In terms of the larger animals, it is difficult to escape the conclusion that the quantities of meat produced by them, as well as the efforts needed to kill and render them, created an opportunity for structured social rituals, such as large scale feasting. (Jacques et al. 2014:23)

Jacques et al. credibly argue, in my view, that this area was a special place, even in the Mesolithic ninth and eighth millennia. They point to the area as being rich in resources, including read algae that colors flints into a bright pink. The implications of these data are vast and we await new reports that will help clarify the historical depth of this most famous Neolithic henge monument area.

The site area of Avebury (Figures 7.45 and 7.46) represents another huge ritualized landscape of the Neolithic. There is of course the huge stone circle more than 100 m in diameter. There are causewayed enclosures, long barrows, chambered barrows such as the Shelving Stone, numerous settlements throughout the phases of the Neolithic, palisaded enclosures, "avenues," the massive conical Silbury Hill, and so forth (Figure 7.44). Organizationally similar to the

Figure 7.45. Henge monument of Avebury. Adapted from Google Earth.

Paracas occupation in Chincha, Peru, the Avebury area can be seen as a place of congregation, procession, exchange, and other highly structured behaviors, most likely timed by astronomical phenomena.

The Western European Neolithic is the time in which the first complex stateless societies developed in the region. The barrow and later henge monuments, located almost always in open areas, literally shout theatrical constructions to attract people in ritually ordered times of the year. The evidence for a ritualized economy from these monuments is strong, complemented by elaborate artificial landscapes designed to attract and retain groups of co-operative people for long periods of time.

The Importance of Astronomical Markers in Special Places

One theme consistently appears in our comparative archaeological analysis of special places constructed by people in many stateless societies. This is the importance of solstice and possibly lunar alignments in the architecture and ground monuments of these sites. Professional archaeologists are very wary of the claims of astronomical alignments for ancient sites. There are a number of reasons for this. The first reason centers on the excesses of the archaeoastronomers themselves in the 1970s and 1980s. These scholars, like all scholars who make new discoveries, became overly enthusiastic and over-interpreted some of what they found. There was far too much speculation, and

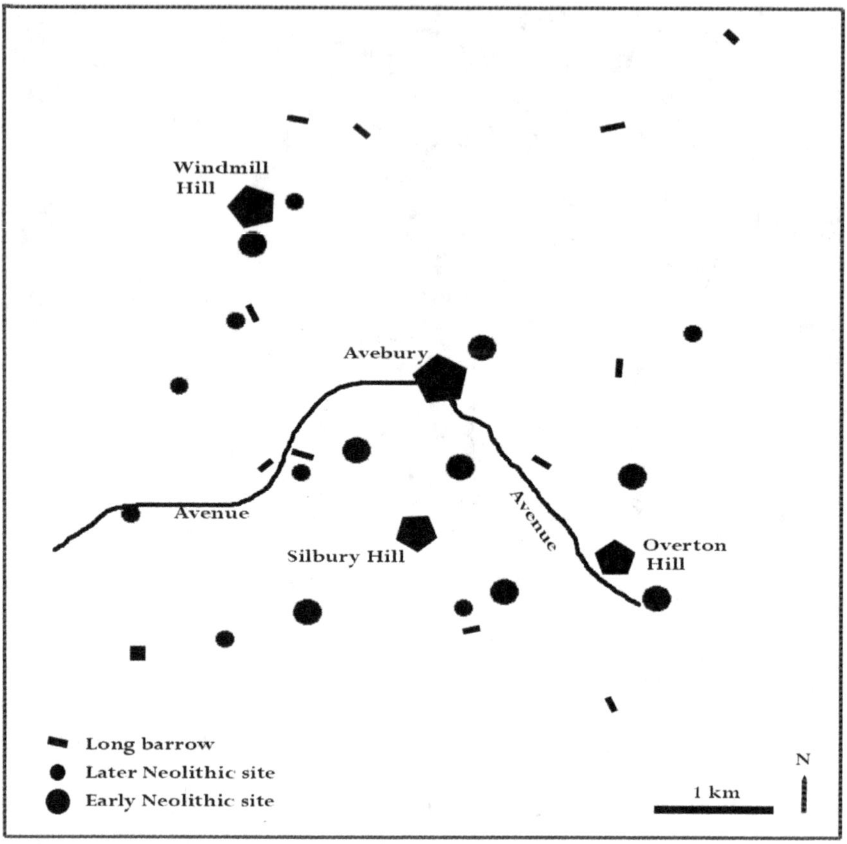

Figure 7.46. Avebury area showing sites and barrows. Adapted from Whittle (1993:Figure 1).

some of the theories proposed argued for levels of complexity in astronomical alignments in the ancient world that were simply not warranted. A second problem is with verification. As many have pointed out, all one needs is two points on the ground or a straight ancient wall and they will most likely line up to something in the night sky. The way in which we ascertain whether an alignment is statistically probable or not is complicated. A final factor is the enthusiastic support of archaeoastronomy by fringe groups such as New Age movements, Druids, the pesky ancient astronaut crowd, and others outside of mainstream scientific thinking who give the field a bad name.

Having noted these objections, it is clear to me that the marking of solstices in the ancient world was quite common. Anyone who spends time outdoors in

June or December, like most field archaeologists, will realize that the solstices are extremely obvious to any observer. It is quite impressive to watch the sun set a little to the left or right each day and then one day watch it "stop" and begin to go back the other direction. As Aveni (2000b:74) tells us, the very word solstice means "sun-stand" or "sun-stopped." It does not require any esoteric knowledge or any sophisticated calculation to mark that moment with a doorway, a line of posts or stone, a wall, or a geoglyph. And it does not take long for any rational person to realize that such a phenomenon that occurs twice a year in the precise time each year is an extraordinarily useful means of marking time.

The ethnographic record is clear that many people in stateless societies had ways to mark cycles of time that were important to them. Many of these involve some use of astronomical phenomena. Erna Gunther tells us that the Klallam people observe the winter solstice: "The old men watch the sunrise and take note of the place where the sun appears. When the sun reaches a certain point and starts back again, they know spring is coming" (Gunther 1927:227). This is, of course, the winter solstice. The ethnographer noted that another group nearby observe that when the sun "rises two days in succession in a certain hollow in the hills, the winter is about to begin." She also writes that the Kwakiutl and Nootka both observed the winter solstice as well (Gunther 1927:227–229).

Malinowski offers some fascinating observations on the nature of time reckoning among the Trobriand Islanders. He says that for everyday activities that require some kind of calendar, the people use an informal calendar based upon seasons and nature. Malinowski describes this as the "meteorological and the cultural" (Malinowski 1927:204). These criteria are somewhat imprecise, based upon bio-indicators following the fishing and climate cycles. In fact, they change with both the vagaries of the seasons and geographical location. However,

> To the garden magicians, and the elders who plan gardening and other events of tribal life, feasts, expeditions, and mortuary ceremonies, as independent scheme of time-reckoning is still necessary and the names of various successive moons are very convenient. It is in this somewhat esoteric and specialized way that the naming of the moon is chiefly used. (Malinowski 1927:211)

In other words, the more complicated calendar used by religious specialists for major events is based upon astronomical observations, in this case the lunar cycle. We can see how places like Stonehenge and Poverty Point encode not local knowledge of the individual group for local purposes but rather the knowledge of the religious specialists and chiefs who are responsible for the major events in their societies.

Douglas Oliver (1974:264) noted that the Maohis of Tahiti calculated annual cycles of time in several ways: (1) the daily rise and fall of the sun; (2) lunar months of 29 nights; (3) annual cycles of 12 to 13 lunar months; (4) annual cycles of breadfruit ripening; (5) seasonal cycles based upon observations of the Pleiades; and (6) annual cycles based upon weather conditions at sea. This was not a necessarily complicated way to reckon time, but it was certainly effective. This was a very practical system, using astronomical and naturally occurring cyclical phenomena.

Patrick Kirch and Roger Green observe, "The rising and settings of the Pleiades were widely observed in many Polynesian societies, where they were used to mark the change in seasons ... and/or to mark the commencement of the 'year'" (Kirch and Green 2001:263).

According to Evans, the Dusun of Borneo reckoned time with two systems. People in the hilly part of the country would use the agricultural season from planting to harvest, a period of six months, while a lowlander would use the wet padi season, which lasted eight or nine months (Evans 1912:395). Frederick Damon tells us that a small group in Papua New Guinea, the Muyuw, uses astronomical phenomena to mark time and schedule work:

> Muyuw do use celestial phenomena to organize production. Tides are most important here. Alternation between spring and neap tide determines when people are likely to be fishing or gardening, and the occurrence of the high and low tide during a 24 hour period decides which part of the day will be devoted to fishing. (Damon 1982:227)

The Mae Enga lived in the western highlands of New Guinea. According to M. J. Meggitt, the Mae Enga use a calendar to organize the inter- and intra-clan festivals. These include a variety of rituals, funerals, house construction, and gatherings in fairs or barter markets. He notes, "For all these activities to be efficiently co-ordinated, they need some means of consistently marking the passage of time; and this exists in their luni-solar calendar" (Meggitt 1958:74). He goes on to describe this system as a "sequence of synodic[30] months" that is correlated with "seasonal activities by adjustment to observed variations of the ecliptic" (Meggitt 1958:74). Meggitt also recorded the tracing of the solstice. His description of the solstice and lunar observances by elders using a mountain range and peaks provides a very compelling analogy for prehistoric sites where clear solar markings are either built into the landscape or architectural features were positioned or aligned with surrounding peaks:

> The elders who are calendar-keepers recognize the variation of the ecliptic, which in these mountains is fairly obviously manifested as differences in the position of sunrise through the year. ... Old Kara men have observed that sunrises "move" from Lupamanda to Taure and back again from the middle of one wet season to the next, that is, from one summer solstice to the next. They have also observed that the sun appears to rest for some days at Lupamanda before starting its return journey along the range. For reasons which are not now explicit, the lunar sequence begins with the first new moon to appear when the sun starts to move north-north-west along the range. It is also recognized that the

sun reaches Taure and moves south-south-east again in the dry seasons months. (Meggit 1958:76)

Here we have a small, sedentary village society utilizing a sophisticated means of keeping track of their calendar for rituals that are important to their own social and economic world.

Edmund Leach (1950) describes Trobriand calendars: "Although the Trobrianders seem to have only ten named months . . . it is still possible to define a Trobriand lunar year. The Trobriand seasonal cycle starts on a full moon (Milamala), so that between the beginning of one seasonal cycle and the next there will always be an exact number of lunations, 12 or 13" (Leach 1950:249). His treatment of the Trobriand calendar is quite extensive and indicates that this stateless society reckoned time in a manner most useful to the society. Most significantly for our discussion, Leach concludes, "We can agree with Hogbin that the need for a calendar arises from the requirements of those who have to arrange festivals and undertakings some distance ahead" (Leach 1950:261). In other words, at least two earlier ethnographers recognized that marking time in stateless societies was intimately linked to ceremonial life.

Evans-Pritchard (1940:95–110) informs us that the Nuer calculated their calendar by the seasons. They have well-defined months, and he notes that most adults can recite the calendar. For the Nuer, time based on ecology is cyclical. T. O. Beidelman tells us that the Kaguru of Tanzania (formerly Tanganyika) "mark the months by the phases of the moon (mwesi) or, to be more exact, Kaguru believe that there is a perpetual succession of different moons, each one marking a new month. At the end of each month the moon dies, is eaten away by the dark, and a new one is then formed, which grows to fullness and then is, in turn, eaten away" (Beidelman 1963:14). They refer to months as moons and have a 13-month calendar. According to Beidelman (1963:15), in the past the months were counted by the use of knotted cords, each with 13 knots. These knots were divided by pegs and represented days.[31]

There are many more ethnographic examples of solstice marking for calendars.[32] Marking the solstice and Pleiades are particularly common. The equinoxes were also commonly used as well. The common feature is that these astronomical phenomena can be seen over a very wide area. They therefore serve as excellent time markers for large, regional cultural events.

From a Local System of Time Reckoning to a Regional One

I argue that creation of the larger special places on the landscape, designed as spaces where people from around a larger region could congregate, involves in many cases a shift from a local system of time reckoning to a regional one.[33] This facilitates participation in events at the regional special places and allows a large number of people over many hundreds of kilometers to keep track of the calendar of events. Not everyone lives near the beach and can observe the waves. Fruit trees or other plants blossom at different times. Virtually all bio-indicators are locally specific. However, certain astronomical phenomena are invariant from the perspective of an observer across a wide landscape. The solstices and equinoxes are examples; their movements and locations are effectively "universal" to viewers in any region. There is very little variation of the azimuth over thousands of kilometers in the solstices because of the enormous distance from the sun to the earth. Likewise, the solstice azimuth effectively does not move over time, at least on a century-long scale. Therefore, two peoples living 1,000 km away from each other could mark the same day with precise accuracy, and this calendar point will not change for very long.

Even if it is cloudy on or near the solstice, the calculation of the days before an event from a sighting a week or so before is easy. Marking solstices is simply a very effective way to structure a calendar of significant events in a stateless society without writing or other forms of formal communication. Marking solstices does not imply a more complex calendrical system, though such a system cannot be ruled out. We see the creation of interregional and interethnic systems of time reckoning that can likewise be built into the architectural landscape. The marking of a solstice is therefore stable and can be used very effectively as a way to permanently anchor a time for a special event that can become a traditional day. It is therefore not surprising that solstice marking is a widespread phenomenon in the social life of stateless societies in the construction of special places on the landscape.

Discussion

We can conceptualize the question of the origin of complex stateless societies as: What were the conditions that led to consistent, multigenerational co-operation between nonkin groups that gave us complex society independently many times in the Old World and the Americas? An important observation is that the processes were similar in independent areas. This means that there is something in the nature of human interaction under similar circumstances that gives rise to emergent complexity. As Flannery and Marcus note in their magisterial work on the origins of social inequality, people in Peru, Mexico, and the Near East all "had a series of recognizable behaviors in common. One of those behaviors was the building of ritual venues" (Flannery and Marcus 2012:121).

The development of complex society begins around 11,000 years ago in the Old World, as represented by sites such as Göbekli Tepe in Turkey, and about 5,000 years ago in the Americas, as seen in settlements such as Watson Brake in Louisiana and Caral in northern Peru. These settlements represent some of the first documented cases in prehistory where people co-operated with each other beyond their kin group on a consistent basis and over a considerable period of time.[34] Following the theoretical framework in this book, this co-operation was transmitted to other people, oftentimes not biologically related, over the generations. Such settlements flourished because people invented new social technologies based upon ritual that permitted increased group co-operation. It was at these times that the appropriate conditions to foster sustained co-operation developed. These co-operative relationships, these social technologies, were then subject to selection like any other tool.

Following the logic of the argument in this book, we can hypothesize a long period of experimentation with different kinds of ritually mediated co-operative organizations. In a context of rising population densities, environmental shifts, and other challenges and opportunities, organizations that were more effective in marshaling resources than others and that provided for sustained co-operation through feasting would survive and be imitated. We would also expect to see most of these organizations collapse under such selective pressures, only to reemerge in different forms and different places over the centuries.

These new and never-before-experienced conditions set up a new evolutionary dynamic in which there was strong selective pressure to develop co-operative labor organizations within a community. In this model, ritual, taboo, and feasting become the means by which these groups can structure their economic co-operation, while initially trade and then eventually war provide the outside resources to fuel the entire system. These stateless societies were able to maintain high levels of co-operation by creating economic rituals and social taboos and by rewarding co-operative agents via feasting and other activities. These rituals have many functions and are almost always associated with economic activities that require coordination and high risk. Leadership arose not as some necessary evil that people were forced to accept but rather as a consensual, managerial style of authority that functioned to schedule labor tasks and redistribute the results of specialized craft production and trade. And these efficient organizations were structured by ritual and rewarded by feasts.

I suggest that the increase in population densities at the end of the Pleistocene altered the costs and benefits of co-operative behavior. In the old Neolithic Revolution model, population pressure was a negative on people's well-being, and it forced people to adapt to this altered circumstance with agriculture, technology, social hierarchy, and eventually war. In this newer perspective, the population densities *allowed* co-operation between groups by drastically altering the cost-benefit landscape. So, in other words, prior to this time (say the first 50,000 years of intellectually modern humans), the least costly behavior when population densities increased in any area was to simply migrate to another area. But once the habitable areas in a region were effectively filled, the simple response of out-migration or fissioning became too costly and the benefits of co-operation became obvious.

A ritualized economy, in the absence of state coercion, regulates the redistribution of co-operatively produced goods in a way that is fair and understandable by the community. The taboos also serve to punish free-riders and allow the community to exercise some order in the way in which work, war, and trade are conducted, without resorting to coercive means, which is extremely costly from an individual and group perspective. In other words, in the absence of a coercive state, ritual, feasting, and taboo serve as the means by which social norms are encoded, and they provide the means to coordinate economic activities among autonomous households. Of course, the coercive state developed later, in the Bronze Age, and here we see an entirely different set of relationships in society.

In this regard, the architecture of these earliest sites makes a lot sense. These are not just simple villages where people decided to add some special buildings. They are places where periodic rituals were performed that brought people together and supported specialized craft production and trade, and where elaborate feasts occurred. It is not a coincidence that many of the sites investigated do not have one central building or pyramid but rather have numerous ones that were replicated across the site and landscape. Whether it is Göbekli Tepe, Caral, Poverty Point, or the numerous stone circles in the early Neolithic of Western Europe, the settlements literally scream of pilgrimage and performance by multiple groups. I argue that each of the groups were related and built their particular section of the site and that co-operation occurred between these groups via the rituals described above.

The empirical record is quite clear on the fact that the construction of ritual and/or ceremonial places is the first indication of complexity. The origins of complex society begin with modest settlements that attract people from afar for some kind of ritual activity that we are just beginning to understand. We also see in the empirical record that parallel to these rituals, people engaged in increased economic specialization and craft production, trade, and feasting. The production of stone objects is a particularly salient characteristic of these early experiments in complexity. Feasting, ritual, and the intensification of economic production developed together in a few places around the globe. Over time, the adoption of agriculture ensured substantially greater economic surplus, which fueled a competitive cycle, including raiding and war, which ultimately led to the development of state societies in some areas.

In true evolutionary fashion, there were many experiments with these new social configurations. We can predict many starts and failures, with intense oscillations between social complexity and a return to a more comfortable mobile way of life. In other words, we predict many instances where people tried unsuccessfully to create complex co-operative social relationships, complete with some kind of nondomestic public architecture, that lasted a short generation or even less. These settlements would develop and recede as people sought to promote in-group co-operation for their mutual benefit. We would expect this pattern to be found over a wide region and to go on for millennia.

The first few thousand years can be therefore considered a kind of experimentation with these new social organizations. Eventually, certain kinds of organizations were more effective at creating and concentrating resources than others. Through a process of economic or cultural selection, the most successful co-operative social configurations either absorbed neighboring groups or were imitated by others. (Empirically, the latter seems to be the most common.) As these forms of social organization spread, a new evolutionary dynamic is created. Selective pressures play a greater and greater role as the most successful groups outcompete others. This new evolutionary dynamic, first given theoretical coherency by Joyce Marcus (2008), is known as chiefly cycling in pre- or nonstate contexts. These successful organizations provided the context for thousands of people to congregate, crystallizing into the great towns of the later Neolithic in the Eastern Hemisphere or the Formative-period towns in the Americas.

States

This book ends with the formation of class-based societies that emerge out of complex stateless ones. These of course are what we customarily understand to be archaic states (Feinman and Marcus 1998). State formation is rare and is not an event but rather a long process that took centuries to occur. As with the emergence of complex stateless societies, there is something emergent in the nature of human interaction in certain contexts that leads in a few instances to the development of coercive, class societies. These cultural contexts, to be sure, include the existence of numerous, competing complex stateless societies in any region. What happens in state formation is the transformation of norms of co-operation to ideologies in which coercion becomes a major factor in maintaining the group. Coercion, conflict, ideologies of power, and so forth emerge as part of the state-building process.

We witness a set of changes in society understandable in anthropological game theory terms. First, there is the obvious change from small to large groups. This fact alone substantially shifts the costs and benefits of co-operation among individuals and groups. This is paralleled by a shift in the autonomy of the household and kin group. In small, stateless societies, individuals and small groups can voluntary associate with leaders. They likewise and can choose to not participate in ceremonial events, particularly the kind that obligate them for future labor or other resources. As population sizes increase and as production is increasingly specialized, the cost of non-co-operation for individuals and small kin groups goes up to the point that they are forced into accepting social relationships that are economically and socially disadvantageous to them. In a quasi-urban setting, for instance, in which many basic subsistence goods are produced in a complex chain of specialized labor, storage, and transport, nonparticipation in specialized labor by individuals is very costly, if not fatal. In a context in which interregional conflict is present, the possibility for out-migration or non-co-operation is practically nil. Therefore, the high-population-density environment of the early state transition is a context in which the bulk of people are indirectly coerced into economically and disadvantageous social organizations.

In game theory terms, we see a shift from simple non-co-operative game strategies to an environment where both non-co-operative as well as co-operative strategies come into play.[35] In the latter, various *coalitions* of groups co-operate among themselves. In the state formation process, coalitions develop out of stateless societies. These coalitions can be classes as understood in the traditional literature and/or can be groups of elites that emerge from the managerial leadership.

Henry Wright (1984:69) suggested that the development of states would be accompanied by a process whereby local ranked groups would merge into "a region-wide chiefly or noble class." Archaeological research in the last 30 years has indeed demonstrated that such local elite groups develop horizontal ties among themselves in a sort of mutual aid phenomenon. The emergence of a regional elite ideology that reinforces leaders' power is part of the process of elite class formation. Wright, in the same paper (Wright 1984:69), refers to this as the generation of "an ideology of chiefly sanctity sufficient to command regular tribute from commoner producers."

From an anthropological game theory perspective, we also see a shift in the ratio of strategies in the evolution of state societies. All forms of prosocial behaviors to enforce co-operation are a mixture of rewards (reciprocity, redistribution, economies of scale, etc.) and punishment (taboo, ostracism, etc.). The second great transformation in human social organization – to coercive states – is the change from the overwhelming use of collective action *rewards* in complex stateless ones to the predominance of *punishment* strategies in the evolution of states. Likewise, the demographic context in which this shift occurs also alters the costs and benefits of such strategies and behaviors. In broadest terms, the rise of the state is accomplished by increasing punishment and coercive mechanisms such as ideologies of power and hierarchy, military and policing power, ecological stress, and so forth.

This will have to be the topic of another book.

8 Epilogue: "No Beans, No Jesus"

> Some 15 years ago a missionary wanted to get members for his congregation. He started in by giving them meat, bread, beans and coffee. It was only a short time until he had his church filled. He found it was costing him more than he could afford and he cut off the meat; a few members dropped out. The expense was still more than he could stand and he cut out the bread. He still couldn't stand the gaff, and he cut out the beans. There was, only one old cripple left. The Missionary asked the old man what was the matter. The old man told him, "No beans, no Jesus."[1]
>
> – Reported by John Wetherill in 1936

This book outlines a way to understand the emergence of complex societies out of our egalitarian ancestors. It is firmly based on the concepts of evolutionary game theory and collective action theory in anthropology. It is informed by economic anthropology and economic theory. I conclude that, in the absence of coercion, self-interested actors create norms of collective behavior that result in greater benefits for all group members. As we know from the analysis of small-group behavior, the danger for sustained co-operation comes from individuals who free-ride on the group. We have seen, from both theory and empirical evidence, that the key to keeping a group co-operating involves a mix of reward and punishment.

I have suggested, again based on theory, observation, and historical data, that people in stateless societies *ritualize* these norms of co-operation as the most successful strategy to keep a group working together. By making these norms easily understandable and predictable for the members of a group, and by backing them with taboo and reward, co-operation is sustainable in the absence of political coercion. The archaeological record is replete with sites from stateless societies that demonstrate these principles. The art, architecture, and layout of complex stateless societies around the world demonstrate how people built special places on the landscape and pursued reward strategies to attract followers. These special places were communal, public, and very much different from ordinary buildings. Solstice time marking seems to be a particularly important component of scheduling feasts and other events in these societies. These special places and the norms associated with them constitute the ritualized nature of economic organization in societies with minimal formal coercion.

In this theoretical framework, ritual, or religion as some would argue, is central in keeping groups together in the face of the collective action pressures. The role that religion and ritual play in society in general remains highly controversial. On one side, neo-Marxists see religion as masking political and economic inequality and, in effect, allowing leaders to manipulate people into accepting class divisions. These are the so-called false consciousness models. In this view, leaders create and perpetuate ideologies of dominance in which they become intermediaries between the divine and the commoner. Elites manipulate commoners and insert themselves as an indispensable component of social, economic, political, and cultural life. It is as if farmers, for instance, genuinely believe that the crops will not grow if the chiefs do not perform their prescribed ritual functions. While such concepts may have utility in extremely complex archaic and secondary state contexts in which a small segment of the population can be manipulated, ethnographic and historical evidence strongly suggests that people in most societies are not fooled by such pretensions. Our ancestors were simply smarter than that, and there is in fact very little evidence to the contrary.

Functionalist theory, in turn, assumes that ritual plays a role in making a society more adaptive or efficient in the face of exogenous pressures or internal demographic stress. In these models, ideology and ritual reinforce the integration of society and decision making, giving such groups an advantage over others less integrated ones. In this framework, most people are assumed to be intelligent and reasonably rational, both socially and economically. They are smart, adaptive agents not so easily fooled into thinking that the elite are required to intervene in their daily lives for economic success. Following Malinowski and two generations of such theory in economic anthropology, I adopt a form of functionalism regarding ritual and religion. But it is necessary to reiterate time and again that that while most religious behavior is ritualized, the majority of ritual behavior that we carry out is not religious in the common understanding of the term. We go to work and we take our place in meetings in unconscious but prescribed ways. We converse, eat, drink, play, take care of our children, interact with others for our mutual benefit, and live life largely without formal rules of social behavior. We constantly code-shift, altering our behavior to conform to the expected norms of the social occasion.

As cultural beings, we implicitly understand the order of social things in our world. We act accordingly to those rules. We do not need explicit instructions for most ordinary interactions. And we were not born with these implicit understandings. Rather, we were born with the capacity to learn behaviors for survival and reproduction, just as we are born with the capacity for language in general but not for Russian or Swahili in particular. The capacity for language is genetic; any specific language is cultural. We learn these instructions through countless interactions in culture from the moment we take our first breath, and as we get older, we transmit these norms to new generations. As with language, the platforms for learning social behaviors are genetic; the specific kinds of social organizations that we create are overwhelmingly cultural.

I hope this book demonstrates that when these implicit, ritualized norms are used to regulate our economic behavior in small groups, we can maintain high levels of co-operation among our fellows. In stateless societies, groups of people who create and reinforce these ritualized norms of economic behavior well will create more resources and make a richer material and social life for all in the group – in these instances, these co-operative norms tend to spread. Norms that are less successful at maintaining co-operation in a competitive environment will disappear over time.

Obviously, religious sentiments have a very broad range of effects on people – from the material and mundane to the ethereal and spiritual. What concerns us here is how nonstate, noncodified, and nonsystematic religious beliefs are given form in ritualized social interactions, specifically economic ones. In this regard, I am the first to admit that the term "ritual" is problematic, and it is only with some hesitation that I have used it in this book. I searched in vain for another word other than "ritual"; in the end, English simply does not provide an adequate term. As my colleague Paula Sabloff once suggested, the term "habituated" might be more appropriate. But this term would have to be qualified to something like "culturally constructed habituation via prosocial costly behavior and various forms of taboo backed by fear of sanction. " The word "ritual" is therefore more parsimonious, and it does indeed convey the general sense that I want. So I retained it for simplicity. "Ritualized" means simply that common, recurring social acts can be internalized as cultural norms and that they can be carried out *implicitly*, without formal rules or guidelines. People internalize behaviors that are useful, and they ultimately reject or modify, in the aggregate, behaviors that are counterproductive and/or that do not work very well.

The danger with the word "ritual" is twofold. First, there is a strong and deep link between the words religion and ritual in English, and these have a common etymological link in classical Latin in the etymon *rītuālis* (*OED* entry for "ritual"). In common usage therefore, ritual is usually linked with religion and there is a vast nonscientific and esoteric literature that uses religion in the ancient world to promote odd, quirky, and even dangerous ideas about non-Western peoples. In that sense, a focus on ritual as religion can easily feed into the misperceptions about stateless peoples being superstitious or overly bound by "irrational" custom and religion, as I described in the beginning of this book.

This book was not written in a vacuum, of course. Over the last generation, there has been a justifiably renewed interest in the role of ritual and religion in non-Western societies in fields such as archaeology, history, and anthropology. Archaeologists have rightly noted many of the ritual features of early settlements. The role of religion and various belief systems is a serious and important field of study. As archaeology has progressed beyond the narrow focus on technology and subsistence, we are enriching our understanding by bringing in ritual behaviors in the past.

Unfortunately, these data have sometimes been interpreted in ways that ironically sound like nineteenth-century scholars who saw primitive peoples everywhere as bound by irrational myth and religion. It is irritatingly ironic that we are seeing some of these prejudices turning up in the serious archaeological literature yet again. In some cases, these interpretations border on variants of very modern and very Western New Age ideologies that are projected onto the ancients. The discoverer of the significance of Göbekli Tepe for instance, the late Klaus Schmidt, provided the world with one of the great archaeological sites of our time. He is to be commended as much as any field archaeologist can be commended. But he also almost immediately pronounced the site a "temple," ignoring evidence of intense lithic production and the other economic and more prosaic features of the site. It was not long before the press jumped all over that. In more than one occasion, the popular press is quick to point out that the site is near Urfa, "the birthplace of Abraham." If a reporter noted that Stonehenge was located near Glastonbury, the birthplace of Saint Arthur of Glastonbury, people would think it absurd. Yet, there seems to be little reaction to linking an 11,000-year-old site with an apocryphal Biblical figure.

As one reads the popular and some academic literature, one sees an attempt to reduce this vastly complex social phenomenon that evolved in millennia of historical context into a place where people went to solely to pray and commune with the gods. It was as if the 50,000-year-old tradition of making a living and enriching one's social life in the Paleolithic suddenly ended with a new revelation about the metaphysical world carved in stone and set in circles, and that this revelation was the spark of civilization. I choose the term "revelation" intentionally.

This New Age inclination is unfortunately common in some archaeological and anthropological circles in the Americas (for a particularly insightful analysis of the intersections of archaeology and religion, see Fowles 2013). Here, North American archaeologists, justifiably reacting to the excesses of some neoevolutionary materialist theory, have countered with extreme forms of postmodern silliness. In a semipopular but serious article about new discoveries in Cahokia, a science writer for the venerable magazine *Archaeology* quotes, "Our point is that religion created Cahokia, and it was intimately part of everything everyone did and thought, enmeshed and entangled so completely in daily living that everything was religion. . . . This Mississippian religion that began at Emerald and Cahokia was a new way of life" (Toner 2015:45).

The article even goes on to suggest that Cahokia sent out religious "missions" to other parts of the upper Midwest. The very concept of a religious mission is such a Western idea, without any real parallel in the non-Western world, that using it as an analogy for Mississippian artifact distribution is dubious at best. And the Christian theological metaphor behind this concept is barely disguised here – a new way of living, a new life, a "revelation" if you will, occurred at Cahokia, leading to the "rebirth" of this culture. The Christian fundamentalist metanarrative underlying this interpretation is barely disguised. We see the same in some semipopular press about Göbekli Tepe that refers to it as the "birth" of religion. The clear implication, of course, is that once some revealed truth dawned on, or was given to these people, civilization was born. From a New Age perspective, the ancients were liberated by religion and myth in a way that we can only hope to achieve in our overly secular and materialist modern world.

A recent article in a highly respected science journal reported that some ancient geoglyphs in Peru were for "prayer walking" (Geddes 2009). The quote probably did not translate well, but we see here this almost instant and unquestioning reference to some kind of esoteric, religious, and profoundly modern New Age behavior on the part of our ancestors. There is no question that the lines were most likely designed to walk on, but again, reducing a complex social phenomenon to what comes across as simple superstition is not backed by ethnographic or historical reality. Football fields in the Western world are walked on considerably, as are lawns on college campuses. The oracle site at Delphi was walked on quite a bit. There is a huge repertoire of behaviors that can be used to explain walkways in the ancient and modern worlds. Why do we seem to instinctively reach for the esoteric, prayerful, contemplative, peaceful, happy peasant model for people in premodern society? Is it because we project our own belief systems onto the past in an uncritical way? Is it because the joyful, religious peasants are no threat to our own economic and political hegemony in the twenty-first century? Whatever the answer, I repeat that there is no empirical basis for such fantasies. This kind of thinking is what I refer to as "phenomenological ethnocentrism." It eschews scientific method as hegemonic while adopting a phenomenological methodology to mask what are in reality simply worn Western projections onto distant people.

Many of us have been misquoted by science writers, and it understandable that these kinds of sentences get in print and are then erroneously attributed to our names. I am certain that our colleagues referred to above did not intend to precisely mean what ultimately came out in the popular press. But as we see in these writings, the idea that the "primitives" were permeated with superstition and religion while we are grounded in a more rational reality lives on as a triumphalist stereotype in the popular Western mind. The only difference from the nineteenth-century colonialists is the modern patronizing twist that the ancients were happier than us because they embraced their deep spirituality. Such modern views are a legacy from our colonialist past when the very concept of the "primitive" was created.

This impulse is unfortunately not confined to the Americas. One rather extreme example from the other hemisphere is a study of the famous Cretan Bronze Age peak sanctuaries. Here, the archaeologists followed ersatz "research" by a New Age practitioner on how to achieve altered states of consciousness by imitating some of the poses in Minoan figurines. The archaeologists then entered such altered states themselves after imitating the figurines on an actual peak sanctuary in Crete. This led them to conclude that there was a "transformative spiritual engagement between the populace and their ritual landscape" (Peatfield and Morris 2012). The fact that one of the authors leads goddess tours to Crete with the New Age guru is probably germane as well.[2] This is truly a crude kind of ethnocentric imposition of a Western spirituality fad onto a people far in the past. Once again, we return to the nineteenth-century concept of the primitive writ large, in this case offered not by some bothersome but charmingly quaint group of amateurs, but by actual professional archaeologists communing in altered states on a mountaintop in rural Crete.[3]

Fogelin's criticism of recent research in anthropological archaeology is particularly cogent here:

> From this [structuralist] perspective, religion seems to be everywhere and all pervasive in nonwestern societies. The problem with this perspective is the growing recognition (noted by Insoll 2004, p. 17) that some societies, even "traditional" societies, have only a limited interest in things religious (Barth 1961, Douglas 1982, Kemp 1995). The separation of church and state may be a modern, western notion, but it is a mistake to assume that people in other societies were or are necessarily ruled by their religious beliefs any more, or less, than are Europeans or Americans. (Fogelin 2007:60)

The venerable Mary Douglas (1973:36) is even more blunt: "The idea that primitive man is by nature deeply religious is nonsense." In fact, the bulk of people's time in stateless society is spent in more prosaic affairs, much like in our own society.[4] It is true that Firth and Malinowski both observed that religion and magic permeated life in the stateless societies they studied. But as I hope that I have demonstrated, these exceptions prove the rule: these ethnographers then proceeded to describe countless observations as to how these beliefs served practical ends, both material and social. Religion for the Trobriand Islanders, the Tikopians, or the Maori, was not the search for the purpose of life in a Western New Age sense, at least as it was described by them and by the ethnographers. Rather, their world was ordered by ritualized practices insofar as they served the purposes of survival and helped maintain a rich and meaningful social life.

We return to title of this epilogue. The sage old man and his profound observation capture a fundamental concept in the social structure of political and economic life in stateless societies: you can have all the religion that you can create, but without a solid economic foundation, no ritually based co-operative group can survive for long. Economic provisioning, in this case meat and beans, was required to create a rich and enduring co-operative group at the church. Feasting by a large and disparate group of people cannot survive without the economic wherewithal that undergirds these complex social interactions.

Likewise, the mere capacity to create surplus cannot sustain co-operation without a strong set of rules, ritualized at the community and supracommunity level, that creates the necessary conditions to sustain that system. The anthropological game theory of the small group teaches us that a set of norms consistent with the concept of conditional co-operation is necessary to create and maintain a vibrant economy in stateless societies. Such a relationship is codeterminative, or "recursive" in contemporary parlance. The sustained interaction of many people in stateless societies requires both a capacity to create the economic surplus that fuels this interaction along with the norms necessary to insure that these relationships do not collapse.

The preacher in our little anecdote built a great following by first providing food and drink for weekly feasts. For a brief while he had a great religious following and his special place, the church, was filled with people. But, like countless other aspiring chiefs in history, his luck ran out when his material resources were depleted – he learned a bit too late the iron clad law of faction-building and intensive co-operation in small groups: "no beans, no Jesus."

Notes

1 THE EVOLUTION OF HUMAN CO-OPERATION

1 The term "irrational" carries too many pejorative implications. I therefore propose the term "social rationality," which covers altruistic and other prosocial behaviors on the part of individuals.

2 As my colleague David Carballo points out, inclusive fitness has been recently criticized by Martin Nowak's group (see, in particular, Allen, Nowak, and Wilson 2013).

3 I thank Dr. Paula Sabloff of the Santa Fe Institute for suggesting the term "habituation" instead of "ritualization." The former term is superior in that it emphasizes the nonreligious aspect of many forms of ritual.

4 In an earlier time, we would have described this as how "man lived in a state of nature."

5 The term "polis" refers to Greek city-states, particularly in some kind of ideal form (see the *Oxford English Dictionary*).

6 Hobbes himself was more nuanced in his writings, which were quite extensive. He recognized that some people in society act in purely ego-directed ways, that some refuse to act except out of fear, that others are socially responsible, and that some are "vainglorious."

7 Explaining co-operation *between* related individuals is, relatively speaking, rather easy. Concepts in evolutionary biology, including kin selection, kin altruism, and inclusive fitness, allow us to understand degrees of co-operation between related individuals. Explaining altruistic behavior between distantly related individuals is substantially more difficult and is a core subject of this book.

8 Gintis nicely defines the distinction between rational actor game theory and evolutionary game theory vis-à-vis the concept of emergent complexity. "That game theory does not stand alone entails denying methodological individualism, a philosophical position asserting that all social phenomena can be explained purely in terms of the characteristics of rational agents, the actions available to them, and the constraints that they face. This position is incorrect because, as we shall see, *human society is a system with emergent properties, including social norms, that can no more be analytically derived from a model of interacting rational agents than the chemical and biological properties of matter can be analytically derived from our knowledge of the properties of fundamental particles*" (Gintis 2009:xiv, emphasis added).

2 ECONOMIC ANTHROPOLOGY OF STATELESS SOCIETIES: THE RISE AND FALL OF HOMO ECONOMICUS

1 "The canonical model in economics considers people to be rational and self-regarding" (Camerer and Fehr 2006:47). The terms "selfish" and "ego directed" are also used to mean "self-regarding." In this light, the economic model was consistent with other dominant scientific ones at the turn of the last century, particularly evolutionary biology. Both of these traditions, at base, required individual actors to be ego directed. We will address aspects of this observation at greater length in subsequent chapters.

2 Georges Fauquet (1941:169) describes the historical progression as beginning with "peddling," soon to be supplanted by regular trade, which becomes primitive markets, which in turn evolve into big fairs. He then argues that "commercial towns arise at the cross-roads of the land and merchant cities grow at the cross-ways of the seas."

3 It is quite fascinating that immediately below this famous aphorism, Smith addresses the surprisingly "modern" issue of whether this propensity is biological or is a product of human creation: "Whether this propensity be one of those original principles in human nature, of which no further account can be given, or whether, as seems more probable, it be the necessary consequence of the faculties of reason and speech, it belongs not to our present subject to inquire." This fascinating aside does betray an important aspect of Smith's theoretical framework. In this view, people can create, through reason and action, characteristics that effectively become part of human nature, and these have the same causal significance as characteristics that are part of our genetic endowment.

4 Both Rousseau and Hobbes espoused different types of social contract theory. The former said that people should subordinate themselves to the common good and implies some kind of altruistic motivations; for the latter, this subordination was done by the necessity to preserve life.

5 Gary Urton and John Papadopoulos (2012:5–6) make the interesting argument that modern economic analysis begins with Plato's *Politics*.

6 This quote is actually an amalgam of two of Spencer's sentences.

7 This concept of evolution is very distinct from that of Charles Darwin. In Spencerian models, evolution is an unfolding of what are essential characteristics. Just as the acorn becomes the oak, cultures unfold over time to express their inherent traits. In this sense, the least evolved societies, such as savages, are incompletely developed forms of more complex societies. This process is similar to what many economists and economic historians adopted vis-à-vis primitive economies in the nineteenth and twentieth centuries, in the sense that these were incompletely developed forms of more advanced economies.

8 In an interesting essay, Horwitz (1995:367) discusses the tension between seeing the production of objects by indigenous Americans as an indication of rational economic behavior and the inclination to "distinguish primitives (natural man) from European Americans (economic man)."

9 At this point we have to make the distinction between economic and colloquial definitions of "rational." Definitions of rational behavior in Western economic theory include terms like "maximizing material resources" and "making logical and prudent decisions" to achieve this goal. In classic or canonical economic theory,

individual decisions are based on the pursuit of material advantage. The colloquial understanding of "rational" is broader and recognizes nonmaterial "utilities" (see Chapter 3) and collective decision making, or at least aggregate decisions made by many individuals. For an in-depth discussion of these issues, see http://plato.stanford.edu/entries/economics/#5.

10 Mitchell is referring to William McDougall, the author of the celebrated book *An Introduction to Social Psychology* (London: Methuen & Company, 1908).

11 In his influential book 1904 *Die entstehung der volkswirtschaft* (*The Emergence of Economies*), Carl Beucher argues that "primitive" peoples did not have any natural tendencies for barter or exchange. These societies, he said, were self-sufficient or autarkic at the household or family level. This is obviously incorrect.

12 The term "substantivist" did not gain general use until the early 1960s.

13 A good case could be made that the "other directed" camp also drew upon Western theological traditions, but this is a subject for another publication.

3 CONDITIONAL CO-OPERATORS: THE EVOLUTIONARY GAME THEORY REVOLUTION

1 At the risk of sounding overly simplistic, one example would be the tradition of the reciprocal gift. Anthropologists have exhaustively studied gift exchange. I side with the camp that understands that gifts are given with the expectation of gifts being received at some future date. There is no pure gift.

2 In the last decade or so, economists have been incorporating this assumption into their models of human behavior. One of many examples: "All this evidence suggests that people are not motivated solely by material self-interest" (Dufwenberg and Kirchsteiger 2004:269).

3 Kin selection is a form of altruism by one individual directed at related individuals. Reciprocal altruism occurs when an individual acts in an altruistic way to another in the expectation of a future reward from that individual.

4 There are some exceptions, such as Wilson and Dugatkin (1997) and Wilson and Sober (1994).

5 Of course, I confine my focus to the Holocene. Prior to this, there was obviously substantial changes in the evolved psychology of humans.

6 The theoretical basis of the PD was developed by M. Flood and M. Dresher in 1950 while working at RAND Corporation. The term "prisoner's dilemma" is attributed to Albert Tucker. For more information, see Tucker (1983).

7 The famous "Dunbar number" of around 150 people with whom one person can maintain meaningful social relationships is probably at the low end. Other work suggests higher numbers, but not much higher. At any rate, we can see how that after a relatively small community size is reached, the underlying assumptions of the IPD will weaken.

8 The idea that costly punishment actually increases payoffs is debated and largely falls along group-versus-individual selection theoretical positions. Dreber et al. (2008), for instance, report on an experimental game and conclude: "Our data show that costly punishment strongly disfavours the individual who uses it and hence it is opposed by individual selection in cooperation games in which direct reciprocity is possible." Egas and Riedl (2008:876) are even more explicit: "Our results show

that the total pay-off in a group is the lowest when altruistic punishment successfully enhances cooperation. . . . In this sense, our result sheds some doubt on group selection models for the evolution of cooperation by altruistic punishment, where sanctioning within a group is supposed to increase cooperation and thereby the competitive strength of that group."

9 There is also a methodological issue here. Conducting UG or dictator games in your own culture avoids certain biases on the part of the participants that stem from cultural misunderstanding about the nature of games and such research in general. A dynamic instantly arises among Westerners and people in different cultures. Unanticipated false expectations by people playing a game may also skew results toward what participants feel they should do, as opposed to acting as they normally do in such social situations. As someone who has worked and interacted with people in non-Western, small-scale societies for over 30 years, I have seen this firsthand on innumerable occasions.

10 Abstract at http://econpapers.repec.org/bookchap/oxpobooks/9780199262052.htm (accessed March 19, 2014). Note that the abstract appears to have a typo. The original sentence makes sense when one removes a "that" where the ellipsis was inserted in the quoted text.

11 It is perhaps a bit ethically troubling that such "primitive" behaviors in others become "reasonable" only after we recognize similar behavior among ourselves.

12 "We report several results. First, our design worked across all stakes levels: the *median offer* was 20 percent of the pie, and the 95 percent confidence interval was 2.5–50 percent of the pie. Interestingly, while the offer proportions are significantly lower in the higher stakes treatments compared to the lowest stakes treatment, the actual amount offered *increases* as stakes increase. Second, among responders we find a considerable effect of stakes: while at low stakes we observe rejections in the range of the extant literature, in the highest stakes condition we observe only a single rejection out of 24 responders" (Andersen et al. 2011:3428).

13 In another study from China, experimenters found, "Focusing first on the effect of prize level, statistical tests indicate that subjects were more risk-averse when real monetary prizes were multiplied by 10" (Kachelmeier and Shehata 1992:1129).

14 The work of Henrich and his colleagues has created this healthy debate in evolutionary psychology that is thankfully tangential to this book. This is the debate as to whether results of experimental games reveal a universal evolved psychology that is expressed in different ways in different cultures or whether cultural norms affect the very nature of the evolution of human cognition? (See Watters [2013] for a readable, nontechnical review of this debate.) As someone who has spent years living and working in small non-Western societies, the second option seems the most obvious to me. However, while fascinating, this debate is moot vis-à-vis our understanding of small group co-operation in stateless societies because regardless of the ontological status of the behavior, the effects on co-operative behavior are the same.

15 For instance, Henrich found that in cultures where gift giving was used to create social debts, respondents in the UG would refuse offers over 50 percent. This is not because they were not acting out of self-interest. Quite to the contrary: they were avoiding debt obligations to the proposers. In cultures where strict debt obligations are not created by generosity, respondents almost always take offers above 40 percent.

4 THE ROLE OF COERCION IN SOCIAL THEORY

1 One could argue that kings and other despotic leaders in state societies were deposed as well. In the case of stateless societies, this occurred on a much more frequent basis and did not require organized violence or rebellion to occur. The chiefs were simply replaced.

2 A good example here would be the Trobriand Islanders as studied by Malinowski and the Trobriand Islanders as studied by later ethnographers. Through the work of ethnographers like Annette Weiner (1988), we can see the changes wrought by colonial and postcolonial administration and globalization.

3 Meggitt provides an extensive observation of Big Man behavior, illustrating how Big Men are not simply aggressive entrepreneurs acting in their own self interest but instead are leaders who negotiate a range of social and political pressures to succeed. To earn prestige and material reward, they must behave in prosocial ways that benefit a very wide swath of people in their group:

> although the Big Man is an obvious figure in Te exchanges in that considerable numbers of valuables enter and leave the clan through his transactions, other men in the group also receive and disburse wealth on these occasions, albeit on a reduced scale. Some men, especially those who aspire to be Big Men, may appear to act along in this whereas others may join forces and pool their resources to make a better showing. Nevertheless many of these men are at the same time likely to contribute to, or to benefit from, the dealings of the Big Man. Moreover, the latter is not only entitled but is also bound to direct clan policy by making statements such as: "It is in the long-term interest of our clan as a whole to ensure that so many pigs go to such-and-such Te partners in clan X rather to men in certain other clans"; and he may use persuasion, promises or threats to achieve this end. Thus, his clansmen, even when they seem to be reaching independent decisions as to the disposition of their valuables, have strong pressures on them to balance these public requirements against their own private desires and so reach compromises more or less compatible with both sets of interests. (Meggitt 1974:189)

4 Hayden has an impressively long list of publications that outline his theory. These can be found in his book (Hayden 2014).

5 As we mentioned briefly in the first chapter, people can be forced into doing things not in their self-interest by exogenous factors – population pressure, environmental degradation, and so forth – or by endogenous ones, by other people in and out of their group. In the latter case, the result usually centers on the involuntary contribution of labour or tribute by one person or group of people to another person or group of people.

6 It is of course a bit more complex than this, but this observation is a central one in social evolutionary theory. I will discuss it at greater length later.

5 THE RITUALIZED ECONOMY: HOW PEOPLE IN STATELESS SOCIETIES CO-OPERATE

1 Alesina's (2013:428) modification of the Weberian position to focus on the effects of labor capital accumulation that resulted from Protestant teachings is quite cogent. This refinement provides specific mechanisms whereby religious beliefs can affect macroeconomic shifts in society: "Lutheran doctrine (promoted by

German reformer Martin Luther) influenced the accumulation of human capital in regions that adopted Protestantism.... [Becker and Woessmann 2009]. The German sociologist Max Weber argued that the teaching of hard work and frugality, and especially the view that economic success in life was a sign of being chosen for a next life, spurred entrepreneurship. Perhaps it was not the religious teaching and moral values of Protestantism that created an economic growth impetus, but rather the latter's emphasis on the accumulation of human capital, of men and women."

2 There is not enough space to discuss the various functionalisms of Malinowski, Radcliffe-Brown, and others. Malinowski (see Firth 1975; Radcliffe-Brown 1949) defined functionalism in terms of what institutions contribute to subsistence and survival. Radcliffe-Brown and the structural functionalists, in contrast, defined functionalism in terms of what a cultural institution contributed to keeping society integrated. The term "functionalism" as used here emphasizes the role that various cultural institutions, such as a ritualized economy, play in the subsistence and social life of a group. Epistemologically, it is always important to remember that the function something plays does not necessarily explain its origin or existence.

3 Weber's ideas continue to provide insight into modern economies. Renneboog and Spaenjers, for instance, investigated the role of religion in the financial decisions of religious and nonreligious households, concluding that there were significant differences in the Dutch households they investigated. A number of other recent studies (Renneboog and Spaenjers 2012:103, note 1) support this correlation.

4 Amanda Cohen (2010) takes Bell's concepts and reworks them for a very interesting archaeological analysis.

5 The observation that "raiding and trading go hand in hand" is a common one in the ethnographic literature.

6 Malinowski was instrumental in bringing the concept of the "pure gift" to the center in economic anthropology. This is a gift "in which an individual gives an object or renders a service without expecting or getting any return" (Malinowski 1922[1966]:177). The celebrated Marcel Mauss (1950) dismissed the concept of the pure gift as consequential in economic exchange theory. The vast bulk of gifts in society serve to discharge a social or economic debt or to obligate others in a series of reciprocal obligations. This is one of the key means by which small groups structure their co-operation.

7 Technically, in really large groups of workers, the payoff can be exponential.

8 Some theorists suggested that economies of scale were also created by an increase in the skills of workers over time. This is theoretically likely but not relevant to most instances in the economies of stateless societies.

9 "The greatest improvement in the productive powers of labour... seem to have been the effects of the division of labour" (Smith 1937 [1776]:volume 1).

10 Vincent Pigott (1998) uses Burton's model to interpret second-millennium BCE copper mining in prehistoric Thailand.

11 In an earlier book (Stanish 1992:18–23), I provide numerous examples in which the household is the basic unit of production and consumption in Andean societies, both in the historical period and the ethnographic present. I also offer a long argument that the household was the basic unit in the distant past from at least the Early Horizon on. Andean peoples developed a very sophisticated norm of reciprocity called by various names – most commonly *waje waje*.

12 My article on household archaeology and a book on a similar theme (Stanish 1989, 1992) extensively develop this argument, and the details need not be repeated here.
13 According to his map, the Tikul Batu Ulu longhouse was approximately 12 × 42 m and contained nine households, with six additional separate households built adjacent to the main structure.
14 Families in this economic sense are not necessarily composed of biologically related individuals. Families also include fictive kin, adoptees, and even captives.
15 The Tsimshian practice avunculocality. This is why each lineage houses maternal nephews of the chief.
16 The term "moiety" is used to refer to the division of a group in two under any principle, including kinship, though most often moieties involve descent groups.
17 Hegmon cites Eggan (1950:109) and Whiteley (1985, 1986) in this quote. Whiteley goes further and does not see as many economic functions residing in the matrilineage as the earlier scholars. He feels that this is an imposition of African structural-functionalist models and quasi-evolutionary worldviews ascribing too much economic control to corporate groups. In his view: "Neither 'clans' nor 'lineages,' therefore, can be regarded as the operative economic units [for animal husbandry] (Whiteley 1985:371). He sees these resources as individually owned by households. Either position supports the view that the Hopi had strong household organizations and larger social institutions that dealt with political and ritual issues.
18 Hämäläinen (1998:490), for instance, writes: "From about the late 1730s to the late 1770s, a peculiar pattern of alternating trade and warfare dominated Comanche-New Mexican relations." Ewers (1966:3) notes, "Intertribal trade was almost as typical of Northern Plains Indian life as was intertribal warfare."
19 Evans was truly a most unsympathetic figure in the history of early-twentieth-century anthropology. His book is full of racist diatribes and reeks of ethnocentrism and race and class superiority. One must use this source with extreme caution. Therefore, I pull from it only these few empirical references to actual behavior. I do not use the heavily biased interpretations or suppositions that permeate his work.
20 The only "nonlocal" element is that the rituals are Islamic in character, but there was no evidence of outside influence when these ethnographies were conducted.
21 Moreno (1987:53) takes the concept of ritualized agricultural cycles a step higher regarding Indian rice cultivation: "While this reproductive view of agriculture is not unique to India, but is found elsewhere in human societies, an agricultural cycle in India could perhaps be more genuinely conceived as a ritualized process, a sacramental rite of passage or samskara effected on the divine body of the Earth."
22 A corollary is that as societies become more hierarchical and coercive, ritual organisation of the labor process shifts considerably. This parallels the shift from complex stateless societies to state organizations permeated by coercive authority. Magic as a means of punishment is less important, though the ideology of reciprocity is oftentimes maintained. This concept is illustrated by a quote from Evans-Pritchard:

> It is clear that the communal garden magic of the Trobrianders is absent from Zande life because the Azande do not cultivate their gardens by joint labor. It is more difficult to see why the Trobriand chief uses magic as part of his machinery of government, whereas the Zande chief does not use this weapon of chastisement, but this difference can readily be understood when the position of the chief is known in both

societies. The Trobriand chief was unable to exercise great executive power as may readily be judged from the almost entire absence of corporal punishment, whereas the mutilations and executions inflicted on their subjects by the Zande chiefs are one of the many signs of their real power. *The Zande chief, therefore, had little need of magic to enforce his rule....* Magic used in communal undertakings such as we find in the Trobriands, in garden work, in trading expeditions, in building canoes, and in other forms of economic enterprise has no counterpart in Zande life. (Evans-Pritchard: 1929:638–639, emphasis added)

23 On one extreme side, scholars like Stephen Reyna and R. E. Downs (1999) see all non-Western conflict as nonlethal or merely as the result of personal violence. They argue that all war seen today in non-Western peoples is due to the negative influence of capitalism. At the other side are people who view war as an inherent feature of at least male-evolved psychology and a major factor in cultural evolution. Most fall somewhere in between, seeing war as common to almost all cultures but highly variant across space and time (Arkush and Stanish 2006; Keeley 1996; LeBlanc 2003; Vayda 1974). What we can say with some assurance is that organized conflict such as raiding has been documented in the vast majority of cultures and as far back as the Epi-Paleolithic in the Old World and the Archaic in the Americas.

24 The majority of participants in raids are young men between 15 and 30 years old, although there is evidence for some women participating as well in some cultures. Sociobiologists downplay evidence for women's participation for obvious reasons. When females are found in archaeological contexts that would suggest female participation in raids, sociobiologists respond that an increase in conflict results in an increase in domestic abuse. Maschner and Reedy-Maschner (1998:23), for instance, argue that "evidence for spouse abuse, in the form of skeletal injuries on females, rises during periods of warfare (Lambert 1994:116–119; Lange 1996). It has also been argued that the stress of warfare, or at least the perception of stress, leads to increasing evidence of spouse abuse (Ember and Ember 1992)." I find this argument a bit specious at best, an attempt to take obvious contradictory data and neutralize them with an inelegant and ad hoc argument.

25 It is useful to distinguish between raiding and warfare. Both are organized forms of violence directed at other people to obtain some kind of material, social, or political advantage. Warfare generally implies territorial aggrandizement, while raiding implies a search for booty.

26 The research for this work was carried out in the 1930s. The investigators were interested in a linguistic analysis of how boys were trained in the art of raiding. The raiding itself took place before the late nineteenth century.

27 Another debate focuses on whether European influence intensified or even caused war among native North American groups. The evidence, in my opinion, is quite strong that patterns of organized violence well predate European migrations (see Bamforth 1994). Raiding was endemic in many stateless societies around the world, going back in fact to early hunter-gatherers in the archaeological record.

28 As we have seen, stealing is another form of exchange (that is, something of value moves from one person or group to another), but one that is not mutually beneficial and therefore highly unstable as a social mechanism.

29 Parts of this section are paraphrased and revised from Stanish and Coben (2013).

30 See, e.g., North (1977) and Cook (1966).

31 Currency is "money actually in physical circulation or capable of being put into circulation" (Einzig 1949:327). The first coinage was developed in the sixth century BCE in Greece and India. Money has many functions in nonmarket societies. It is used as a unit of value, a unit of account, a unit of payment, a medium of exchange, and a symbol of status, and it encodes history of use (like Kula "money").

32 Barter exchange is also extremely common in all noncapitalist economies and is found in many nominally capitalist societies that have large populations outside of urban centers. We also find forms of it in developed Western economies as well, though it is substantially reduced in significance. An example would be birthdays, which have customary rules of reciprocity. While the objects are obtained in price-making markets, they are subsequently exchanged as rough equivalencies in a complex system of deferred reciprocities mediated by complex social rules.

33 Chapman's article is a masterpiece of opaque and at times incomprehensible prose. In spite of this, it is in my opinion a very important contribution to economic anthropology literature. It has been ignored due to the structure of her argument and the style of writing, not the substance of the argument itself. She presents data and makes comments that seem to contradict some of Polanyi's most fundamental points, yet she constantly returns to a defense of his position. When she says, for instance, "Barter is distinguished from other types of direct exchange, because it is a purely economic transaction, involving no mutual obligation between the partners" (Chapman 1980:39), she directly contradicts the substantivist notion of the embedded economy. I happen to agree with her that barter is indeed not "embedded" in the sense suggested by Polanyi. Nevertheless, her basic point about barter is crucial to understanding how stateless societies exchange goods at a regional level.

34 One reason economic anthropologists have avoided much discussion of barter fairs is an implicit assumption that these are a Western and European phenomenon. As such, they would not be a good analogy for non-Western societies. In reality, we find that the behavior of people from different villages or ethnic groups meeting in a common, predetermined place to exchange goods and other valuables is virtually a cultural universal in complex hunter-gatherer and settled village societies. From an anthropological game-theoretical perspective, it does not matter where the behavior takes place. As long as the socioeconomic and political context is similar – small groups in preindustrial environments, for instance – the behavior can be a useful analogy for any people who create strategies in a similar social and natural environment.

35 Some scholars believe that such trade centers were not indigenous but rather reactions to European expansion. However, archaeological and historical data suggest otherwise. Dobyns's view is held by most scholars: "Europeans could establish trading relationships with Native Americans only because the latter already engaged in international trade on a significant scale and well understood the principles of commodity exchange" (Dobyns 1984:23).

36 And see Schneider (2002:66) for a different interpretation of this "neoformalist" assumption.

37 David Graeber (2011:30) argues that barter occurs only between groups of strangers "who might otherwise be enemies." It is somewhat odd that in a 500+ page book on

anthropological economics he does not mention the Trobriand Islanders, the undisputed champions of non-Western barter exchange economies. The Trobriand Kula system was a massive, complex, and autochthonous barter system conducted over long distances by decidedly nonstrangers who formed partnerships for life and even passed them on to heirs.

38 Annette Weiner (1988) describes the Kula after the introduction of speedboats and market economies. The traditional Kula exchange adapted to these changed conditions.

39 From North (1977:712), citing a reanalysis of Malinowski's data in Belshaw (1965).

40 As an archaeologist, I have spent up to 12 weeks at one stretch living in isolated villages in a beautiful part of the Andes. After only a few weeks, it becomes abundantly clear how important are the structured social occasions where new people are met, exotic foods are consumed, new friendships are created, and so forth.

41 *Pechara* or *bechara:* "a lawsuit, decision in a suit, v. to hold court, decide a case" (Howell and Bailey 1900:119).

42 Western economists have recognized nonmaterial motivations in modern society. The concept of "hedonic utility" is important here (Reichhardt 2006; Wolpert 2010). It can be defined as the amount of satisfaction (utility) one receives from experiencing pleasurable sensations. It is a concept surrounded by some controversy because it is difficult to measure, for one thing. On the other hand, it is difficult to dispute the fact that people are motivated by pleasurable experiences, and it is incumbent upon economists to deal with this messy cultural and psychological phenomenon.

43 The Pathan Hill tribes described in the 1970s were effectively stateless societies living in the boundaries of modern but utterly weak states.

44 As the economist Jon Elster (1989:107) observed: "While useful, the [act of] ostracism is not rational."

45 The authors cite Zahavi (1975), Grafen (1990), and Johnstone (1997) in this quote.

46 Of course, the 21 percent that did not have these special places demonstrates the variability in the kinds of organizations created by stateless societies. Such variation is the core of evolutionary selection, described in the following chapter.

47 See http://southernsudan.prm.ox.ac.uk/details/1967.26.187/ for more information.

6 AN ANTHROPOLOGICAL GAME THEORY MODEL FOR THE EVOLUTION OF RITUALIZED ECONOMIES

1 This of course describes a methodology in science that emphasizes empirical observation (induction) as the key component of the scientific process. Arthur describes this quite elegantly:

> In problems of complication then, we look for patterns; and we simplify the problem by using these to construct temporary internal models or hypotheses or schemata to work with. We carry out localized deductions based on our current hypotheses and act on them. As feedback from the environment comes in, we may strengthen or weaken our beliefs in our current hypotheses, discarding some when they cease to perform, and replacing them as needed with new ones. In other words, when we cannot fully reason or lack full definition of the problem, we use

simple models to fill the gaps in our understanding. Such behavior is inductive. (Arthur 1994:406–407)

2 Hill et al. (2011) list a range of 5.8 to 81.6 adults, with a mean of around 25 for the "mean experienced band size" in a study of 32 contemporary and/or historic hunter-gatherer societies.
3 Richerson, Boyd, and Bettinger (2001:389) use the term "competitive ratchet" to define the establishment of a new evolutionary dynamic such as this. This is a useful term. It captures the reality that strategies employed by groups have to shift in response to changing social and environmental factors.
4 Hill et al. note that anthropologists assume that most people in hunter-gatherer band societies are related. However, their review of 32 societies indicates that a surprisingly high percentage of individuals are not related. Therefore, they conclude, "these patterns produce large interaction networks of unrelated adults and suggest that inclusive fitness cannot explain extensive cooperation in hunter-gatherer bands" (Hill et al. 2011:1286). This observation supports the theoretical position of this book focusing on the learned capacity of agents to develop strategies of co-operation among both kin and nonkin.
5 The capacity to alter our environments in such marked ways is another feature that distinguishes humans from other social animals. The recent development of niche construction theory promises to refine our understanding of this process.
6 There are many instances in nature where nonhuman agents alter their environment. While it is certainly a question of degree and not kind, the degree is so substantially huge when humans are involved that it is virtually a moot, semantic argument about the role of humans in changing their environment.
7 Or to be precise: Jean-Baptiste Pierre Antoine de Monet, Chevalier de Lamarck (August 1, 1744–December 18, 1829). For more details, see http://en.wikipedia.org/wiki/Jean-Baptiste_Lamarck (accessed December 21, 2014).
8 The third entry for "evolution" in the OED is "The process of unrolling, opening out, or revealing." The etymology is from the classical Latin *evolution*, or "the action of unrolling a scroll (Cicero)."
9 Ironically, Darwin himself did not like the term "evolution" due to the implication of "progress" that its own etymology implies.
10 One could make a distinction here between "biological" and "economic" fitness. The former refers to the reproductive success of an individual and his/her kin group; the latter refers to the resource success of an individual and his/her kin group.
11 The entire group selection debate began a generation ago with a number of papers. David S. Wilson and Elliot Sober (1994) argue, for instance, that humans can replicate their group strategies through extragenetic means, specifically the inheritance of successful group strategies.
12 Such a position is supported by work reported by Marsh et al. (2014), in which intense altruistic co-operators exhibit neural mechanisms that are opposite that of psychopaths.
13 "*Homo reciprocans* could have evolved during the Paleolithic era in a similar way: reciprocators could have proliferated in a population, even if their fitness was reduced by upholding norms, so long as reciprocators were more likely to find themselves with other reciprocators in well ordered groups capable of surviving bouts of material scarcity and attacks that norm-flouting groups could not endure" (see, describing Bowles and Gintis's work, Shalizi 1999:19).

7 THE EVOLUTION OF RITUALIZED ECONOMIES: THE ARCHAEOLOGICAL EVIDENCE

1 The figure of 200 years is the approximate inherent error (2 sigma) in most carbon dates, unless they are statistically analyzed using Bayesian or other kinds of techniques (see Levine and Stanish 2014).

2 I use the dating convention of the research area that is discussed. Therefore, dates are provided in cal BP, cal BCE or in uncalibrated dates depending upon the published materials.

3 This is the cultural area from roughly Ecuador in the north to Antofagasta, Chile, in the south.

4 "alrededor de 3400 a.C. (calib.), en el Primer Edificio de Sechín Bajo se utiliza un tipo particular de adobes en forma de pirámide trunca alargada, prensada con los dedos, junto con adobones bolsiformes."

5 These dates are uncalibrated as reported.

6 This section is modified from Stanish and Haley (2005:63–64).

7 Some data discussed here were presented by Dr. Shady at the conference "Avances en la investigacíon de Perú prehispánico, un estado de la cuestíon," sponsored by the Ministry of Culture–Cusco in April 2015.

8 This information was presented at a conference sponsored by the Ministry of Culture in Pisac, Peru, in April 2015.

9 The means by which the community on the Island of the Sun established a fair division of the work was quite ingenious. The construction of a community building was slated for a hillside with a very uneven surface. The community leaders effectively negotiated and eyeballed the volume of earth and then marked out rectangular sections with chalk on the ground. The width of each rectangle correlated to the slope of the hill – a higher slope meant a more narrow area that each group had to move and hence an equivalent volume for each section of the community. Curiously, it did not correspond to the size of the group, but each group was responsible for an equal amount of ground. Once the work began, there was a very friendly competition to see who could finish first.

10 Also called sunken courts, sunken temples, or sunken plazas.

11 Using OxCal 4.2.4 at a 95.4 percent probability.

12 Arkush and Tung (2013) in their article use "at least 1500 BCE" as the date for the art on the Cerro Sechín site. As described in the book, I use the date from the base of the "temple" as recorded by the excavators. It is possible that these carvings were later – it is extremely difficult to determine dates of monumental architecture that was around for a many centuries. However, the date used in this section is the best approximation that we have.

13 The carbon dates for the site (Nuñéz 1984:132), using modern calibrations at the 95.4 percent confidence level (OxCal 4.2), are quite broad. They range from 780 to 388 BCE, 407 to 766 CE, and 20 BCE to 321 CE.

14 Many other geoglyphs and petroglyphs are most likely later in date, in the Late Intermediate period and Late Horizon (Zori and Brandt 2012). Others in the region may be earlier, based upon stylistic similarities to Formative-period traditions.

15 Zori and Brandt (2012) concentrate on the post-Formative periods; we suggest that data presented by Nuñéz from Pircas indicates that this process occurred in the Formative as well.

16 The above discussion is modified from Stanish et al. (2014).
17 The southernmost site is now destroyed but was documented by a team in the 1980s (Canziani 1992).
18 The site was first described by Dwight Wallace (1971) and later by José Canziani (1992).
19 Other sites in the region have unequivocal evidence of habitation. These domestic structures are easy to discern in this environment.
20 Recent research makes it likely that even earlier complex sites will be found in East Asia. An example would be pottery-bearing levels at a site in Hunan province, China, that has associated carbon dates of as early as circa 21,000 BP. Several other recently discovered or reworked sites in other areas of the country have dates at or before 15,000 BP (see Boaretto et al. 2009; Kuzmin 2006; Wu et al. 2012). If these early dates are even remotely accurate, then it is highly likely that the poorly studied regions of China will contain complex settlements millennia earlier than those in Anatolia.
21 This was reported at the 2013 Shanghai Archaeology Forum by Professor Colin Renfrew and the late Dr. Klaus Schmidt.
22 It is also of course possible that the ore was valued as a raw material itself, though this seems a bit unlikely.
23 One of the common mistakes made by modern archaeologists is to deny a defensive function for a walled settlement if the wall does not completely encircle a site, or if there are “too many” doorways. A brief study of military history reveals, in fact, that these are misconceptions (see Arkush and Stanish 2005).
24 J. Saunders (2012:32) notes that the Conly site in Louisiana may be as early as 6000 BCE, but much work remains to be done in this period.
25 As cited from Lauro and Lehmann (1982:28).
26 These conclusions are based upon a 20-cm contour base map provided by the Center for Archaeology at Tulane and Google Earth. Calculations of the position of the sunsets in 1700 BCE were made with Starry Night Pro software. There is an imperceptible variation in the positions over the last 4,000 years.
27 Tick island was renamed to the Harris Creek site.
28 North American archaeologists have recognized the significance of feasting in these early special places: “Finally, feasting, possibly combined with the completion of construction projects or as periodic ritual where commodities are exchanged, is an activity that serves to cement social relationships and reinforce social roles” (Lindauer and Blitz 1997:185). “Additional evidence ties on-mound food consumption to ritual activities. For example, at the late platform mound site of Osceola, the remains of maize and tobacco seeds are confined to mound contexts, lending support to the hypothesis (Rose et al., 1984) that ‘maize was adopted as an elite or ceremonial food in the Lower Mississippi Valley’ (Fritz and Kidder, 1993, p. 9)” (Lindauer and Blitz 1997:186).
29 A barrow is a mound of earth or stone erected over a grave (OED).
30 “Synodic period, the time required for a body within the solar system . . . to return to the same or approximately the same position relative to the Sun as seen by an observer on the Earth” (Encyclopaedia Britannica, “synodic period,” www.britannica.com/EBchecked/topic/578497/synodic-period).
31 Similar observations were made about the Hopi lunar calendar and solstice shrines, as well as the Pawnee earth lodge as an observatory (detailed in McCluskey 1977, 1982, 2015).

32 Aveni (2000b) is an excellent source for more examples, particularly the many instances recorded among the Maya of Central America.
33 Timothy Pauketat (2012) discusses a number of solstice and other astronomical alignments in Mississippian cultures of North America.
34 Of course, research progresses all the time, and new sites that alter this picture will be discovered.
35 I gratefully acknowledge my conservations with David Wolpert of the Santa Fe Institute in this discussion.

8 EPILOGUE: "NO BEANS, NO JESUS"

1 As reported in http://archive.org/stream/southwesternmonu1936depa/southwesternmonu1936depa_djvu.txt. The quote by Wetherill on page 117 is even more charming since he ends the story with, "I think that is the matter with my report 'No beans, no report.' (Or a darn poor excuse of a report.)" The famous Wetherill in fact worked for the Bureau of Indian Affairs at the Navajo National Monument, as well as running his famous trading post.

This story also appears in at least one other version suggesting that it may have been an apocryphal legend at the time:

> The Padre in charge of one Mission was rather disappointed at the attendance at the Sunday morning service. He thought about it and finally came up with an idea. He would supply some kind of breakfast for his flock after the service each Sunday. As the word got around, the attendance increased, and after a few Sundays, he had a very good gathering each time. It was quite a bit of work and some expense. So after a couple of months, the Padre, being convinced that he had finally established the true religion in the hearts and minds of his Indians, began to cut back on the breakfast. He then noticed that the attendance began to fall off. Finally, as he reached the point of no breakfast, only one old Indian came. After the service the Padre talked to the old man and asked him why the other Indians didn't come. The old Indian looked at him and said, "No beans – no Jesus." (Papers from the estate of Charles de Young Elkus, http://researcharchive.calacademy.org/research/library/elkus/stories/frelign.htm).

2 For additional information, see www.robinettekennedy.com (accessed June 11, 2015).
3 There are certain depictions of ritual behavior on ceramic art in the Andean culture of Moche that should not be replicated by anyone, professional or amateur.
4 I base this statement on my read of ethnographies and my three decades of observing people in village societies in the Andes.

References

Adler, Michael A., and Richard H. Wilshusen. 1990. Large-scale integrative facilities in tribal societies: Cross-cultural and southwestern US examples. *World Archaeology* 22(2): 133–146.

Aiello, Leslie C., and R. I. M. Dunbar. 1993. Neocortex size, group size, and the evolution of language. *Current Anthropology* 34(2): 184–193.

Aldenderfer, Mark S. 1989. The archaic period in the South-Central Andes. *Journal of World Prehistory* 3(2): 117–158.

Alesina, Alberto. 2013. Women, fertility, and the rise of modern capitalism. *Science* 342(6157): 427–428.

Allen, Benjamin, Martin A. Nowak, and Edward O. Wilson. 2013. Limitations of inclusive fitness. *Proceedings of the National Academy of Sciences* 110: 20135–20139.

Allen, Michael. 1984. Elders, chiefs, and Big Men: Authority legitimation and political evolution in Melanesia. *American Ethnologist* 11(1): 20–41.

Alvard, Michael S., and David A. Nolin. 2002. Rousseau's whale hunt? Coordination among big-game hunters. *Current Anthropology* 43(4): 533–559.

Andersen, Steffen, Seda Ertaç, Uri Gneezy, Moshe Hoffman, and John A. List. 2011. Stakes matter in ultimatum games. *American Economic Review* 101(7): 3427–3439.

Anderson, David G., Michael Russo, and Kenneth E. Sassaman. 2007. Mid-Holocene cultural dynamics in southeastern North America. In *Climate Change and Cultural Dynamics: A Global Perspective on Mid-Holocene Transitions*, edited by David G. Anderson, Kirk A. Maasch, and Daniel H. Sandweiss, pp. 457–489. Elsevier, London.

Andreoni, James, and John H. Miller. 1993. Rational cooperation in the finitely repeated prisoner's dilemma: Experimental evidence. *Economic Journal* 103(418): 570–585.

Ankermann, Bernhard. 1929. *Das Eingeborenenrecht. Ostafrika*. Strecker and Schröder, Stuttgart.

Appadurai, Arjun. 1994. Commodities and the politics of value. In *Interpreting Objects and Collections*, edited by S. Pearce, pp. 76–91. Routledge, London.

Arkush, Elizabeth, and C. Stanish. 2005. Interpreting conflict in the ancient Andes: Implications for the archaeology of warfare. *Current Anthropology* 46(1): 3–28.

Arkush, Elizabeth, and Tiffiny A. Tung. 2013. Patterns of war in the Andes from the Archaic to the Late Horizon: Insights from settlement patterns and cranial trauma. *Journal of Archaeological Research* 21(4): 307–369.

Arnold, Jeanne E. 1996. The archaeology of complex hunter-gatherers. *Journal of Archaeological Method and Theory* 3(1): 77–126.

Asad, Talal. 1972. Market model, class structure and consent: a reconsideration of Swat political organisation. *Man* 7(1): 74–94.

Aten, Lawrence. 1999. Middle Archaic ceremonialism at Tick Island, Florida: Ripley P. Bullen's 1961 excavation at the Harris Creek site. *Florida Anthropologist* 52(3): 131–200.

Atkinson, Richard J. C., C. M. Piggott, and Nancy K. Sandars. 1951. *Excavations at Dorchester, Oxon.* Ashmolean Museum, Oxford.

Arthur, W. Brian. 1994. Inductive reasoning and bounded rationality. *The American Economic Review* 84(2): 406–411.

Aveni, Anthony F. 1990. An assessment of previous studies of the Nazca Lines. In *The Lines of Nazca*, edited by A. Aveni, pp. 1–40. American Philosophical Society 183, Philadelphia.

2000a. *Between the Lines: The Mystery of the Giant Ground Drawings of Ancient Nasca, Peru.* University of Texas Press, Austin.

2000b. *Empires of Time: Calendars, Clocks and Cultures.* I. B. Tauris, London.

2003. Archaeoastronomy in the ancient Americas. *Journal of Archaeological Research* 11(2): 149–191.

Axelrod, Robert. 1980. Effective choice in the Prisoner's Dilemma. *Journal of Conflict Resolution* 24(1): 3–25.

1997. *The Complexity of Cooperation: Agent-Based Models of Competition and Collaboration.* Princeton University Press, Princeton.

Axelrod, Robert, and William D. Hamilton. 1981. The evolution of cooperation. *Science* 211(4489): 1390–1396.

Bamforth, Douglas B. 1994. Indigenous people, indigenous violence: Precontact warfare on the North American Great Plains. *Man* 29(1): 95–115.

Bandy, Matthew S. 2004. Fissioning, scalar stress, and social evolution in early village societies. *American Anthropologist* 106(2): 322–333.

2005. New World settlement evidence for a two-stage Neolithic Demographic Transition. *Current Anthropology* 46(S):S109–S115.

Banning, Edward B. 2011. So fair a house. *Current Anthropology* 52 (5): 619–660.

Bar-Yosef, Ofer. 1986. The walls of Jericho: An alternative interpretation. *Current Anthropology* 27: 157–162.

1998. The Natufian culture in the Levant, threshold to the origins of agriculture. *Evolutionary Anthropology: Issues, News, and Reviews* 6(5): 159–177.

Bar-Yosef, Ofer, and Anna Belfer-Cohen. 1991. From sedentary hunter-gatherers to territorial farmers in the Levant. In *Between Bands and States*, edited by Susan Gregg, pp. 181–202. Southern Illinois University Center for Archaeological Investigations, no. 9, Carbondale, IL.

Barnes, John A. 1966. Durkheim's Division of Labour in Society. *Man (New Series)* 1(2): 158–175.

Barth, Frederik. 1961. *Nomads of South Persia: The Basseri Tribe of the Khamseh Confederacy.* Little, Brown, Boston.

Bean, Lowell John, and Florence Connolly Shipek. 1978. Luiseño. In *Handbook of North American Indians*, vol. 5, *California*, edited by R. F. Heizer and W. C. Sturtevant, pp. 550–563. Smithsonian Institution, Washington, DC.

Becker, Sascha O., and Ludger Woessmann. 2009. Was Weber wrong? A human capital theory of Protestant economic history. *Quarterly Journal of Economics* 124 (2): 531–596.

Beidelman, Thomas O. 1963. Kaguru time reckoning: An aspect of the cosmology of an East African People. *Southwestern Journal of Anthropology* 19(1): 9–20.

Bell, Catherine. 1997. *Ritual: Perspectives and Dimensions*. Oxford University Press, Oxford.

Bellwood, Peter, and Marc Oxenham. 2008. The expansions of farming societies and the role of the neolithic demographic transition. In *The Neolithic Demographic Transition and Its Consequences*, edited by P. Bocquet-Appel and O. Bar-Yosef, pp. 13–34. Springer, New York.

Belshaw, Cyril. 1965. *Traditional Exchange and Modern Markets*. Prentice Hall, Englewood Cliffs, NJ.

Benchley, Elizabeth. 1970. Mississippian calendars: Some suggestions. *Proceedings of the Annual Meetings of the American Anthropological Association*, San Diego.

Benfer, Robert A. 2011. Giant Preceramic animal effigy mounds in South America? Project Gallery article. *Antiquity* 329. www.antiquity.ac.uk/projgall/benfer329/ (accessed June 5, 2015).

2012. Monumental architecture arising from an early astronomical-religious complex in Peru, 2200–1750 BC. In *Early New World Monumentality*, edited by R. Burger and R. M. Rosenswig, pp. 313–363. University of Florida Press, Gainesville.

Benz, Marion, and Alexander Gramsch. 2006. Zur sozio-politischen Bedeutung von Festen. Eine Einführung anhand von Beispielen aus dem Alten Orient und Europa. *Ethnographisch-archaologische Zeitschrift* 47(4): 417–437.

Berger, Roger, Heiko Rauhut, Sandra Prade, and Dirk Helbing. 2010. Bargaining over time in ultimatum game experiments. CCSS Working Paper Series. Swiss Federal Institute of Technology. http://web.sg.ethz.ch/wps/CCSS-10–002 (accessed March 23, 2015).

Bettencourt, Luis, and Geoffrey West. 2010. A unified theory of urban living. *Nature* 467: 912–913.

Bischof, Henning. 2012. Los periodos Arcaico Tardío, Arcaico Final y Formativo Temprano en el valle de Casma: evidencias e hipótesis. *Boletín de Arqueología PUCP* 13: 9–54.

Boaretto, Elisabetta, Xiaohong Wu, Jiarong Yuan, Ofer Bar-Yosef, Vikki Chu, Yan Pan, Kexin Liu et al. 2009. Radiocarbon dating of charcoal and bone collagen associated with early pottery at Yuchanyan Cave, Hunan Province, China. *Proceedings of the National Academy of Sciences* 106 (24): 9595–9600.

Boas, Franz. 1966. *Kwakiutl Ethnography*. University of Chicago Press, Chicago.

Bocquet-Appel, Jean-Pierre, and Ofer Bar-Yosef. 2008. Prehistoric demography in a time of globalization. In *The Neolithic Demographic Transition and Its Consequences*, edited by P. Bocquet-Appe and O. Bar-Yosef, pp. 1–10. Springer, New York.

Boehm, Christopher 1993. Egalitarian behavior and reverse dominance hierarchy [and comments and reply]. *Current Anthropology* 34(3): 227–254.

Boserup, Esther. 1965. *The Conditions of Agricultural Growth: The Economics of Agrarian Change under Population Pressure*. Aldine, Chicago.

Bowles, Samuel. 2008. Being human: Conflict: Altruism's midwife. *Nature* 456(7220): 326–327.

Bowles, Samuel, Robert Boyd, Ernst Fehr, and Herbert Gintis. 1997. Homo reciprocans: A research initiative on the origins, dimensions, and policy implications of reciprocal fairness. www.umass.edu/preferen/gintis/homo.pdf (accessed June 5, 2015).

Bowles, Samuel, and Herbert Gintis. 2008. The evolutionary basis of collective action. In *The Oxford Handbook of Political Economy*, edited by Donald A. Wittman and Barry R. Weingast, pp. 951–967. Oxford Handbooks Online.

2011. *A Cooperative Species: Human Reciprocity and its Evolution*. Princeton University Press, Princeton, NJ.

Boyd, Matthew. 1998. Interdependence and power: Complexity in hunter-gatherer/farmer exchanges. *Plains Anthropologist* 43(165): 311–319.

Boyd, Rob, Herbert Gintis, Samuel Bowles, and Peter J. Richerson. 2003. The evolution of altruistic punishment. *Proceedings of the National Academy of Sciences* 100(6): 3531–3535.

Boyd, Robert, and Sarah Mathew 2012. Comment on "The false allure of group selection." http://edge.org/conversation/the-false-allure-of-group-selection#rbsm.

Boyd, Robert, and Peter J. Richerson. 1992. Punishment allows the evolution of cooperation (or anything else) in sizable groups. *Ethology and Sociobiology* 13(3): 171–195.

2002. Group beneficial norms can spread rapidly in a structured population. *Journal of Theoretical Biology* 215(3): 287–296.

Brecher, Kenneth, and William G. Haag. 1983. Astronomical alignments at Poverty Point. *American Antiquity* 48(1): 161–163.

Briones, Luis. 2006. The geoglyphs of the north Chilean desert: An archaeological and artistic perspective. *Antiquity* 80(307): 9–24.

Bromley, R. J., and Richard Symanski. 1974. Marketplace trade in Latin America. *Latin American Research Review* 9(3): 3–38.

Burger, Richard. 1995. *Chavin and the Origins of Andean Civilization*. Thames Hudson, New York.

Burger, Richard, and Lucy Burger. 1980. Ritual and religion at Huaricoto. *Archaeology* 33(6): 26–32.

Burger, Richard, and Robert Rosenswig, eds. 2012. *Early New World Monumentality*. University Press of Florida, Gainesville.

Burl, Aubrey. 2000. *The Stone Circles of Britain, Ireland, and Brittany*. Yale University Press, New Haven, CT.

Burton, John. 1984. Quarrying in a tribal society. *World Archaeology* 16(2): 234–247.

Byock, Jesse L. 1986. Governmental order in early Medieval Iceland. *Viator* 17(1): 19–34.

2003. Feuding in Viking Age Iceland's great village. In *Conflict in Medieval Europe: Changing Perspectives on Society and Culture*, edited by Warren Brown and Piotr Górecki, pp. 229–241. Ashgate, Aldershot, UK.

Camerer, Colin F., and Ernst Fehr. 2006. When does "economic man" dominate social behavior? *Science* 311(5757): 47–52.

Cameron, Lisa A. 1999. Raising the stakes in the ultimatum game: Experimental evidence from Indonesia. *Economic Inquiry* 37(1): 47–59.

Canziani, José A. 1992. Arquitectura y urbanismo del período Paracas en el Valle de Chincha. *Gaceta Arqueológica Andina* 6: 87–117.

2009. *Ciudad y territorio en los Andes: Contribuciones a la historia del urbanismo prehispánico*. Fondo Editorial de la Pontificia Universidad Católica del Perú, Lima, Perú.

Carballo, David M. 2013. Cultural and evolutionary dynamics of cooperation in archaeological perspective. In *Cooperation and Collective Action: Archaeological Perspectives*, edited by D. M. Carballo, pp. 3–33. University Press of Colorado, Boulder.

Carballo, David M., Luis Barba, Agustín Ortiz, Jorge Blancas, Nicole Cingolani, Jorge H. Toledo Barrera, David Walton, Isabel Rodríguez López, and Lourdes Couoh 2014. Suprahousehold consumption and community ritual at La Laguna, Mexico. *Antiquity* 88(339): 141–159.

Carballo, David M., Paul Roscoe, and Gary M. Feinman. 2014. Cooperation and collective action in the cultural evolution of complex societies. *Journal of Archaeological Method and Theory* 21(1): 98–133.

Carneiro, Robert L. 1970. A theory of the origin of the state. *Science* 169(3947): 733–738.

1988. The circumscription theory. *American Behavioral Scientist* 31(4): 497–511.

2000. The transition from quantity to quality: A neglected causal mechanism in accounting for social evolution. *Proceedings of the National Academy of Sciences* 97(23): 12926–12931.

Carpenter, Jeffrey, Eric Verhoogen, and Stephen Burks. 2005. The effect of stakes in distribution experiments. *Economics Letters* 86: 393–398.

Carrier, James G. 2009. Simplicity in economic anthropology: Persuasion, form, and substance. In *Economic Persuasions*, edited by S. Gudeman, pp. 15–30. Berghahn Books, New York.

Caughey, John Lyon, III. 1970. *Cultural Values in a Micronesian Society*. PhD dissertation, Department of Anthropology, University of Pennsylvania, Philadelphia.

Çelik, Bahattin. 2011. Karahan Tepe: A new cultural centre in the Urfa area in Turkey. *Documenta Praehistorica* XXXVIII: 241–253.

Chacón, María Isabel, Barbara Pickersgill, and Daniel Debouck. 2005. Domestication patterns in common bean (*Phaseolus vulgaris* L.) and the origin of the Mesoamerican and Andean cultivated races. *Theoretical and Applied Genetics* 110(3): 432–444.

Chagnon, Napoleon. 1988. Life histories, blood revenge, and warfare in a tribal population. *Science* 239: 985–992.

2012. *The Yanomamo*. Cengage Learning, Independence, KY.

Chapman, Anne. 1980. Barter as a universal mode of exchange. *L'Homme* 20(3): 33–83.

Chaudhuri, Ananish 2011. Sustaining cooperation in laboratory public goods experiments: A selective survey of the literature. *Experimental Economics* 14(1): 47–83.

Choi, Jung-Kyoo, and Samuel Bowles. 2007. The coevolution of parochial altruism and war. *Science* 318(5850): 636–640.

Christensen, Jonas. 2004. Warfare in the European Neolithic. *Acta Archaeologica* 75(2): 129–156.

Claassen, Cheryl. 2010. *Feasting with Shellfish in the Southern Ohio Valley: Archaic Sacred Sites and Rituals*. University of Tennessee Press, Knoxville.

Clare, Lee, Eelco J. Rohling, Bernhard Weninger, and Johanna Hilpert. 2008. Warfare in Late Neolithic/Early Chalcolithic Pisidia, southwestern Turkey: Climate induced social unrest in the late 7th millennium calBC. *Documenta Prehistorica* 35: 65–92.

Clark, John E. 2004. Surrounding the sacred: Geometry and design of early mound groups as meaning and function. In *Signs of Power: The Rise of Cultural Complexity in the South-East*, edited by Jon L. Gibson and Philip J. Carr, pp. 162–213. University of Alabama Press, Tuscaloosa.

Clark, Kenneth, and Martin Sefton. 2001. The sequential prisoner's dilemma: Evidence on reciprocation. *Economic Journal* 111(468): 51–68.

Codere, Helen. 1957. Kwakiutl society: Rank without class. *American Anthropologist* 59: 473–486.

Cohen, Amanda Beth. 2010. *Ritual and Architecture in the Titicaca Basin: The Development of the Sunken Court Complex in the Formative Period*. PhD dissertation, Department of Anthropology, UCLA.

Cohen, Mark N. 1975. Archaeological evidence for population pressure in pre-agricultural societies. *American Antiquity* 40(4): 471–475.

Cohen, Percy S. 1967. Economic analysis and economic man. In *Themes in Economic Anthropology*, edited by R. Firth, pp. 91–117. Routledge, London.

Collard, Mark, Kevan Edinborough, Stephen Shennan, and Mark G. Thomas. 2010. Radiocarbon evidence indicates that migrants introduced farming to Britain. *Journal of Archaeological Science* 37(4): 866–870.

Collocott, Ernest, and Edgar Vyvyran. 1927. Kava ceremonial in Tonga. *The Journal of the Polynesian Society* 36(1): 21–47.

Conklin, William. 1996 La arquitectura de la Huaca de Los Reyes. *Arinka* 10: 88–100.

Connolly, Robert P. 2006. An assessment of radiocarbon age results from the Poverty Point site. *Louisiana Archaeology* 27: 1–14.

Cook, Scott. 1966. The obsolete "anti-market" mentality: A critique of the substantive approach to economic anthropology. *American Anthropologist* 68(2, Part 1): 323–345.

Cowgill, George L., and Edwin N. Wilmsen. 1975. Population pressure as a non-explanation [with discussion]. *Memoirs of the Society for American Archaeology* 30: 127–133.

Craddock, Paul T., Michael R. Cowell, Morven N. Leese, and Michael J. Hughes. 1983. The trace element composition of polished flint axes as an indicator of source. *Archaeometry* 25(2): 135–163.

Crocker, Christopher. 1969. Men's house associates among the Eastern Bororo. *Southwestern Journal of Anthropology* 25(3): 236–260.

Curry, Andrew. 2008. Seeking the roots of ritual. *Science* 319: 278–280.

Dalton, George. 1969. Theoretical issues in economic anthropology. *Current Anthropology* 10(1): 63–102.

1977. Aboriginal economies in stateless societies. In *Exchange Systems in Prehistory*, edited by T. Earle and J. Ericson, pp. 191–212. Academic Press, New York.

Damon, Frederick H. 1982. Calendars and calendrical rites on the northern side of the Kula Ring. *Oceania* 52(3): 221–239.

de Laguna, Frederica. 1952. Some dynamic forces in Tlingit society. *Southwestern Journal of Anthropology* 8(1): 1–12.

Dietler, Michael. 1990. Driven by drink: The role of drinking in the political economy and the case of early Iron Age France. *Journal of Anthropological Archaeology* 9(4): 352–406.

Dietler, Michael, and Ingrid Herbich. 2001. Feasts and labor organization: Dissecting a fundamental economic practice. In *Feasts*, edited by M. Dietler and B. Hayden, pp. 240–264. Smithsonian Institution, Washington, DC.

Dietrich, Oliver, Manfred Heun, Jens Notroff, Klaus Schmidt, and Martin Zarnkow. 2012. The role of cult and feasting in the emergence of Neolithic communities: New evidence from Gobekli Tepe, south-eastern Turkey. *Antiquity* 86(333): 674–695.

Dobyns, Henry F. 1984. Trade centers: The concept and a rancherian culture area example. *American Indian Culture and Research Journal* 8(1): 23–35.

Douglas, Mary. 1954. Foreword. In M. Mauss, *The Gift: Forms and Functions of Exchange in Archaic Societies*. Norton, New York.

1973. *Natural Symbols: Exploration in Cosmology*. Random House, Vintage Books, New York.

1982. *Natural Symbols: Explorations in Cosmology*. Pantheon, New York.

Dove, Michael R. 1983. Theories of swidden agriculture, and the political economy of ignorance. *Agroforestry Systems* 1(2): 85–99.

1985. *Swidden Agriculture in Indonesia: The Subsistence Strategies of the Kalimantan Kant*. Vol. 43. Mouton, Amsterdam.

Dreber, Anna, David G. Rand, Drew Fudenberg, and Martin A. Nowak. 2008. Winners don't punish. *Nature* (452): 348–351.

Dufwenberg, Martin, and Georg Kirchsteiger. 2004. A theory of sequential reciprocity. *Games and Economic Behavior* 47(2): 268–298.

Durkheim, Emile. 1912. *The Elementary Forms of the Religious Life*. George Allen and Unwin, London.

Dye, David H. 2013. Trends in cooperation and conflict in Native eastern North America. In *War, Peace, and Human Nature: The Convergence of Evolutionary and Cultural Views*, edited by Douglas Fry, pp. 132–150. Oxford University Press, Oxford.

Dye, Tom. 1994. Population trends in Hawai'i before 1778. *Hawaiian Journal of History* 28: 1–20.

Earle, Timothy. 1977. A reappraisal of redistribution: Complex Hawaiian chiefdoms. In *Exchange Systems in Prehistory*, edited by T. Earle and J. Ericson, pp. 213–229. Academic Press, New York.

1978. *Economic and Social Organization of a Complex Chiefdom: The Halelea District, Kaua'i, Hawaii*. Anthropological Papers, Museum of Anthropology, no. 63. University of Michigan, Ann Arbor.

1997. *How Chiefs Come to Power: The Political Economy in Prehistory*. Stanford University Press, Palo Alto, CA.

Easley, David, and Jon Kleinberg. 2010. *Networks, Crowds, and Markets: Reasoning about a Highly Connected World*. Cambridge University Press, Cambridge.

Edel, Matthew. 1970. Karl Polanyi's concept of non-market trade: Discussion. *Journal of Economic History* 30(1): 127–130.

Edel, May Mandelbaum. 1937. The Bachiga of East Africa. In *Cooperation and Competiton among Primitive Peoples*, edited by M. Mead, pp. 127–152. Transaction, New Brunswick, NJ.

Egas, Martijn, and Arno Riedl. 2008. The economics of altruistic punishment, and the maintenance of cooperation. *Proceedings of the Royal Society* 275(1637): 871–878.

Eggan, Fred. 1950. *Social Organization of the Western Pueblos*. University of Chicago Press, Chicago.

Einzig, Paul. 1949. *Primitive Money: In Its Ethnological, Historical and Economic Aspects*. Pergamon Press, London.

Ellis, William. 1826. *Narrative of a Tour through Hawaii, or, Owhyhee: With Remarks on the History, Traditions, Manners, Customs, and Language of the Inhabitants of the Sandwich Islands*. Fisher and Jackson, London.

Elster, Jon. 1989. Social norms and economic theory. *Journal of Economic Perspectives* 3(4): 99–117.

Ember, Carol, and Melvin Ember. 1992. Resource unpredictability, mistrust, and war. *Journal of Conflict Resolution* 36: 242–262.

Engel, Frédéric A. 2010. *Arqueología Inedita de la Costa Peruana*. Asamblea Nacional de Rectores, Lima.

Engels, Frederick. 1972. *The Origin of the Family, Private Property and the State in the Light of the Researches of Lewis H. Morgan*. International, New York.

Evans, Ivor H. N. 1912. Notes on the religious beliefs, superstitions, ceremonies and tabus of the Dusuns of the Tuaran and Tempassuk Districts, British North Borneo. *Journal of the Anthropological Institute of Great Britain and Ireland* 42: 380–396.

1922. *Among Primitive Peoples in Borneo: A Description of the Lives, Habits and Customs of the Piratical Head-hunters of North Borneo, with an Account of Interesting Objects of Prehistoric Antiquity Discovered in the Island*. Seeley, Service, London.

Evans-Pritchard, Edward. 1929. The morphology and function of magic. *American Anthropologist* 31(4): 619–641.

1933. Zande blood-brotherhood. *Africa* 6(4): 369–401.

1940. *The Nuer*. Vol. 940. Clarendon, Oxford.

Ewers, John C. 1966. Plains Indian reactions to the Lewis and Clark expedition. *Montana: Magazine of Western History* 16(1): 2–12.

Faris, Ellsworth. 1934. Review of "Emile Durkheim on the Division of Labor in Society." *American Journal of Sociology* 40(3): 376–377.

Fauquet, Georges. 1941. The place of co-operation in economic and social life. *Annals of Public and Cooperative Economics* 17(1): 168–175.

Fehr, Ernst, and Urs Fischbacher. 2003. The nature of human altruism. *Nature* 425: 785–791.

Fehr, Ernst, and Simon Gächter. 2002. Altruistic punishment in humans. *Nature* 415: 137–140.

Feinman, Gary. 2000. Cultural evolutionary approaches and archaeology: Past, present, and future. In *Cultural Evolution, Contemporary Viewpoints*, edited by G. Feinman and L. Manzanilla, pp. 3–12. Kluwer Academic, New York.

Feinman, Gary M., and Joyce Marcus, eds. 1998. *Archaic States*. School for Advanced Research, Santa Fe, NM.

Feinman, Gary, and Jill Neitzel. 1984. Too many types: An overview of sedentary prestate societies in the Americas. *Advances in Archaeological Method and Theory* 7: 39–102.

Feldman, Robert. 1980. *Aspero, Peru: Architecture, Subsistence Economy and Other Artifacts of a Preceramic Maritime Chiefdom*. PhD dissertation, Department of Anthropology, Harvard University.

1985. Preceramic corporate architecture: Evidence for the development of non-egalitarian social systems in Perú. In *Early Ceremonial Architecture in the Andes*, edited by C. Donnan, pp. 71–92. Dumbarton Oaks, Washington, DC.

Fennell, Lee Anne. 2011. Ostrom's law: Property rights in the commons. *International Journal of the Commons* 5(1): 9–27.

Fenton, William N. 1965. The Iroquois Confederacy in the twentieth century: A case study of the theory of Lewis H. Morgan in ancient society. *Ethnology* 4(3): 251–265.

Finlayson, Bill, Steven J. Mithen, Mohammad Najjar, Sam Smith, Darko Maričević, Nick Pankhurst, and Lisa Yeomans. 2011. Architecture, sedentism, and social complexity at Pre-Pottery Neolithic A WF16, southern Jordan. *Proceedings of the National Academy of Sciences* 108(20): 8183–8188.

Firth, Raymond. 1929. *Primitive Economics of the New Zealand Maori*. Routledge, London.

1959. *Economics of the New Zealand Maori*. Government Printer, Wellington.

1965. *Primitive Polynesian Economy*. Norton, New York.

1967. *The Work of the Gods in Tikopia*. Humanities Press, New York.

1970. *Rank and Religion in Tikopia: A Study in Polynesian Paganism and Conversion to Christianity*. George Allen and Unwin, London.

1972. Methodological issues in economic anthropology. *Man* 7(3): 467–475.

1975. An appraisal of modern social anthropology. *Annual Review of Anthropology* 4(1): 1–26.

Flannery, Kent. 1972a. The origins of the village as a settlement type in Mesoamerica and the Near East: A comparative study. In *Man, Settlement, and Urbanism*, edited by R. J. Ucko, R. Tringham, and G. W. Dimbleby, pp. 23–53. Duckworth, London.

1972b. The cultural evolution of civilizations. *Annual Review of Ecology and Systematics* 3: 399–426.

1995. Prehistoric social evolution. In *Research Frontiers in Anthropology*, edited by C. R. Ember and M. Ember, pp. 1–26. Prentice Hall, Englewood Cliffs, NJ.

2002. The origins of the village revisited: From nuclear to extended households. *American Antiquity* 67(3): 417–433.

Flannery, Kent V., and Joyce Marcus. 2012 *The Creation of Inequality: How Our Prehistoric Ancestors Set the Stage for Monarchy, Slavery, and Empire*. Harvard University Press, Cambridge, MA.

Flannery, Kent V., Joyce Marcus, and Robert G. Reynolds. 1989. *The Flocks of the Wamani: A Study of Llama Herders on the Punas of Ayacucho, Peru*. Academic Press, San Diego.

Fogelin, Lars. 2007. The archaeology of religious ritual. *Annual Review of Anthropology* 36: 55–71.

Ford, Richard I. 1983. Inter-Indian exchange in the Southwest. In *Handbook of North American Indians*, vol. 9, *Southwest*, edited by A. Ortiz, pp. 711–722. Smithsonian Institution, Washington, DC.

Fortes, Meyer. 1962. Ritual and office in tribal society, In *Essays on the Ritual of Social Relations*, edited by M. Gluckman, pp. 53–88. Manchester University Press, Manchester.

Fowles, Severin. 2013. *An Archaeology of Doings: Secularism and the Study of Pueblo Religion*. School of Advanced Research, Santa Fe, NM.

Freeman, Derek. 1974. The evolutionary theories of Charles Darwin and Herbert Spencer [and comments and replies]. *Current Anthropology* 15(3): 211–237.

Fuchs, Peter R., Renate Patzschke, Claudia Schmitz, Germán Yenque, and Jesús Briceño. 2012. Investigaciones arqueológicas en el sitio de Sechín Bajo, Casma. *Boletín de Arqueología PUCP* 10: 111–135.

Gavrilets, Sergey. 2015. Collective action and the collaborative brain. *Journal of the Royal Society Interface* 12: 20141067. doi:10.1098/rsif.2014.1067.

Geddes, Linda. 2009. Ancient Peruvian geoglyphs might have been prayer walks. *New Scientist*, 201(2692): 12.

Geertz, Hildred. 1975. An anthropology of religion and magic. *Journal of Interdisciplinary History* 6(1): 71–89.

Ghezzi, Ivan, and Clive Ruggles. 2007. Chankillo: A 2300-year old solar observatory in Coastal Peru. *Science* 315: 1239–1243.

2011. The social and ritual context of horizon astronomical observations at Chankillo. International Symposium on Archaeoastronomy. *Proceedings IAU Symposium* 278: 144–153.

Gibson, Alex. 2014. Space and episodic ritual at the monumental Neolithic round mound of Duggleby Howe, North Yorkshire, England. In *Préhistoires Méditerranéennes [En ligne], Colloque | 2014, mis en ligne le 29 octobre 2014, consulté le 17 mai 2015*. http://pm.revues.org/924.

Gibson, Jon L. 2000. *The Ancient Mounds of Poverty Point: Place of Rings*. University Press of Florida, Gainesville.

Gibson, Jon L., and Mark Melancon. 2004. In the beginning: Social contexts of first pottery in the lower Mississippi Valley. In *Early Pottery, Technology, Function, Style, and Interaction in the Lower Southeast*, edited by R. Saunders and C. T. Hays, pp. 169–192. University of Alabama Press, Tuscaloosa.

Gignoux, Christopher R., Brenna M. Henn, and Joanna L. Mountain. 2011. Rapid, global demographic expansions after the origins of agriculture. *Proceedings of the National Academy of Sciences* 108(13): 5154–5162.

Gintis, Herbert. 2000a. Beyond "Homo economicus": Evidence from experimental economics. *Ecological Economics* 35(3): 311–322.

2000b. *Game Theory Evolving: A Problem-Centered Introduction to Modeling Strategic Behavior*. Princeton University Press, Princeton, NJ.

2000c. Strong reciprocity and human sociality. *Journal of Theoretical Biology* 206(2): 169–179.

2009. *The Bounds of Reason: Game Theory and the Unification of the Behavioral Sciences*. Princeton University Press, Princeton, NJ.

2012. The false allure of group selection. http://edge.org/conversation/the-false-allure-of-group-selection#rbsm (accessed June 4, 2015).

Gintis, Herbert, Samuel Bowles, Robert Boyd, and Ernst Fehr. 2005. Moral sentiments and material interests: Origins, evidence, and consequences. In *Moral Sentiments and Material Interests: The Foundations of Cooperation in Economic Life*, edited by Samuel Bowles, Robert Boyd, and Ernst Fehr, pp. 3–39. MIT Press, Boston.

Gintis, Herbert, Eric Alden Smith, and Samuel Bowles. 2001. Costly signaling and cooperation. *Journal of Theoretical Biology* 213(1): 103–119.

Glencross, Bonnie, and Başak Boz. 2013. Representing violence in Anatolia and the Near East during the transition to agriculture. In *The Routledge Handbook of Bioarchaeology of Human Conflict*, edited by C. Knüsel and M. Smith, pp. 90–108. Routledge, Oxon.

Goette, Lorenz, and Stephan Meier. 2011. Can integration tame conflicts? *Science* 334: 1356–1357.

Goldin, Liliana R., and Brent Metz. 1991. An expression of cultural change: Invisible converts to Protestantism among highland Guatemala Mayas. *Ethnology* 30(4): 325–338.

Goldman, Irving. 1937a. The Ifugao of the Philippine Islands. In *Cooperation and Competition among Primitive Peoples*, edited by Margaret Mead, pp. 153–179. McGraw-Hill, New York.

1937b. The Bathonga of South Africa. In *Cooperation and Competition among Primitive Peoples*, edited by Margaret Mead, pp. 354–381. McGraw-Hill, New York.

Goodfellow, David Martin. 1939. *Principles of Economic Sociology: The Economics of Primitive Life as Illustrated from the Bantu Peoples of South and East Africa*. Routledge, London.

Goody, Jack. 1972. The evolution of the family. In *Household and Family in Past Time*, edited by P. Laslett and R. Wall, pp. 103–124. Cambridge University Press, Cambridge.

Gould, Stephen J. 1991. *Bully for Brontosaurus: Reflections in Natural History*. Norton, New York.

Graeber, David. 2011. *Debt: The First 5,000 Years*. Melville House, Brooklyn, NY.

Grafen, Alan. 1990. Biological signals as handicaps. *Journal of Theoretical Biology* 144: 517–546.

Graffam, Gray, Mario Rivera, and Alvaro Carevič. 1996. Ancient metallurgy in the Atacama: Evidence for copper smelting during Chile's early Ceramic period. *Latin American Antiquity* 7(2): 101–113.

Granovetter, Mark. 1985. Economic action and social structure: The problem of embeddedness. *American Journal of Sociology* 91(3): 481–510.

Griffin, Arthur F. 2011. Emergence of fusion/fission cycling and self-organized criticality from a simulation model of early complex polities. *Journal of Archaeological Science* 38(4): 873–883.

Griffin, Arthur F., and Charles Stanish. 2007. An agent-based model of prehistoric settlement patterns and political consolidation in the Lake Titicaca Basin of Peru and Bolivia. *Structure and Dynamics* 2(2). https://escholarship.org/uc/item/2zd1t887 (accessed June 5, 2015).

Griffin, Laura. 2013. Spatial linkages on the Peruvian south coast in the Early Horizon: A view from Cerro del Gentil. Master's thesis, Archaeology, University of California, Los Angeles.

Guerrero, Emma, Stephan Naji, and Jean-Pierre Bocquet-Appel. 2008. The signal of the Neolithic demographic transition in the Levant. In *The Neolithic Demographic*

Transition and Its Consequences, edited by P. Bocquet-Appel and O. Bar-Yosef, pp. 57–80. Springer, New York.

Guffroy, J. 1989 Un centro ceremonial formativo en el Alto Piura. *Bulletin, Institut Francais d'etudes Andines* 18(2): 161–207.

Gunther, Erna. 1927. Klallam ethnography. *University of Washington Publications in Anthropology* 1(5): 171–314.

Güth, Werner, Rolf Schmittberger, and Bernd Schwarze. 1982. An experimental analysis of ultimatum bargaining. *Journal of Economic Behavior and Organization* 3(4): 367–388.

Haas, Jonathan, Winifred Creamer, and Alvaro Ruiz. 2004. Dating the Late Archaic occupation of the Norte Chico region in Peru. *Nature* 432(7020): 1020–1023.

Håkansson, N. Thomas. 1998. Rulers and rainmakers in precolonial South Pare, Tanzania: Exchange and ritual experts in political centralization. *Ethnology* 37(3): 263–283.

Hallum, Anne Motley. 2003. Taking stock and building bridges: Feminism, women's movements, and Pentecostalism in Latin America. *Latin American Research Review* 38(1): 169–186.

Hämäläinen, Pekka. 1998. The western Comanche trade center: Rethinking the Plains Indian trade system. *Western Historical Quarterly* 29(4): 485–513.

Hammond, Dorothy. 1970. Magic: A problem in semantics. *American Anthropologist* 72(6): 1349–1356.

Harris, Marvin. 1984. A cultural materialist theory of band and village warfare: The Yanomamo test. In *Warfare, Culture, and Environment*, edited by R. B. Ferguson, pp. 111–140. Academic Press, New York.

Harris, Sandra. 2001. Being politically impolite: Extending politeness theory to adversarial political discourse. *Discourse and Society* 12(4): 451–472.

Hastorf, Christine A. 2005. The upper (middle and late) Formative in the Titicaca region. In *Advances in Titicaca Basin Archaeology* 1, edited by C. Stanish, A. Cohen, and M. Aldenderfer, pp. 65–94. Cotsen Institute of Archaeology, University of California Press, Los Angeles.

Hayden, Brian. 1990. Nimrods, piscators, pluckers, and planters: The emergence of food production. *Journal of Anthropological Archaeology* 9: 31–69.

1992. Models of domestication. In *Transitions to Agriculture in Prehistory*, edited by T. D. Price and A. B. Gebauer, pp. 11–19. Monographs in World Archaeology, vol. 4. Prehistory Press, Madison, WI.

1994. Competition, labor, and complex hunter-gatherers. In *Key Issues in Hunter-Gatherer Research*, edited by Ernest Burch Jr. and Linda J. Ellanna, pp. 223–239. Berg, Oxford.

1996. Feasting in prehistoric and traditional societies. In *Food and the Status Quest: An Interdisciplinary Perspective*, edited by P. Wiessner and W. Schiefenhövel, pp. 127–147. Berghahn Books, Oxford.

2014. *The Power of Feasts: From Prehistory to the Present*. Cambridge University Press, Cambridge.

Hayden, Bryan, and Ron Adams. 2004. Ritual structures in transegalitarian communities. In *Complex Hunter-Gatherers: Evolution and Organization of Prehistoric Communities on the Plateau of Northwestern North America*, edited by W. Prentis and I. Kujit, pp. 84–102. University of Utah Press, Salt Lake City.

Hegmon, Michelle. 1989. Social integration and architecture. In *The Architecture of Social Integration in Prehistoric Pueblos*, edited by W. D. Lipe and M. Hegmon, pp. 5–14. Occasional Papers, no. 1. Crow Canyon Archaeological Center, Cortez, CO.

1991. The risks of sharing and sharing as risk reduction: Interhousehold food sharing in egalitarian societies. In *Between Bands and States*, edited by Susan Gregg, pp. 309–329. Southern Illinois University Center for Archaeological Investigations, no. 9, Carbondale.

Henrich, Joseph. 2000. Does culture matter in economic behavior? Ultimatum Game bargaining among the Machiguenga of the Peruvian Amazon. *American Economic Review* 90(4): 973–979.

2009. The evolution of costly displays, cooperation and religion: Credibility enhancing displays and their implications for cultural evolution. *Evolution and Human Behavior* 30(4): 244–260.

n.d. Evolutionary links between cooperation, religion, and ritual: The evolution of inferentially potent displays and their implications for cultural evolution. www2.psych.ubc.ca/~henrich/Website/Papers/Evolution_of_Costly_Displays05.pdf (accessed June 3, 2015).

Henrich, Joseph, Robert Boyd, Samuel Bowles, Colin Camerer, Ernst Fehr, and Herbert Gintis. 2004. *Foundations of Human Sociality: Economic Experiments and Ethnographic Evidence from Fifteen Small-Scale Societies*. Oxford University Press, Oxford.

Henrich, Joseph, Robert Boyd, Samuel Bowles, Colin Camerer, Ernst Fehr, Herbert Gintis, and Richard McElreath. 2001. In search of Homo economicus: Behavioral experiments in 15 small-scale societies. *American Economic Review* 91: 73–78.

Henrich, Joseph, Jean Ensminger, Richard McElreath, Abigail Barr, Clark Barrett, Alexander Bolyanatz, Juan Camilo Cardenas et al. 2010c. Markets, religion, community size, and the evolution of fairness and punishment. *Science* 327(5972): 1480–1484.

Henrich, Joseph, Steven Heine, and Ara Norenzayan. 2010a. Most people are not WEIRD. *Nature* 466(7302): 29.

2010b. The weirdest people in the world? *Behavioral and Brain Sciences* 33(2–3): 61–83.

Henrich Joseph, and Natalie Henrich. 2004. Fairness without punishment: Behavioral experiments in the Yasawa Islands, Fiji. www2.psych.ubc.ca/~henrich/pdfs/HenrichFijiGamesDraft02.pdf (accessed June 3, 2015).

Henrich, Joseph, Carolyn Lesorogol, Frank Marlowe, David Tracer, and John Ziker. 2006. Costly punishment across human societies. *Science* 312(5781): 1767–1770.

Herskovits, Melville. 1968[1940]. Economizing and rational behavior. In *Economic Anthropology*, edited by E. LeClair and H. Schneider, pp. 40–55. Holt, Reinhart, and Winston, New York.

Higginbottom, Gail, and Roger Clay. 2014. Origins of megalithic astronomy in Britain. arXiv preprint arXiv:1402.1338. http://papers.ssrn.com/sol3/papers.cfm?abstract_id=2382345 (accessed June 3, 2015).

Hill, Kim R., Robert S. Walker, Miran Božičević, James Eder, Thomas Headland, Barry Hewlett, A. Magdalena Hurtado, Frank Marlowe, Polly Wiessner, and Brian Wood.

2011. Human social structure co-residence patterns in hunter-gatherer societies show unique. *Science* 331(6022): 1286–1289.
Hindess, Barry, and Paul Q. Hirst. 1975. *Pre-Capitalist Modes of Production.* Routledge, London.
Hirth, Kenneth G. 1996. Political economy and archaeology: Perspectives on exchange and production. *Journal of Archaeological Research* 4(3): 203–239.
1998. The distributional approach: A new way to identify marketplace exchange in the archaeological record. *Current Anthropology* 39(4): 451–376.
Hirth, Kenneth, and Joanne Pillsbury. 2013. *Merchants, Markets, and Exchange in the Pre-Columbian World.* Dumbarton Oaks Research Library and Collection, Washington, DC.
Hobbes, Thomas. 1651. Leviathan. Traité de la matière, de la forme et du pouvoir de la république ecclésiastique et civile. Traduction originale de M. Philippe Folliot, 2002. www.catallaxia.free.fr/Hobbes%20-%20leviathan.pdf.
Hodder, Bromwell W. 1965. The distribution of markets in Yorubaland. *Scottish Geographical Magazine* 81: 48–58.
Hodder, Ian. 1982. *Symbols in Action: Ethnoarchaeological Studies of Material Culture.* Cambridge University Press, Cambridge.
Hodgson, Geoffrey M. 2000. What is the essence of institutional economics? *Journal of Economic Issues* 34(2): 317–329.
Hogbin, H. Ian. 1938. Social advancement in Guadalcanal, Solomon Islands. *Oceania* 8(3): 289–305.
Hollingsworth, J. Rogers, and Robert Boyer. 1997. Coordination of economic actors and social systems of production. In *Contemporary Capitalism*, edited by J. R. Hollingsworth and R. Boyer, pp. 1–48. Cambridge University Press, Cambridge.
Hommon, Robert. 2013. *The Ancient Hawaiian State.* Oxford University Press, Oxford.
Hooper, Paul L., Hillard S. Kaplan, and James L. Boone. 2010. A theory of leadership in human cooperative groups. *Journal of Theoretical Biology* 265(4): 633–646.
Horwitz, Howard. 1995. Selling relics, preserving antiquities: The Professor's House and the narrative of American anthropology. *Configurations* 3(3): 353–389.
Hose, Charles, and William McDougall. 1912. The pagan tribes of Borneo. *Project Gutenberg.* www.searchengine.org.uk/pdfs/8/930.pdf (accessed October 23, 2014).
Howell, William, and Demetrius James Sandford Bailey. 1900. *A Sea Dyak Dictionary, in Alphabetical Parts, with Examples and Quotations Shewing the Use and Meaning of Words.* American Mission Press, Singapore.
Humphreys, Paul. 1986. Causation in the social sciences: An overview. *Synthese* 68(1): 1–12.
Humphreys, S. C. 1969. History, economics, and anthropology: The work of Karl Polanyi. *History and Theory* 8(2): 165–212.
Imamuddin, Sayyid Muhammad. 1981. *Muslim Spain: 711–1492 AD: A Sociological Study.* Vol. 2. Brill, Leiden, Netherlands.
Insoll, Timothy. 2004. *Archaeology, Ritual, Religion.* Routledge, London.
Jacques, David, and Tom Phillips. 2014. Mesolithic settlement near Stonehenge: Excavations at Blick Mead, Vespasian's Camp, Amesbury. *Wiltshire Archaeological & Natural History Magazine* 107: 7–27.

Janetski, Joel C. 2002. Trade in Fremont society: Contexts and contrasts. *Journal of Anthropological Archaeology* 21 (3): 344–370.

Jenks, Albert Ernest. 1902. Economic man – a definition. *American Anthropologist* 4: 201–206.

Jensen, Hans E. 1987. The theory of human nature. *Journal of Economic Issues* 21(3): 1039–1073.

Johansen, Bruce Elliott, and Barbara Alice Mann, eds. 2000. *Encyclopedia of the Haudenosaunee (Iroquois Confederacy)*. Greenwood, Westport, CT.

Johnson, Allen W., and Timothy K. Earle. *The Evolution of Human Societies: From Foraging Group to Agrarian State*. Stanford University Press, Palo Alto, CA.

Johnson, Douglas H. 1982. Ngundeng and the "Turuk": Two narratives compared. *History in Africa* 9: 119–139.

Johnson, D., D. Proulx, and S. Mabee. 2002. The correlation between geoglyphs and subterranean water resources in the Río Grande de Nasca drainage. In *Andean Archaeology II: Art, Landscape and Society*, edited by H. Silverman and W. Isbell, pp. 307–332. Kluwer Academic/Plenum, Oxford.

Johnson, Jay K. 2000. Beads, microdrills, bifaces, and blades from Watson Brake. *Southeastern Archaeology* 19(2): 95–104.

Johnstone, Rufus A. 1997. The evolution of animal signals. In *Behavioural Ecology: An Evolutionary Approach*, edited by J. R. Krebs and N. B. Davies, pp. 155–178. Blackwell, Oxford.

Jones, Dan. 2010. A WEIRD view of human nature skews psychologists' studies. *Science* 328(5986): 1627.

Junker, Laura Lee. 1999. *Raiding, Trading, and Feasting: The Political Economy of Philippine Chiefdoms*. University of Hawaii Press, Honolulu.

Kaberry, Phyllis M. 1941a. The Abelam tribe, Sepik District, New Guinea: A preliminary report. *Oceania* 11(3): 233–258.

1941b. The Abelam tribe, Sepik District, New Guinea: A preliminary report (continued). *Oceania* 11(4): 345–367.

Kaberry, Phyllis M. 1965. Political organization among the Northern Abelam. *Anthropological Forum* 1(3–4): 334–372.

Kachelmeier, Steven J., and Mohamed Shehata. 1992. Examining risk preferences under high monetary incentives: Experimental evidence from the People's Republic of China. *American Economic Review* 18(5): 1120–1141.

Kalberg, Stephen. 2002. Introduction. *The Protestant Ethic and the Spirit of Capitalism* by Max Weber. Roxbury, Los Angeles, CA.

Kantner, John, and Kevin Vaughn. 2012. Pilgrimage as costly signal: Religiously motivated cooperation in Chaco and Nasca. *Journal of Anthropological Archaeology* 31(1): 66–82.

Kato, Tsuyoshi. 1988. Agricultural rituals and rice cultivation in Negeri Sembilan: A reconstruction from oral history. *Southeast Asian Studies* 26(2): 109–131.

Keasbey, Lindley M. 1903. Economic anthropology. *Science* 17(425): 286–290.

Keeley, Lawrence H. 1988. Hunter-gatherer economic complexity and "population pressure": A cross-cultural analysis. *Journal of Anthropological Archaeology* 7(4): 373–411.

1996. *War before Civilization*. Oxford University Press, Oxford.

Keeley, Lawrence H., Marisa Fontana, and Russell Quick. 2007. Baffles and bastions: The universal features of fortifications. *Journal of Archaeological Research* 15(1): 55–95.

Kemp, Barry J. 1995. How religious were the ancient Egyptians? *Cambridge Archaeological Journal* 5: 25–54.

Kessell, John L. 1995. *Kiva, Cross and Crown: The Pecos Indians and New Mexico, 1540–1840.* Western National Parks Association, Tucson, AZ.

Kidder, Tristram R., Anthony L. Ortmann, and Lee J. Arco. 2008. Poverty Point and the archaeology of singularity. *SAA Archaeological Record* 8(5): 9–12.

Kinambo, Isaria N. 1969. *A Political History of the Pare of Tanzania c. 1500–1900.* East African Publication House, Nairobi.

Kirch, Patrick V. 1994. The pre-Christian ritual cycle of Futuna, Western Polynesia. *Journal of the Polynesian Society* 101(3): 255–298.

1995. *The Wet and the Dry: Irrigation and Agricultural Intensification in Polynesia.* University of Chicago Press, Chicago.

2001. Polynesian feasting in ethnohistoric, ethnographic, and archaeological contexts: A comparison of three societies. In *Feasts: Archaeological and Ethnographic Perspectives on Food, Politics, and Power*, edited by M. Dietler and B. Hayden, pp. 168–184. Smithsonian Institution Press, Washington, DC.

Kirch, Patrick Vinton, and Roger C. Green. 2001. *Hawaiki, Ancestral Polynesia: An Essay in Historical Anthropology.* Cambridge University Press, Cambrige.

Knight, Vernon James. 2001. Feasting and the emergence of platform mound ceremonialism in eastern North America. In *Feasts: Archaeological and Ethnographic Perspectives on Food, Politics, and Power*, edited by M. Dietler and B. Hayden, pp. 311–333. Smithsonian Institution Press, Washington, DC.

Kohler, Stefan. 2013. More fair play in an Ultimatum Game after resettlement in Zimbabwe: A field experiment and a structural model. *PLoS ONE* 8(5): e64791. doi:10.1371/journal.pone.0064791.

Kolb, Michael J. 1997. Labor mobilization, ethnohistory, and the archaeology of community in Hawai'i. *Journal of Archaeological Method and Theory* 4(3–4): 265–285.

Kornienko, Tatiana V. 2009. Notes on the cult buildings of northern Mesopotamia in the Aceramic Neolithic period. *Journal of Near Eastern Studies* 68(2): 81–102.

Kuijt, Ian. 2000. Life in Neolithic farming communities, an introduction. In *Life in Neolithic Farming Communities: Social Organization, Identity, and Differentiation*, edited by I. Kuijt, pp. 3–13. Springer, New York.

2001. Lithic inter-assemblage variability and cultural-historical sequences: A consideration of the Pre-Pottery Neolithic A occupation of Dhra,' Jordan. *Paleórient* 27: 107–125.

Kurtz, Donald V. 1974. Peripheral and transitional markets: The Aztec case. *American Ethnologist* 1(4): 685–705.

Kuzmin, Yaroslav V. 2006. Chronology of the earliest pottery in East Asia: Progress and pitfalls. *Antiquity* 80(308): 362–371.

Lambers, Karsten, and Martin Sauerbier. 2006. GIS-based visibility studies of the Nasca geoglyphs at Palpa, Peru. In *Recording, Modeling and Visualization of Cultural Heritage*, edited by E. Baltasavias, A. Gruen, L. van Gool, and M. Pateraki, pp. 249–261. Taylor and Francis, London.

Lambert, Patricia Marie. 1994. *War and Peace on the Western Front: A Study of Violent Conflict and its Correlates in Prehistoric Hunter-Gatherer Societies of Coastal Southern California*. PhD dissertation, Department of Anthropology, University of California, Santa Barbara.

Lange, C. 1996. Silent victims: Women and children in warfare. Paper presented in the symposium Writing Horror: Archaeologies of Violence, the Eighteenth Annual Conference of the Theoretical Archaeology Group, University of Liverpool, December 16–18.

Lauro, James, and Geoffrey R. Lehmann. 1982. *The Slate Site: A Poverty Point Lapidary Industry in the Southern Yazoo Basin, Mississippi*. Mississippi Department of Archives and History, Jackson.

Leach, Edmund R. 1950. Primitive calendars. *Oceania* 20(4): 245–262.

Leach, Jerry. 1983. *The Kula*. Cambridge University Press, Cambridge.

LeBlanc, Steven A. 2003. *Constant Battles: The Myth of the Peaceful, Noble Savage*. Macmillan, New York.

LeClair, Edward. 1962. Economic theory and economic anthropology. *American Anthropologist* 64(6): 1179–1203.

Le Clair, Edward Emil, and Harold K. Schneider, eds. 1968. *Economic Anthropology: Readings in Theory and Analysis*. Holt, Rinehart and Winston, New York.

Lee, Richard B. 1990. Primitive communism and the origin of social inequality. In *The Evolution of Political Systems: Sociopolitics in Small-Scale Sedentary Societies*, edited by S. Upham, pp. 225–246. Press Syndicate of the University of Cambridge, Cambridge.

Levine, Abigail, and Charles Stanish. 2014. The importance of multiple 14C dates from significant archaeological contexts. *Journal of Archaeological Method and Theory* 21(4): 824–836.

Levine, David K., and Wolfgang Pesendorfer. 2002. The evolution of cooperation through imitation. IMBS Colloquium. UC Irvine: Institute for Mathematical Behavioral Sciences. https://escholarship.org/uc/item/45g5k560 (accessed June 5, 2015).

Levine, Frances. 1991. Economic perspectives on the Comanchero trade. In *Farmers, Hunters, and Colonists*, edited by K. Spielmann, pp. 155–169. University of Arizona Press, Tucson.

Levine, Frances, and Anna LaBauve. 1997. Examining the complexity of historic population decline: A case study of Pecos Pueblo, New Mexico. *Ethnohistory* 44(1): 75–112.

Lima, Steven L. 1989. Iterated Prisoner's Dilemma: An approach to evolutionarily stable cooperation. *American Naturalist* 134(5): 828–834.

Lindauer, Owen, and John H. Blitz. 1997. Higher ground: The archaeology of North American platform mounds. *Journal of Archaeological Research* 5(2): 169–207.

Lindstrom, Lamont. 1981. Big Man: A short terminological history. *American Anthropologist* 83(4): 900–905.

Lienhardt, Godfrey. 1958. The Western Dinka. In *Tribes without Rulers: Studies in African Segmentary Systems*, edited by John Middleton and David Tait, pp. 97–135. Routledge and Kegan Paul, London.

Lipo, Carl P., Timothy D. Hunt, and Robert C. Dunnell. 2012. Formal analyses and functional accounts of groundstone plummets from Poverty Point, Louisiana. *Journal of Archaeological Science* 39(1): 84–91.

Lipson, Ephraim. 1937. *The Economic History of England*. Black, London.

Lowie, Robert Harry. 1920. *Primitive Society*. Boni and Liveright, New York.

Lubbock, John. 1882. *Origins of Civilisation and the Primitive Condition of Man*. 4th ed. Longmans, Green, London.

Lumbreras, Luis Guillermo. 2001. Uhle y los asentamientos de Chincha en el siglo XVI. *Revista del Museo Nacional* 49: 13–87.

Luning, Sabine. 2005. Ritual territories and dynamics in the annual bush fire practices of Maane, Burkina Faso. Patrimoines naturels, territoires et identités. Éditions de l'IRD, colloques et séminaires, 443–473. http://horizon.documentation.ird.fr/exl-doc/pleins_textes/divers10-07/010037555.pdf (accessed June 3, 2015).

MacDonald, Kevin. 2002[1994]. *A People That Shall Dwell Alone: Judaism as a Group Evolutionary Strategy with Diaspora Peoples*. iUniverse.

Macfarlan, Shane J., Robert S. Walker, Mark V. Flinn, and Napoleon A. Chagnon. 2014. Lethal coalitionary aggression and long-term alliance formation among Yanomamö men. *Proceedings of the National Academy of Sciences* 111(47): 16662–16669.

Macneil, Ian R. 1986. Exchange revisited: Individual utility and social solidarity. *Ethics* 96(3): 567–593.

Mahdi, Niloufer Qasim. 1986. Pukhtunwali: Ostracism and honor among the Pathan hill tribes. *Ethology and Sociobiology* 7(3): 295–304.

Malinowski, Bronislaw. 1918. Fishing in the Trobriand Islands. *Man* 18: 87–92.

1920a. Kula: The circulating exchange of valuables in the archipelagoes of Eastern New Guinea. *Man* 20: 97–105.

1920b. War and weapons among the natives of the Trobriand Islands. *Man* 20: 10–12.

1921. The primitive economics of the Trobriand Islanders. *Economic Journal* 31(121): 1–16.

1927. Lunar and seasonal calendars in the Trobriands. *Journal of the Anthropological Institute of Great Britain and Ireland* 57: 203–215.

1948. *Magic, Science and Religion and Other Essays*. Beacon Press, Boston.

1966[1922]. *Argonauts of the Western Pacific*. G. Routledge, London.

Malville, J. McKim. 2014. The Casma Valley of Peru: A cradle of pre-Inca astronomy. *Mediterranean Archaeology and Archaeometry* 14(3): 169–178.

Man, Edward Horace. 1975[1883]. *The Aboriginal Inhabitants of the Andaman Islands*. Sanskaran Prakashak, Delhi.

Marcus, Joyce. 2008. The archaeological evidence for social evolution. *Annual Review of Anthropology* 37: 251–266.

Marcus, Joyce, and Kent V. Flannery. 1996. *Zapotec Civilization: How Urban Society Evolved in Mexico's Oaxaca Valley*. Thames and Hudson, New York.

2004. The coevolution of ritual and society: New 14C dates from ancient Mexico. *Proceedings of the National Academy of Sciences* 101(52): 18257–18261.

Marsh, Abigail A., Sarah A. Stoycos, Kristin M. Brethel-Haurwitz, Paul Robinson, John W. VanMeter, and Elise M. Cardinale. 2014. Neural and cognitive characteristics of extraordinary altruists. *Proceedings of the National Academy of Sciences* 111(42): 15036–15041.

Martinoia, Rozenn. 2003. That which is desired, which pleases, and which satisfied: Utility according to Alfred Marshall. *Journal of the History of Economic Thought* 25(3): 349–364.

Maryanski, Alexandra. 2010. WEIRD societies may be more compatible with human nature. *Behavioral and Brain Sciences* 33(2–3): 103–104.

Maschner, Herbert D. G., and Katherine L. Reedy-Maschner. 1998. Raid, retreat, defend (repeat): The archaeology and ethnohistory of warfare on the North Pacific rim. *Journal of Anthropological Archaeology* 17: 19–51.

Mathew, Sarah, and Robert Boyd. 2011. Punishment sustains large-scale cooperation in prestate warfare. *Proceedings of the National Academy of Sciences* 108(28): 11375–11380.

Mauss, Marcel. 1950. *Essai sur le don. Forme et raison de l'échange dans les sociétés archaïques*. Presses universitaires de France, Paris.

McCluskey, Stephen C. 1977. The astronomy of the Hopi Indians. *Journal for the History of Astronomy* 8: 174–195.

1982. Historical archaeoastronomy: The Hopi example. In *Archaeoastronomy in the New World*, vol. 1, edited by A. Aveni, pp. 11–58. Cambridge University Press, Cambridge.

2015. Hopi and Puebloan ethnoastronomy and ethnoscience. In *Handbook of Archaeoastronomy and Ethnoastronomy*, edited by C. Ruggles, pp. 649–658. Springer, New York.

McDougall, William. 1908. *An Introduction to Social Psychology*. Methuen, London.

McNeill, Desmond. 1990. Alternative interpretations of Aristotle on exchange and reciprocity. *Public Affairs Quarterly* 4(1): 55–68.

Mead, Margaret, ed. 1937. *Cooperation and Competition among Primitive Peoples*. McGraw-Hill, New York.

Meggitt, Mervyn J. 1958. Mae Enga time reckoning and calendar. *Man* 58(87): 74–77.

1967. The pattern of leadership among the Mae-Enga of New Guinea 1. *Anthropological Forum* 2(1): 20–35.

1974. "Pigs are our hearts!" The Te exchange cycle among the Mae Enga of New Guinea. *Oceania* 44(3): 165–203.

Mensforth, Robert P. 2001. Warfare and trophy taking in the Archaic period. In *Archaic Transitions in Ohio and Kentucky Prehistory*, edited by Olaf Prufer, Sara E. Pedde, and Richard S. Meindl, pp. 110–138. Kent State University Press, Kent, OH.

Mercer, Roger, and Frances Healy. 2008. *Hambledon Hill, Dorset, England: Excavation and Survey of a Neolithic Monument Complex and Its Surrounding Landscape*. English Heritage, London.

Miles, Douglas. 1964. The Ngadju longhouse. *Oceania* 35(1): 45–57.

Mill, John Stuart. 1874[1844]. *Essays on Some Unsettled Questions of Political Economy*. Longmans, London.

Miller, Bruce G., and Daniel L. Boxberger. 1994. Creating chiefdoms: The Puget Sound case. *Ethnohistory* 41(2): 267–293.

Mirsky, Jeannette. 1937. The Eskimo of Greenland. In *Cooperation and Competition among Primitive Peoples*, edited by M. Mead, pp. 51–86. McGraw-Hill, New York.

Mitchell, Wesley C. 1910. The rationality of economic activity. *Journal of Political Economy* 18(3): 197–216.

Mitchell, William P. 1976. Irrigation and community in the central Peruvian highlands. *American Anthropologist* 78(1): 25–44.

Mithen, Steven, and Bill Finlayson. 2000. WF16, a new PPNA site in southern Jordan. *Antiquity* 74(283): 11–12.

Mithen, Steven J., Bill Finlayson, Sam Smith, Emma Jenkins, Mohammed Najjar, and Darko Maricevic. 2011. An 11,600 year-old communal structure from the Neolithic of southern Jordan. *Antiquity* 85(328): 350–364.

Monaghan, John. 1990. Reciprocity, redistribution, and the transaction of value in the Mesoamerican fiesta. *American Ethnologist* 17(4): 758–774.

1996a. The archaeology of plazas and the proxemics of ritual: Three Andean traditions. *American Anthropologist* 98(4): 789–802.

1996b. *Architecture and Power in the Ancient Andes: The Archaeology of Public Buildings*. Cambridge University Press, Cambridge.

2007a. Early fifth millennium BC Prehispanic architecture from far northern Peru. *Antiquity* 81(312). www.antiquity.ac.uk/projgall/moore1 (accessed September 13, 2013).

2007b. Proyecto Arqueológico Tumbes: 2006 Annual Report. Annual Report to National Science Foundation – Submitted March 16, 2007. Proyecto Arqueológico Tumbes. www.cla.csudh.edu/project/exreports.pdf (accessed September 13, 2013).

Morgan, Lewis Henry. 1962[1851]. *League of the Iroquois*. Corinth Books, New York.

Morgenstern, Oskar. 1976. The collaboration between Oskar Morgenstern and John von Neumann on the theory of games. *Journal of Economic Literature* 14(3): 805–816.

Moreno, Manual. 1987. Agriculture as a sacrament: A new approach to the cycle of rice in south India. http://soar.wichita.edu/handle/10057/1794 (accessed October 23, 2014).

Morin, Jesse. 2006. Non-domestic architecture in prehistoric complex hunter-gatherer communities: An example from Keatley Creek, on the Canadian Plateau of British Columbia. MA thesis, Department of Anthropology, University of British Columbia.

Moseley, Michael E. 1975. Prehistoric principles of labor organization in the Moche Valley, Peru. *American Antiquity* 40(2): 191–196.

1992. *The Incas and Their Ancestors: The Archaeology of Peru*. Thames and Hudson, New York.

Munier, Bertrand, and Costin Zaharia. 2002. High stakes and acceptance behavior in Ultimatum bargaining: A contribution from an international experiment. *Theory and Decision* 53(3): 187–207.

Myers, Milton. 1983. *The Soul of Modern Economic Man: Ideas of Self-Interest, Thomas Hobbes to Adam Smith*. University of Chicago Press, Chicago.

Naveh, Danny. 2003. PPNA Jericho: A socio-political perspective. *Cambridge Archaeological Journal* 13(1): 83–96.

Nettle, Daniel, and Robin I. M. Dunbar. 1997. Social markers and the evolution of reciprocal exchange. *Current Anthropology* 38(1): 93–99.

Newcomb, William W. 1950. A re-examination of the causes of Plains warfare. *American Anthropologist* 52(3): 317–330.

Newlander, Khori S. 2012. *Exchange, Embedded Procurement, and Hunter-Gatherer Mobility: A Case Study from the North American Great Basin*. PhD dissertation, Department of Anthropology, Hamilton College, New York.

Norenzayan, Ara, and Azim F. Shariff. 2008. The origin and evolution of religious prosociality. *Science* 322(5898): 58–62.

North, Douglass C. 1977. Markets and other allocation systems in history: The challenge of Karl Polanyi. *Journal of European Economic History* 6(3): 703–716.

1989. Institutions and economic growth: An historical introduction. *World Development* 17(9): 1319–1332.

1990. A transaction cost theory of politics. *Journal of Theoretical Politics* 2(4): 355–367.

Novakova, Julie, and Jaroslav Flegr. 2013. How much is our fairness worth? The effect of raising stakes on offers by proposers and minimum acceptable offers in dictator and ultimatum games. *PLoS ONE* 8(4): e60966. doi:10.1371/journal.pone.0060966.

Nowak, Martin A., Karen Page, and Karl Sigmund. 2000. Fairness versus reason in the Ultimatum Game. *Science* 289: 1773–1775.

Nuñéz, Lautaro. 1984. El asentamiento Pircas: Nuevas evidencias de tempranas ocupaciones agrarias en el norte de Chile. *Estudios Atacameños* 7: 117–134.

Oberg, Kalervo. 1943. A comparison of three systems of primitive economic organization. *American Anthropologist* 45(4): 572–587.

Oliver, Douglas L. 1949. *Studies in the Anthropology of Bougainville, Solomon Islands*. Peabody Museum of American Archaeology and Ethnology, Cambridge, MA.

1974. *Ancient Tahitian Society*. Vol. I. University Press of Hawaii, Honolulu.

Onuki, Yoshio. 2012. El Periodo Arcaico en Huánuco y el concepto del Arcaico. *Boletín de Arqueología PUCP* 3: 325–333.

Opler, Morris E., and Harry Hoijer. 1940. The raid and war-path language of the Chiricahua Apache. *American Anthropologist* 42(4): 617–634.

Ortmann, Anthony L. 2007. *The Poverty Point Mounds: Analysis of the Chronology, Construction History, and Function of North America's Largest Hunter-Gatherer Monuments*. PhD dissertation, Department of Anthropology, Tulane University, New Orleans, LA.

Overing, Joanna. 1993. The anarchy and collectivism of the "primitive other": Marx and Sahlins in the Amazon. In *Socialism: Ideals, Ideologies, and Local Practice*, edited by C. M. Hann, pp. 20–37. Routledge, Oxon.

Parker Pearson, Mike, Ros Cleal, Peter Marshall, Stuart Needham et al. 2007. The age of Stonehenge. *Antiquity* 81(313): 617–639.

Parker Pearson, Mike, Josh Pollard, Colin Richards, Julian Thomas, Christopher Tilley, Kate Welham, and Umberto Albarella. 2006. Materializing Stonehenge: The Stonehenge riverside project and new discoveries. *Journal of Material Culture* 11(1–2): 227–261.

Pauketat, Timothy R. 2012. *An Archaeology of the Cosmos: Rethinking Agency and Religion in Ancient America*. Routledge, New York.

Pearsall, Deborah M. 2008. Plant domestication and the shift to agriculture in the Andes. In *The Handbook of South American Archaeology*, edited by H. Silverman and W. Isbell, pp. 105–120. Springer, New York.

Peatfield, Alan A. D., and Christine Morris. 2012. Dynamic spirituality on Minoan peak sanctuaries. In *Archaeology of Spiritualities*, edited by K. Rountree, C. Morris, and A. Peatfield, pp. 227–245. Springer, New York.

Pels, Peter. 1997. The anthropology of colonialism: Culture, history, and the emergence of Western governmentality. *Annual Review of Anthropology* 26: 163–183.

Perelman, Michael. 2000. *The Invention of Capitalism: Classical Political Economy and the Secret History of Primitive Accumulation*. Duke University Press, Durham, NC.

Peros, Matthew C., Samuel E. Munoz, Konrad Gajewski, and André E. Viau. 2009. Prehistoric demography of North America inferred from radiocarbon data. *Journal of Archaeological Science* 37(3): 656–664.

Perrot, Nicolas. 1911. *The Indian Tribes of the Upper Mississippi Valley and Region of the Great Lakes*. Arthur H. Clark, Cleveland, OH.

Persky, Joseph. 1995. The ethology of Homo economicus. *Journal of Economic Perspectives* 9(2): 221–231.

Peters, Joris, and Klaus Schmidt. 2004. Animals in the symbolic world of Pre-Pottery Neolithic Göbekli Tepe, south-eastern Turkey: A preliminary assessment. *Anthropozoologica* 39(1): 179–218.

Petersen, Glenn. 1982. *One Man Cannot Rule a Thousand: Fission in a Ponapean Chiefdom*. University of Michigan Press, Ann Arbor.

Peterson, Jean Treloggen. 1978. Hunter-gatherer/farmer exchange. *American Anthropologist* 80(2): 335–351.

Pfeiffer, Thomas, Claudia Rutte, Timothy Killingback, Michael Taborsky, and Sebastian Bonhoeffer. 2005. Evolution of cooperation by generalized reciprocity. *Proceedings of the Royal Society B: Biological Sciences* 272(1568): 1115–1120.

Piatek, Bruce John. 1994. The Tomoka mound complex in northeast Florida. *Southeastern Archaeology* 13(2): 109–118.

Phillips, Ruth B. 1978. Masking in Mende Sande society initiation rituals. *Africa: Journal of the International African Institute* 48(3): 265–277.

Pigott, Vincent C. 1998. Prehistoric copper mining in the context of emerging community craft specialisation in northeast Thailand. In *Social Approaches to an Industrial Past: The Archaeology and Anthropology of Mining*, edited by E. W. Herbert, A. Bernard Knapp, and V. C. Pigott, pp. 205–225. Routledge, London.

Pinker, Steven. 2012. The false allure of group selection. http://edge.org/conversation/the-false-allure-of-group-selection#rbsm.

Plattner, Stuart. 1989. Introduction. In *Economic Anthropology*, edited by S. Plattner, pp. 1–20. Stanford University Press, Palo Alto, CA.

Polanyi, Karl. 1944. *The Great Transformation*. Reinhart, New York.

Posey, Darrell A. 1982. Time, space, and the interface of divergent cultures: The Kayapó Indians of the Amazon face the future. *Revista de Antropología* 25: 89–104.

Powers, Simon T., and Laurent Lehmann. 2014. An evolutionary model explaining the Neolithic transition from egalitarianism to leadership and despotism. *Proceedings of the Royal Society B: Biological Sciences*, http://classic.rspb.royalsocietypublishing.org/content/281/1791/20141349.full.

Pozorski, Shelia, and Thomas Pozorski. 2008. Early cultural complexity on the coast of Peru. In *The Handbook of South American Archaeology*, edited by H. Silverman and W. Isbell, pp. 607–631. Springer, New York.

Price, Neil. 2002. *The Viking Way: Religion and War in Late Iron Age Scandinavia.* Uppsala University, Humanistisk-samhällsvetenskapliga vetenskapsområdet, Faculty of Arts, Department of Archaeology and Ancient History, Archaeology, Uppsala, Sweden.

Price, T. Douglas, and Ofer Bar-Yosef. 2011. The origins of agriculture: New data, new ideas. *Current Anthropology* 52(S4): S163–S174.

Provinse, John H. 1937. Cooperative ricefield cultivation among the Sian Dyaks of Central Borneo. *American Anthropologist* 39(1): 77–102.

Purrington, Robert D. 1983. Supposed solar alignments at Poverty Point. *American Antiquity* 48(1): 157–161.

Quilter, Jeffrey. 1985. Architecture and chronology at El Paraíso, Peru. *Journal of Field Archaeology* 12(3): 279–297.

1991. Late Preceramic Peru. *Journal of World Prehistory* 5(4): 387–438.

2013. *The Ancient Central Andes.* Routledge World Archaeology, Routledge, London.

Quilter, Jeffrey, and Terry Stocker. 1983. Subsistence economies and the origins of Andean complex societies. *American Anthropologist* 85(3): 545–562.

Quinn, Rhonda L., Bryan D. Tucker, and John Krigbaum. 2008. Diet and mobility in Middle Archaic Florida: Stable isotopic and faunal evidence from the Harris Creek archaeological site (8Vo24), Tick Island. *Journal of Archaeological Science* 35(8): 2346–2356.

Rabin, Matthew. 1993. Incorporating fairness into game theory and economics. *American Economic Review* 83(5): 1281–1302.

Radcliffe-Brown, Alfred R. 1949. Functionalism: A protest. *American Anthropologist* 51(2): 320–323.

Rand, David G., Corina E. Tarnita, Hisashi Ohtsuki, and Martin A. Nowak. 2013. Evolution of fairness in the one-shot anonymous Ultimatum Game. *Proceedings of the National Academy of Sciences* 110(7): 2581–2586.

Randall, Asa. 2008. Archaic shell mounds of the St. Johns River, Florida. *SAA Archaeological Record* 8(5): 13–17.

Rankin, Daniel J., and Michael Taborsky. 2009. Assortment and the evolution of generalized reciprocity. *Evolution* 63(7): 1913–1922.

Rappaport, Roy A. 1967. Ritual regulation of environmental relations among a New Guinea people. *Ethnology* 6(1): 17–30.

Rashid, Salim. 1985. Review of "The Soul of Modern Economic Man: Ideas of Self-Interest, Thomas Hobbes to Adam Smith" by Milton Myers. *History of Political Economy* 17(2): 330–332.

Raulet, Harry M. 1974. Review of "Population Growth: Anthropological Implications," edited by Brian Spooner. *Contemporary Sociology* 3(4): 371–373.

Reay, Marie. 1959. *The Kuma: Freedom and Conformity in the New Guinea Highlands.* Melbourne University Press, Melbourne.

Redman, Deborah A. 1997. *The Rise of Political Economy as a Science: Methodology and the Classical Economists.* MIT Press, Boston.

Reiche, María. 1949. *The Mystery on the Desert.* Privately printed, Lima.

1993. *Contribuciones a la geometría y astronomía en el antiguo Perú. Asociación María Reiche para las Líneas de Nasca.* Epígrafe Editores S. A., Lima.

Reichhardt, Tony. 2006. Well-being research: A measure of happiness. *Nature* 444: 418–419.

Reinhard, Johan. 1986. *The Nazca Lines: A New Perspective on Their Origin and Meaning*. Editorial los Piños, Lima.

Renfrew, Colin. 1983. The social archaeology of megalithic monuments. *Scientific American* 249: 152–163.

1994. The archaeology of religion. In *The Ancient Mind: Elements of Cognitive Archaeology*, edited by C. Renfrew and Ezra B. W. Zubrow, pp. 47–54. Cambridge University Press, Cambridge.

2008. Neuroscience, evolution and the sapient paradox: The factuality of value and of the sacred. *Philosophical Transactions of the Royal Society B: Biological Sciences* 363 (1499): 2041–2047.

2013. The sanctuary at Keros: Questions of materiality and monumentality. *Journal of the British Academy* 1: 187–212.

Renfrew, Colin, Michael Boyd, and C. Bronk Ramsey. 2012. The oldest maritime sanctuary? Dating the sanctuary at Keros and the Cycladic Early Bronze Age. *Antiquity* 86(331): 144–160.

Renfrew, Colin, Judson T. Chesterman, and Martin Jim Aitken. 1979. *Investigations in Orkney*. Society of Antiquaries of London, Vol. 38. Distributed by Thames and Hudson, London.

Renneboog, Luc, and Christophe Spaenjers. 2012. Religion, economic attitudes, and household finance. *Oxford Economic Papers* 64(1): 103–127.

Reyna, Stephen, and R. E. Downs. 1999. *Deadly Developments: Capitalism, States and War*. Gordon and Breach, Amsterdam.

Richards, Colin. 1996. Henges and water: Towards an elemental understanding of monumentality and landscape in Late Neolithic Britain. *Journal of Material Culture* 1(3): 313–336.

Richerson, Peter J., Robert Boyd, and Robert L. Bettinger. 2001. Was agriculture impossible during the Pleistocene but mandatory during the Holocene? A climate change hypothesis. *American Antiquity* 66(3): 387–411.

Riehl, Simone, Mohsen Zeidi, and Nicholas J. Conard. 2013. Emergence of agriculture in the foothills of the Zagros Mountains of Iran. *Science* 341(6141): 65–67.

Rodríguez, Aurelio. 1997. *Los campos de geoglífos en la costa central del Perú*. Cuadernos de Investigacíon 9. Pontifica Universidad Catolica del Perú, Lima.

Roe, Lori M. 2010. *Social Complexity and Mound Ceremony in the Coles Creek Culture: Research at the Raffman Mound Center in Madison Parish, Louisiana*. PhD dissertation, Department of Anthropology, Tulane University, New Orleans, LA.

Roscoe, Paul. 2009. Social signaling and the organization of small-scale society: The case of contact-era New Guinea. *Journal of Archaeological Method and Theory* 16(2): 69–116.

2012. Before elites: The political capacities of Big Men. In *Before Elites: Alternatives to Hierarchical Systems in Modelling Social Formations*, Vol. 1, edited by T. L. Kienlin and A. Zimmerman, pp. 41–54. Universitätsforschungen zur prähistorischen Archäologie, Rudolph Habelt, Bonn.

Rosman, Abraham, and Paula Rubel. 1971. *Feasting with Mine Enemy*. Columbia University Press, New York.

2013. Colonialism and the efflorescence of warfare: The New Ireland case. In *Deadly Developments, Capitalism, States and War*, edited by S. Reyna and R. E. Downs, pp. 69–85. Gordon and Breach, Amsterdam.

Roth, Henry Ling. 1968[1896]. *The Natives of Sarawak and British North Borneo*. 2 vols. University of Malaya Press, Kuala Lampur.

Rother, P. 1906. Wie es von der Zeit der deutschen Herrschaft in Pare aussah. *Evangelisch Lutherisches Missionsblatt* 117–122.

Rothman, Mitchell S. 2000. The commoditization of goods and the rise of the state in ancient Mesopotamia. In *Commodities and Globalization: Anthropological Perspectives*, edited by A. Haugerud, M. Priscilla Stone, and P. D. Little, pp. 163–178. Society for Economic Anthropology, Rowman and Littlefield, New York.

Rousseau, Jean-Jacques. 1755[2011]. *Discourse on the Origin and Foundations of Inequality among Men*. Translated, edited, and with an introduction by Helena Rosenblatt. Bedford/St. Martins, New York.

Rowan, Yorke M. 2011. Beyond belief: The archaeology of religion and ritual. *Archeological Papers of the American Anthropological Association* 21(1): 1–10.

Ruggles, Clive. 2014. Geoglyphs of the Peruvian Coast. In *Handbook of Archaeoastronomy and Ethnoastronomy*, edited by C. Ruggles, pp. 821–830. Springer, New York.

Ruggles, Clive, and Nicolas Saunders. 2012. Desert labyrinth: Lines, landscape and meaning at Nazca, Peru. *Antiquity* 86(334): 1126–1140.

Russo, Michael. 1994. Why we don't believe in Archaic ceremonial mounds and why we should: The case from Florida. *Southeastern Archaeology* 13(2): 93–109.

2008. Late Archaic shell rings and society in the Southeast US. *SAA Archaeological Record* 8(5): 18–22.

Sahlins, Marshall. 1972. *Stone Age Economics*. Aldine, Chicago.

Salomon, Frank. 1987. A north Andean status trader complex under Inka rule. *Ethnohistory* 34(1): 63–77.

Samaniego, Lorenzo, Enrique Vergara, and Henning Bischof. 1985. New evidence on Cerro Sechín, Casma Valley, Peru. In *Early Ceremonial Architecture in the Andes*, edited by C. Donnan, pp. 165–190. Dumbarton Oaks, Washington, DC.

Sanderson, Stephen K. 2001. *The Evolution of Human Sociality: A Darwinian Conflict Perspective*. Rowan and Littlefield, Lanham, MD.

Sanfey, Alan G., James K. Rilling, Jessica A. Aronson, Leigh E. Nystrom, and Jonathan D. Cohen. 2003. The neural basis of economic decision-making in the Ultimatum Game. *Science* 300(5626): 1755–1758.

Sassaman, Kenneth E. 1995. The cultural diversity of interactions among mid-Holocene societies of the American Southeast. In *Native American Interactions*, edited by M. Nassaney and K. Sassaman, pp. 174–204. University of Tennessee Press, Knoxville.

2005. Poverty Point as structure, event, process. *Journal of Archaeological Method and Theory* 12(4): 335–364.

2008. The new Archaic: It ain't what it used to be. *SAA Archaeological Record* 8(5): 6–8.

Sassaman, Kenneth E., Meggan E. Blessing, and Asa R. Randall. 2006. Stallings Island revisited: New evidence for occupational history, community pattern, and subsistence technology. *American Antiquity* 71(3): 539–565.

Sassaman, Kenneth E., and Michael J. Heckenberger. 2004. Crossing the symbolic Rubicon in the Southeast. In *Signs of Power: The Rise of Cultural Complexity in the Southeast*, edited by J. L. Gibson and P. J. Carr, pp. 214–233. University of Alabama Press, Tuscaloosa.

Sassaman, Kenneth E., and Asa R. Randall. 2012. Shell mounds of the middle St. Johns Basin, northeast Florida. In *Early New World Monumentality*, edited by Richard Burger and Robert Rosenswig, pp. 53–72. University Press of Florida, Gainesville.

Saunders, Joe. 2012. Early mounds in the lower Mississippi Valley. In *Early New World Monumentality*, edited by R. Burger and R. Rosenswig, pp. 25–52. University Press of Florida, Gainesville.

Saunders, Joe W., Rolfe D. Mandel, C. Garth Sampson, Charles M. Allen, E. Thurman Allen, Daniel A. Bush, James K. Feathers et al. 2005. Watson Brake, a Middle Archaic mound complex in northeast Louisiana. *American Antiquity* 70(4): 631–668.

Saunders, Joe W., Rolfe D. Mandel, Roger T. Saucier, E. Thurman Allen, C. T. Hallmark, Jay K. Johnson, Edwin H. Jackson, Charles M. Allen, Gary L. Stringer, Douglas S. Frink, James K. Feathers, Stephen Williams, Kristen J. Gremillion, Malcolm F. Vidrine, and Reca Jones. 1997. A mound complex in Louisiana at 5400–5000 years before the present. *Science* 277: 1796–1799.

Saunders, Rebecca. 2004. Spatial variation in Orange culture pottery. In *Early Pottery: Technology, Function, Style, and Interaction in the Lower Southeast*, edited by Rebecca Saunders and Christopher Hays, pp. 40–62. University of Alabama Press, Tuscaloosa.

Saunders, Rebecca, and Michael Russo. 2011. Coastal shell middens in Florida: A view from the Archaic period. *Quaternary International* 239(1): 38–50.

Scaglion, Richard. 1996. Chiefly models in Papua New Guinea. *The Contemporary Pacific* 8(1): 1–31.

Schmidt, Klaus. 2000. Göbekli Tepe, southeastern Turkey: A preliminary report on the 1995–1999 excavations. *Paleórient* 26(1): 45–54.

2003. The 2003 Campaign at Göbekli Tepe (Southeastern Turkey). *Neolithics* 2(3): 3–7.

2006. *Sie bauten die ersten Tempel: das rätselhafte Heiligtum der Steinzeitjäger: die archäologische Entdeckung am Göbekli Tepe*. Beck, Munich.

Schneider, Jane. 2002. World markets: Anthropological perspectives. In *Exotic No More: Anthropology on the Front Lines*, edited by J. MacClancy, pp. 64–85. University of Chicago Press, Chicago.

Schwartz, Douglas W. 1957. Climate change and culture history in the Grand Canyon region. *American Antiquity* 22(4): 372–377.

Seaton, S. Lee. 1974. The Hawaiian kapu abolition of 1819. *American Ethnologist* 1(1): 193–206.

Shady Solis, Ruth. 2000. Sustento socioeconómico del estado pristina de Supe-Perú: las evidencias de Caral-Supe. *Arqueología y Sociedad* 13: 49–66.

Shady Solis, Ruth, Daniel Cáceda Guillén, Aldemar Crispín Balta, Marco Machacuay Romero, Pedro Novoa Bellota, and Edna Quispe Loayza. 2009. *Caral, La civilización más Antigua de las Américas: 15 Años Revelendo su Historia*. Instituto Nacional de Cultura, Lima.

Shady Solis, Ruth, Jonathan Haas, and Winifred Creamer. 2001. Dating Caral, a preceramic site in the Supe Valley on the central coast of Peru. *Science* 292: 723–726.

Shady Solis, Ruth, and Christopher Kleihege. 2008. *Caral. La primera civilización de América*. Universidad de San Martín de Porres, Lima.

Shady Solis, Ruth, and Carlos Leyva, eds. 2003. *La Ciudad Sagrada de Caral-Supe*. Instituto Nacional de Cultura, Lima.

Shalizi, Cosma Rohilla. 1999. Homo reciprocans: Political economy and cultural evolution. *Bulletin of the Santa Fe Institute* 14(2): 16–20.

Shields, Christopher J. 2007. *Aristotle*. Routledge, London.

Shorter, Aylward. 1972. *Chiefship in Western Tanzania: A Political History of the Kimbu*. Clarendon Press, Oxford.

Sigmund, Karl, Christoph Hauert, and Martin A. Nowak. 2001. Reward and punishment. *Proceedings of the National Academy of Sciences* 98(19): 10757–10762.

Sikkink, Lynn. 1997. Water and exchange: The ritual of "yaku cambio" as communal and competitive encounter. *American Ethnologist* 24(1): 170–189.

Silverman, Helaine. 1990. The early Nasca pilgrimage center of Cahuachi: Archaeological and anthropological perspectives. In *The Lines of Nazca*, edited by A. Aveni, pp. 209–244. American Philosophical Society, Philadelphia.

2002a. *Ancient Nasca Settlement and Society*. University of Iowa Press, Iowa City.

2002b. Nasca settlement and society on the hundredth anniversary of Uhle's discovery of the Nasca style. In *Andean Archaeology I: Variations in Sociopolitical Organization*, edited by W. Isbell and H. Silverman, pp. 121–159. Kluwer Academic/Plenum, New York.

Silverman, Helaine, and Donald Proulx. 2002. *The Nasca*. Blackwell, Malden, MA.

Simmons, Marc. 1979. History of Pueblo-Spanish relations to 1821. In *Handbook of North American Indians*, vol. 9, *Southwest*, edited by A. Ortiz, pp. 178–193. Smithsonian Institution, Washington, DC.

Slovak, Nicole, and Real Alto. 2003. Heilbrunn timeline of art history. Metropolitan Museum of Art. www.metmuseum.org/toah/hd/real/hd_real.htm (accessed April 21, 2015).

Smith, Adam. 1937[1776]. *The Wealth of Nations*. Modern Library, New York.

Smith, Eric A. 2003. Perspectives from behavioral ecology. In *Genetic and Cultural Evolution of Cooperation*, edited by P. Hammerstein, pp. 401–427. MIT Press, Boston.

Smith, Eric A., and R. Bliege Bird. 2005. Costly signaling and cooperative behavior. In *Moral Sentiments and Material Interests: The Foundations of Cooperation in Economic Life*, edited by H. Gintis, S. Bowels, R. Boyd, and E. Fehr, pp.115–148. MIT Press, Boston.

Smith, Michael E. 1980. The role of the marketing system in Aztec society and economy: Reply to Evans. *American Antiquity* 45(4): 876–883.

2006. How do archaeologists compare early states? *Reviews in Anthropology* 35(1): 5–35.

Smith, Philip E. L. 1972. Changes in population pressure in archaeological explanation. *World Archaeology* 4(1): 5–18.

Sosis, Richard, and Eric Bressler. 2003. Cooperation and commune longevity: A test of the costly signaling theory of religion. *Cross-Cultural Research* 37(2): 211–239.

Sosis, Richard, Howard C. Kress, and James S. Boster. 2007. Scars for war: Evaluating alternative signaling explanations for cross-cultural variance in ritual costs. *Evolution and Human Behavior* 28(4): 234–247.

Spawforth, Anthony J. S. 1989. Agonistic festivals in Roman Greece. *Bulletin of the Institute of Classical Studies* 36(S55): 193–197.

Spencer, Charles S. 1993. Human agency, biased transmission, and the cultural evolution of chiefly authority. *Journal of Anthropological Archaeology* 12: 41–74.

1997. Evolutionary approaches in archaeology. *Journal of Archaeological Research* 5(3): 209–264.

Spencer, Charles S., and Elsa M. Redmond. 1992. Prehispanic chiefdoms of the western Venezuelan Ilanos. *World Archaeology* 24(1): 134–157.

Spencer, Herbert. 1900. *First Principles*. 6th ed. D. Appleton, New York.

Speth, John D., and Katherine A. Spielmann. 1983. Energy source, protein metabolism, and hunter-gatherer subsistence strategies. *Journal of Anthropological Archaeology* 2(1): 1–31.

Spielmann, Katherine A. 1983. Late prehistoric exchange between the Southwest and Southern Plains. *Plains Anthropologist* 28(102, Part 1): 257–272.

2002. Feasting, craft specialization, and the ritual mode of production in small-scale societies. *American Anthropologist* 104(1): 195–207.

2008. Crafting the sacred: Ritual places and paraphernalia in small-scale societies. *Research in Economic Anthropology* 27: 37–72.

Spooner, Brian, ed. 1972. *Population Growth: Anthropological Implications*. MIT Press, Cambridge, MA.

Stanish, C. 1989. Household archaeology: Testing models of zonal complementarity in the south central Andes. *American Anthropologist* 91: 7–24.

1992 *Ancient Andean Political Economy*. University of Texas Press, Austin.

2001. The origins of the state in South America. *Annual Review of Anthropology* 30: 41–64.

2003. *Ancient Titicaca: The Evolution of Complex Society in Southern Peru and Northern Bolivia*. University of California Press, Berkeley.

2004. The evolution of chiefdoms: An economic anthropological model. In *Archaeological Perspectives on Political Economies*, edited by G. Feinman and L. Nichols, pp. 7–24. University of Utah Press, Salt Lake City.

2006. Prehispanic strategies of agricultural intensification in the Titicaca Basin of Peru and Bolivia. In *Agricultural Strategies*, edited by J. Marcus and C. Stanish, pp. 364–397. Cotsen Institute of Archaeology Press, Los Angeles, CA.

2008. Explanation in archaeology, overview. In *Encyclopedia of Archaeology*, edited by D. Pearsall, pp. 1358–1364. Academic Press, New York.

2009. The evolution of managerial elites in intermediate societies. In *The Evolution of Leadership: Transitions in Decision Making from Small-Scale to Middle-Range Societies*, edited by K. Vaughn, J. Eerkens, and J. Kanter, pp. 97–119. School of American Research, Santa Fe, NM.

2010. Labor taxes, market systems and urbanization in the Prehispanic Andes: A comparative perspective. In *Archaeological Approaches to Market Exchange in Ancient Societies*, edited by C. Garraty and B. Stark, pp. 185–205. University of Colorado Press, Boulder.

2012. The revaluation of landscapes in the Inca Empire as Peircean Replication. In *Construction of Value in the Ancient World*, edited by J. Papadopoulos and G. Urton, pp. 80–88. Cotsen Institute of Archaeology Press, Los Angeles, CA.

2013a. The ritualized economy and cooperative labor in intermediate societies. In *Cooperation and Collective Action: Archaeological Perspectives*, edited by D. M. Carballo, pp. 83–92. University Press of Colorado, Boulder.

2013b. What was Tiwanaku? In *Visions of Tiwanaku*, edited by A. Vranich and C. Stanish, pp. 151–166. Cotsen Institute of Archaeology Press, Los Angeles, CA.

Stanish, Charles, and Lawrence Coben. 2013. Barter markets in the pre-Hispanic Andes. In *Merchants, Markets, and Exchange in the Pre-Columbian World*, edited by K. G. Hirth and J. Pillsbury, pp. 419–434. Dumbarton Oaks, Washington, DC.

Stanish, Charles, and Kevin Haley. 2005. Power, fairness, and architecture: Modeling early chiefdom development in the central Andes. In *Power in the Prehispanic Andes*, edited by C. Conlee, D. Ogburn, and K. Vaughn, pp. 53–70. American Anthropological Association Monographs in Archaeology, Washington, DC.

Stanish, Charles, and Abigail Levine. 2011. War and early state formation in the northern Titicaca Basin, Peru. *Proceedings of the National Academy of Sciences* 108(34): 13901–13906.

Stanish, Charles, and Lee H. Steadman. 1994. *Archaeological Research at Tumatumani, Juli, Peru.* Fieldiana Anthropology, Field Museum of Natural History, Chicago.

Stanish, Charles, Henry Tantaleán, Benjamin Nigra, and Laura Griffin. 2014. A 2,300-year-old architectural and astronomical complex in the Chincha Valley, Peru. *Proceedings of the National Academy of Sciences* 11(20): 7218–7223. doi:10.1073/pnas.1406501111.

Strother, Zoë S. 2004. Architecture against the state: The virtues of impermanence in the Kibulu of Eastern Pende chiefs in central Africa. *Journal of the Society of Architectural Historians* 63(3): 272–295.

Sweet, Louise E. 1965. Camel raiding of North Arabian Bedouin: A mechanism of ecological adaptation. *American Anthropologist* 67(5): 1132–1150.

Swezey, Sean L., and Robert F. Heizer. 1977. Ritual management of salmonid fish resources in California. *Journal of California Anthropology* 4(1): 7–29.

Tambiah, Stanley Jeyaraja. 1985. *Culture, Thought, and Social Action: An Anthropological Perspective.* Harvard University Press, Cambridge, MA.

Tait, David. 1958. The territorial pattern and lineage system of Konkomba. In *Tribes without Rulers: Studies in African Segmentary Systems*, edited by J. Middleton and D. Tait, pp. 167–202. Routledge and Kegan Paul, London.

Tantaleán, Henry, Charles Stanish, Michiel Zegarra, Kelita Pérez, and Ben Nigra. 2014. Paracas en el Valle de Chincha: Nuevos Datos y Explicaciones. *Boletín de Arqueología PUCP* 17: 31–57.

Tello, Julio César. 1956. *Arqueología del Valle de Casma: culturas: Chavín, Santa o Huaylas Yunga y Sub-Chimú.* Vol. 1. Editorial San Marcos, Lima.

Thaler, Richard H. 1988. Anomalies: The Ultimatum Game. *Journal of Economic Perspectives* 2(4): 195–206.

Thomas, David Hurst. 1972. Western Shoshone ecology: Settlement patterns and beyond. In *Great Basin Cultural Ecology: A Symposium*, edited by D. D. Fowler, pp. 135–154. Desert Research Institute Publications in the Social Sciences No. 8, Reno, NV.

Tomasello, Michael, Malinda Carpenter, Josep Call, Tanya Behne, and Henrike Moll. 2005. Understanding and sharing intentions: The origins of cultural cognition. *Behavioral and Brain Sciences* 28(5): 675–691.

Toner, Mike. 2015. City of the Moon. *Archaeology* 68(2): 40–45.

Tooker, Elisabeth. 1964. An ethnography of the Huron Indians, 1615–1649. *Bulletin, Smithsonian Institution Bureau of American Ethnology*, 190. U.S. Government Printing Office, Washington, DC.

Trigger, Bruce Graham. 1963. Order and freedom in Huron society. *Anthropologica* 5(2): 151–169.

Tucker, Albert. 1983. The mathematics of Tucker: A sampler. *Two-Year College Mathematics Journal* 14(3): 228–232.

Turner, Billie L., and Marina Fischer-Kowalski. 2010. Ester Boserup: An interdisciplinary visionary relevant for sustainability. *Proceedings of the National Academy of Sciences* 107(51): 21963–21965.

Turner, Billie L., II, Robert Q. Hanham, and Anthony V. Portararo. 1977. Population pressure and agriculture intensity. *Annals of the Association of American Geographers* 67(3): 384–396.

Twiss, Katheryn C. 2008. Transformations in an early agricultural society: Feasting in the southern Levantine Pre-Pottery Neolithic. *Journal of Anthropological Archaeology* 27(4): 418–442.

Tykot, Robert H., and John E. Staller. 2002. The importance of early maize agriculture in coastal Ecuador: New data from La Emerenciana. *Current Anthropology* 43(4): 666–677.

Uberoi, J. P. Singh. 1971. *Politics of the Kula Ring: An Analysis of the Findings of Bronisław Malinowski*. Manchester University Press, Manchester.

Unkel, I., B. Kromer, M. Reindel, L. Wacker, and G. Wagner. 2007. A Chronology of the Pre-Columbian Paracas and Nasca cultures in South Peru based on AMS 14C Dating. *Radiocarbon* 49(2): 551–564.

Unkel, Ingmar, Markus Reindel, Hermann Gorbahn, Johny Isla Cuadrado, Bernd Kromer, and Volker Sossna. 2012. A comprehensive numerical chronology for the pre-Columbian cultures of the Palpa valleys, south coast of Peru. *Journal of Archaeological Science* 39(7): 2294–2303.

Urton, Gary. 1990. Andean social organization and the maintenance of the Nazca Lines. In *The Lines of Nazca*, edited by A. Aveni, pp. 173–206. American Philosophical Society, Philadelphia.

Urton, Gary, and John Papadopoulos. 2012. Introduction: The construction of value in the ancient world. In *The Construction of Value in the Ancient World*, edited by G. Urton and J. Papadopoulos, pp. 1–47. Cotsen Institute of Archaeology Press, Los Angeles, CA.

Van Gijseghem, H., and K. J. Vaughn. 2008. Regional integration and the built environment in middle-range societies: Paracas and early Nasca houses and communities. *Journal of Anthropological Archaeology* 27(1): 111–130.

Vaughn, Kevin J. 2004. Households, crafts, and feasting in the ancient Andes: The village context of early Nasca craft consumption. *Latin American Antiquity* 15(1): 61–88.

Vayda, Andrew P. 1971. Phases of the process of war and peace among the Marings of New Guinea. *Oceania* 42(1): 1–24.

1974. Warfare in ecological perspective. *Annual Review of Ecology and Systematics* 5: 183–193.

Von Hagen, Adriana, and Craig Morris. 1998. *The Cities of the Ancient Andes*. Thames and Hudson, New York.

von Neumann, John. 1928. Zur theorie der gesellschaftsspiele. *Mathematische Annalen* 100(1): 295–320.

von Rueden, Christopher, Michael Gurven, Hillard Kaplan, and Jonathan Stieglitz. 2014. Leadership in an egalitarian society. *Human Nature* 25(4): 538–566.

Wahl, Joachim, and H. G. König. 1987. Anthropologisch-traumatologische Untersuchung dermenschlichen Skelettreste aus dem bandkeramischen Massengrab bei Talheim, Kreis Heilbronn. *Fundberichte aus Baden-Württemberg* 12: 65–193.

Wallace, Dwight. 1971. *Valles de Chincha y de Pisco*. Arqueológicas 13. Museo Nacional de Antropología y Arqueología, Lima.

1985. Paracas in Chincha and Pisco: A reappraisal of the Ocucaje sequence. In *Recent Studies in Andean Prehistory and Protohistory*, edited by P. Kvietok and D. Sandweiss, pp. 67–94. Cornell University Latin American Studies Program, Ithaca, NY.

Wallerstein, Immanuel. 1984. *The Politics of the World-Economy: The States, the Movements and the Civilizations*. Cambridge University Press, Cambridge.

Watson, Aaron. 2001. Composing Avebury. *World Archaeology* 33(2): 296–314.

Watters, Ethan. 2013. We aren't the world. *Pacific Standard Magazine*. www.psmag.com/magazines/magazine-feature-story-magazines/joe-henrich-weird-ultimatum-game-shaking-up-psychology-economics-53135/ (accessed March 19, 2014).

Webb, Clarence H. 1968. The extent and content of Poverty Point culture. *American Antiquity* 33(3): 297–321.

1971. Archaic and Poverty Point zoomorphic locust beads. *American Antiquity* 36(1): 105–114.

1977. *The Poverty Point Culture*.Geoscience and Man, vol. 17, Louisiana State University School of Geoscience.

Webster, Hutton. 1908. *Primitive Secret Societies: A Study in Early Politics and Religion*. Macmillan, New York.

Wedekind, Claus. 1998. Game theory enhanced: Give and ye shall be recognized. *Science* 280: 2070–2071.

Wedekind, Claus, and Manfred Milinski. 1996. Human cooperation in the simultaneous and the alternating Prisoner's Dilemma: Pavlov versus generous Tit-for-Tat. *Proceedings of the National Academy of Sciences* 93(7): 2686–2689.

Weibull, Jörgen W. 1998. What have we learned from evolutionary game theory so far? IUI Working Paper No. 487. www.econstor.eu/bitstream/10419/94851/1/wp487.pdf (accessed June 4, 2015).

Weiner, Annette. 1988. *The Trobrianders of Papua New Guinea*. Thomson, Belmont, CA.

Wells, E. Christian, and Karla L. Davis-Salazar, eds. 2007. *Mesoamerican Ritual Economy: Archaeological and Ethnological Perspectives*. University Press of Colorado, Boulder.

Wendorf, Fred, and Romulad Schild. 2004. Late Paleolithic warfare in Nubia: The evidence and causes. *Adumata: A Semi-Annual Archaeological Refereed Journal on the Arab World* 10: 7–28.

Werner, Dennis. 1980. *The Making of a Mekranoti Chief: The Psychological and Social Determinants of Leadership in a Native South American Society*. PhD dissertation, Department of Anthropology City University of New York, New York.

1982. Chiefs and presidents. *Ethos* 10(2): 136–148.

White, Luise. 1994. Blood brotherhood revisited: Kinship, relationship, and the body in East and central Africa. *Africa* 64(3): 359–372.

Whiteley, Peter M. 1985. Unpacking Hopi "clans": Another vintage model out of Africa? *Journal of Anthropological Research* 41(4): 359–374.

1986. Unpacking Hopi "clans," II: Further questions about Hopi descent groups. *Journal of Anthropological Research* 42(1): 69–79.

Whittle, Alasdair. 1993. The Neolithic of the Avebury area: Sequence, environment, settlement and monuments. *Oxford Journal of Archaeology* 12(1): 29–53.

Whittle, Alasdair, Frances Healy, and Alex Bayliss. 2011. *Gathering Time: Dating the Early Neolithic Enclosures of Southern Britain and Ireland*. Oxbow, Oxford.

Widmer, Randolph J. 1994. The structure of southeastern chiefdoms. In *The Forgotten Centuries: Indians and Europeans in the American South 1521–1704*, edited by Charles M. Hudson and Carmen Chaves Tesser, pp. 125–155. University of Georgia Press, Athens.

Wiessner, Polly. 2001. Of feasting and value: Enga feasts in a historical perspective (Papua New Guinea). In *Feasts: Archaeological and Ethnographic Perspectives on Food, Politics, and Power*, edited by M. Dietler and B. Hayden, pp. 115–143. Smithsonian Institution Press, Washington, DC.

2002. The vines of complexity: Egalitarian structures and the institutionalization of inequality among the Enga. *Current Anthropology* 43(2): 233–269.

Wild, Eva Maria, Peter Stadler, Annemarie Häußer, Walter Kutschera, Peter Steier, Maria Teschler-Nicola, Joachim Wahl, and Helmut J. Windl. 2004. Neolithic massacres: Local skirmishes or general warfare in Europe? *Radiocarbon* 46: 377–385.

Willer, Robb. 2009. Groups reward individual sacrifice: The status solution to the collective action problem. *American Sociological Review* 74(1): 23–43.

Wilson, David John. 1983. The origins and development of complex Prehispanic society in the Lower Santa Valley, Peru: Implications for theories of state origins. *Journal of Anthropological Archaeology* 2(3): 209–276.

1988. *Prehispanic Settlement Patterns in the Lower Santa Valley, Peru: A Regional Perspective on the Origins and Development of Complex North Coast Society*. Smithsonian Institution Press, Washington, DC.

Wilson, David S., and Lee A. Dugatkin. 1997. Group selection and assortative interactions. *American Naturalist* 149(2): 336–351.

Wilson, David S., and Eliott Sober. 1994. Reintroducing group selection to the human behavioral sciences. *Behavioral and Brain Sciences* 17(4): 585–608.

Winiata, Maharaia. 1956. Leadership in pre-European Maori society. *Journal of the Polynesian Society* 65(3): 212–231.

Winter, Edward. 1958. The aboriginal political structure of Bwamba. In *Tribes without Rulers: Studies in African Segmentary Systems*, edited by J. Middleton and John D. Tait, pp. 136–166. Routledge and Kegan Paul, London.

Wohlt, Paul Brandin. 1978. *Ecology, Agriculture and Social Organization: The Dynamics of Group Composition in the Highlands of Papua New Guinea*. PhD dissertation, Department of Anthropology, University of Minnesota.

Wolpert, David H. 2010. Why income comparison is rational. *Games and Economic Behavior* 69(2): 458–474.

Woodburn, James. 1998. Sharing is not a form of exchange: An analysis of property-sharing in immediate-return hunter-gatherer societies. In *Property Relations: Renewing the Anthropological Tradition*, edited by C. M. Hann, pp. 48–63. Cambridge University Press, Cambridge.

Wrangham, Richard W., and Luke Glowacki. 2012. Intergroup aggression in chimpanzees and war in nomadic hunter-gatherers. *Human Nature* 23(1): 5–29.

Wright, Henry T. 1984. Prestate political formations. In *On the Evolution of Complex societies: Essays in Honor of Harry Hoijer*, edited by T. Earle, pp. 41–77. Undena, Malibu, C A.

Wright, Katherine, and Andrew Garrard. 2003. Social identities and the expansion of stone bead-making in Neolithic Western Asia: New evidence from Jordan. *Antiquity* 77(296): 267–284.

Wu, Joshua B., Peter W. Hom, Lois E. Tetrick, Lynn M. Shore, Liangding Jia, Chaoping Li, and Lynda Jiwen Song. 2006. The norm of reciprocity: Scale development and validation in the Chinese context. *Management and Organization Review* 2(3): 377–402.

Wu, Xiaohong, Chi Zhang, Paul Goldberg, David Cohen, Yan Pan, Trina Arpin, and Ofer Bar-Yosef. 2012. Early pottery at 20,000 years ago in Xianrendong Cave, China. *Science* 336(6089): 1696–1700.

Xygalatas, Dimitris, Panagiotis Mitkidis, Ronald Fischer, Paul Reddish, Joshua Skewes, Armin W. Geertz, Andreas Roepstorff, and Joseph Bulbulia. 2013. Extreme rituals promote prosociality. *Psychological Science* 24(8): 1602–1605. doi:10.1177/0956797612472910.

Yang, Mayfair. 1989. The gift economy and state power in China. *Comparative Studies in Society and History* 31: 25–54.

1994. *Gifts, Favors and Banquets: The Art of Social Relationships in China*. Cornell University Press, Ithaca, NY.

Yate, William. 1835. *An Account of New Zealand*. Seeley and Burnside, London.

Young, Michael W. 1971. *Fighting with Food: Leadership, Values and Social Control in a Massim Society*. Cambridge University Press, Cambridge.

Zaggl, Michael A. 2014. Eleven mechanisms for the evolution of cooperation. *Journal of Institutional Economics* 10: 197–230.

Zahavi, Amotz. 1975. Mate selection: A selection for handicap. *Journal of Theoretical Biology* 53: 205–214.

Zent, Stanford. 1999. The quandary of conserving ethnoecological knowledge: A Piaroa example. In *Ethnoecology: Knowledge, Resources and Rights*, edited by T. Gragson and B. Blount, pp. 90–124. University of Georgia Press, Athens.

Zheng, Hong-Xiang, Shi Yan, Zhen-Dong Qin, and Li Jin. 2012. MtDNA analysis of global populations support that major population expansions began before Neolithic time. *Scientific Reports* 2: Article 745. doi:10.1038/srep00745.

Ziegler, Rolf. 1990. The Kula: Social order, barter, and ceremonial exchange. In *Social Institutions: Their Emergence, Maintenance, and Effects*, edited by Michael

Hechter, Karl-Dieper Opp, and Reinhard Wippler, pp. 141–157. Aldine de Gruyter, Hawthorne, NY.

Zori, Colleen. 2011. *Metals for the Inka: Craft Production and Empire in the Tarapacá Valley, Northern Chile*. PhD dissertation, Department of Anthropology, University of California, Los Angeles, CA.

Zori, Colleen, and Erika Brandt. 2012. Managing the risk of climatic variability in late prehistoric northern Chile. *Journal of Anthropological Archaeology* 31(3): 403–421.

Index

Abelam peoples, 71, 72, 163
Adler, Michael, 91, 161
Africa, 18
aggrandizers, 2, 69, 70, 71, 75, 151
aggrandizing behavior, 34, 40, 41, 68
Ain Mallaha archaeological site, 239
Allen, Michael, 73
altruism, 5, 15, 35, 38, 49, 75, 126, 170, 182
  in group selection, 39
altruistic behavior, 13, 38, 70, 76, 155
  as theoretical problem, 35, 38, 181
Alvard, Michael, 97
Amba leaders, 64
Andaman Islands, 67
Andersen, Steffen, 54
Anderson, David, 254
Andreoni, James, 48
anthropological game theory, 5, 36, 55, 57, 76, 80, 81, 82, 86, 90, 103, 104, 126, 129, 143, 145, 148, 156, 158, 176, 178, 267, 268, 275
  and population growth, 173
  and population pressure, 173
  coercion defined, in, 77
  defined, 55
  economy of scale, 95
  role of punishment in, 155
  to explain barter, potlatch, and fairs, 155
appropriate conditions
  for co-operation to emerge, 2, 3, 5, 14, 15, 16, 35, 53, 62, 82, 126, 173, 265
archaic states, 57, 267
Aristotle, 10, 22
  and co-operation, 10
Arkush, Elizabeth, 213
Arthur, Brian, 166
Asad, Talal, 162
Asia, 18
Aspero archaeological site, 198, 203
astronomical alignments
  at Caral, 204
  at Chankillo, 233
  at Pampa del Gentil, 233
  at Poverty Point, 245
  function of in stateless societies, 259
  henge monuments, 258
astronomical markers
  in special places, 259
Atacama Desert, 215, 220
Aten, Lawrence, 251
autarkic household, 20, 278
Avebury archaeological site, 258
Aveni, Anthony, 204, 209, 246
Axelrod, Robert, 13, 45, 56
ayllu, 110

Bachiga people, 145
Bandy, Matthew, 177, 190
Banning, E.B., 235
barter exchange, 96, 132, 135, 145
  and marginal utility, 136
  competitive strategies in, 143
  defined, 135
  establishing value in, 141
  ethnographic examples, 137
  in stateless societies, 93, 132
  in the Andes, 142
  price determination in, 143
  regulated by ritual, 81
  rules and rituals associated with, 144
  setting prices in, 143
  strategies for success in, 145
barter fairs, 20, 93, 112, 133, 134, 139, 141, 206
  actors in, 143
  establishing value in, 137
  exchange values, 145
Bar-Yosef, Ofer, 96, 185, 239
basic unit of production, 105, 106, 194
  defined, 105
Bathonga peoples, 91
Bedouin society
  nonlethal raiding, 128
  raiding in, 128

Belfer-Cohen, Anna, 96
Bell, Catherine, 86, 87
Bellwood, Peter, 187
Benfer, Robert, 206
Bentham, Jeremy, 60
Benz, Marion, 237
Big Men, 63, 69, 70, 71, 72, 125, 150, 163, 280
 and competitive feasting, 74
 and feasts, 74
 as leaders, 59
 as leaders in Te society, 280
 in Enga society, 73
 lacking coercive power, 71
 roles in Abelam society, 72
 status available to many, 73
Bird, Rebecca, 158
Bischof, Henning, 196
Blick Mead archaeological site, 258
blood brotherhood
 among Bachiga people, 145
 among the Zande people, 145
 enhances exchange, 146
blood feuds, 7, 157
 at Jebel Sahaba, 130
 in Middle and Late Archaic, 253
 in Nuer society, 64
Boas, Franz, 27, 150, 153
Bocquet-Appel, Jean-Pierre, 185
Boehm, Christopher, 156
Borneo, 26, 92, 106, 107, 116, 119, 123, 148, 163
Boserup, Esther, 175, 176
Boster, James, 126
Bowles, Sam, 3, 43, 126, 158
Boxberger, 108
Boyd, Robert, 51, 57, 126
Boz, Basac, 240
Bradley, Richard, 257
Brandt, Erika, 219
Brazil, 183
Brecher, Kenneth, 245
Bressler, Eric, 159
Briones, Luis, 208
British Columbia, 108, 164
Bromley, R.J., 133
Burger, Richard, 207, 208, 211
Burkina Faso, 92
Burl, Aubrey, 255
Burton, John, 66, 100, 101
Byock, Jesse, 66

Cahokia archaeological site, 273
canonical model, economic theory, 21, 27, 52, 277
 criticized, 28, 33
 defined, 14
 expectations met, 54
 expectations not met, 45, 53
Canziani, Jose, 229
Caral archaeological site, 185, 198, 200, 201, 203, 204, 206, 207, 214, 220
 ritual objects at, 204
Carballo, David, 7, 46, 203
Carevič, Alvaro, 219
Carneiro, Robert, 175, 179
Caroline Islands, 67
Caserones archaeological site, 216, 219
Casma Valley, 195, 206, 210, 216, 220
Caughey, John, 67
causality
 in the social sciences, 171
causewayed enclosures, 255
Çayönü Tepesi archaeological site, 239
Çelik, Bahattin, 237
Cerro del Gentil archaeological site, 226, 229, 233
 ritual objects in, 227
 sunken patios at, 227
Cerro Sechín archaeological site, 210
 evidence of lethal raiding in, 211, 214
Chagnon, Napoleon, 177, 178
Chankillo archaeological site
 solstice time marking at, 233
Chapman, Anne, 135
chiefly cycling, 181
chiefs, 71
 as coordinator of specialized labor, 100
 authority of in Trobriand Islands, 63
 Western misperceptions, 63
China, 84
Chincha Valley, 220, 223
 Chococota archaeological sites, 228
 map, 221
 Paracas occupation in, 223
Chiricahua Apache
 raiding practices in, 129
Chococota archaeological sites, 228
Chogha Golan archaeological site, 187
Choi, Jung-Kyoo, 126
chulpas, 224
Claassen, Cheryl, 241
Clark, Kenneth, 49
coalitions, 126, 181
 develop in early state societies, 268
coercion, social, 15, 46, 57
 chief, authority of in Eastern Pende, 65
 defined, 77
 derived from group, 40
 ethnographic cases, 62
 forms of in states, 77

coercion, social (*cont.*)
in evolutionary theory, 60, 68, 77, 78
informal, 2
lack of among Abelam leaders, 72
lack of among Amba people, 64
lack of among Andaman chiefs, 67
lack of among in Eastern Pende chiefs, 65
lack of among Kimbu chiefs, 65
lack of among Konkomba leaders, 64
lack of among Kuma leaders, 74
lack of among Mae-Enga leaders, 73, 74
lack of among Mekranoti leaders, 66
lack of among Nuer chiefs, 64
lack of among Piaroa leaders, 66
lack of among Ponapean chiefs, 66
lack of among Solomon Island leaders, 152
lack of among Truk Island chiefs, 67
lack of among Tungei big-men, 66
present in state societies, 2, 6, 40, 49, 57, 61, 62, 71, 84
rare in stateless societies, 2, 3, 6, 9, 10, 34, 60, 62
shift in nature of in evolution of state societies, 62
state-enforced
described, 77
state-enforced versus small group, 77
transformation to ideologies of power in state societies, 267
Collard, Mark, 187
collective action problem, 2, 3, 7, 8, 41, 49, 76, 97, 103, 127, 132, 270
definition of, 7
collective action theory, 76
Columbia River drainage
aboriginal trade fair in, 139
Comanche peoples, 140
competition
in strategy selection, 180
competitive feasts, 93, 147, 149, 153, 239
in Solomon Islands, 150
complex stateless societies, 53, 62, 69, 71, 78, 170, 220
definition of, 1
lack of coercion, 34
successful, 2
conditional co-operation, 3, 6, 15, 16, 35, 43, 275
definition of, 5
co-operation. *See* conditional co-operation, raiding promotes cooperation, collective action problem
among ego-directed actors, 45
and costly signaling, 158
and small groups, 5
cost of, 9, 15
develops among ego-directed actors, 3, 5, 7, 78
economic basis of, 8
problems of. *See* free-rider problem
coresidential group. *See* household
costly punishment, 4, 50
as socially rational, 47
costly signaling, 75, 81, 155, 158, 159, 231
intensified in stateless societies, 160
to promote co-operation, 158
costs and benefits
from demographic context, 268
of co-operation, 267
of co-operative behavior, 265
Cowgill, George, 175
cultural transmission, 7
in strategy selection, 181
Cupisnique architectural tradition, 208
Curry, Andrew, 234
Czech Republic, 54

Damon, Frederick
and Muyuw astronomical time marking, 262
dance feasts, 153
Darwin, Charles, 180
Davis-Salazar, Karla L., 85
de Laguna, Federica, 177
debt obligation
as a strategy, 78, 154
debt obligations, 155
deferred debts, 146
deferred reciprocity, 103
demographic pressure, 174
deterministic versus probabilistic logic, 172
Dictator Games, 54
Dietler, Michael, 146, 150
Dietrich, Oliver, 237
Dinka peoples, 177, 178
division of labor. *See* specialized labor
Dobyns, Henry, 139
on aboriginal North American barter fairs, 138
Douglas, Mary, 94, 274
Dove, Michael, 106
Duggleby Howe archaeological site, 256, 257
Dunbar number, 278
Dunbar, Robin, 45
Durkheim, Emile, 86
Durrington Walls archaeological site, 257
Dusun peoples, 119
Dyak people
feasts schedule agricultural cycle, 116

Dyak peoples, 150, 163
  feasts, 148
Dye, Tom, 192

Earle, Timothy, 59
eastern Ambae people, 73
Eastern Pende people, 65
economic anthropology, 4, 20, 21, 26, 28, 32, 34, 132, 269, 270. *See* formalist school, substantivist school
  and game theory, 35
  formalism, 26
  production and exchange in, 93
economic man, 20, 23, 24, 25, 26
  as inherently maximising, 24
  criticized, 29, 36
Economic Man. *See* Homo economicus
economy of scale, 8, 49, 52, 111, 155, 235
  costs of, 103
  defined in classical economics, 96
  economic advantages, 99
  example, 8
  example among the Tungei, 101
  in stateless societies, 95
  links households into co-operative groups, 97
  material basis of co-operation, 97
  Type 1 defined, 97, 98
  Type 2 defined, 102
economy of scale in stateless society
  cost, 95
  payoff, 95
Edel, May Mandelbaum, 145
Edmonds, Mark, 257
egalitarian ethos
  in stateless societies, 156
El Paraíso archaeological site, 207
El Porvenir archaeological site, 196
embeddedness concept
  Karl Polanyi, 31, 133
  Malinowski, 31
emergent properties, 4, 7, 15, 68, 88, 170, 267, 276
Engels, Friedrich, 32, 60
Evans, Ivor, 119
Evans-Pritchard, E.E., 7, 63, 64, 145, 161
  and Nuer time reckoning, 263
evolution
  of complex societies, 6, 17
evolutionary game theory, 2, 3, 14, 15, 16, 48, 58, 159
  defined, 14
evolutionary game theory, 13
evolved psychology, 41, 42, 43, 56, 57
exchange. *See* barter exchange, market exchange

Faction building
  in barter, 145
fairness, 4, 5, 46, 47, 48, 56, 95, 104, 124, 202
  as socially rational, 53
  favored by natural selection, 48
  outcome in Ultimatum Game, 45
  required in specialized labor, 95
  requisite for co-operation, 46
fairness, norms of
  to promote co-operation, 81
fairs, 93, 112, 131, 137, 138, 140, 141, 155, 206, 219, 230, 231, 233, 262, 277
  as location for barter exchange, 135, 137
  common in the non-Western world, 138
  common in the Western world, 138
  definition of, 134
  regulated by ritual, 81
fairs, Medieval England
  linked to religious rites, 137
feasting
  at Gobekli Tepe, 237
feasts
  collective, 239
feasts and feasting, 9, 10, 19, 69, 70, 71, 73, 74, 75, 81, 84, 86, 87, 109, 112, 117, 118, 146, 151, 152, 198, 203, 231, 239, 265, 275
  among Paracas, 233
  among the Dyak, 150
  among the Sea Dyaks, 148
  as economic utility, 86
  at Cerro del Gentil, 227
  at Gobekli Tepe, 237
  at Harris Creek, 251
  in special places, 160
Feinman, Gary, 7, 147
Fennel, Lee Anne, 166
Fiji, 53
Firth, Raymond, 31, 32, 36, 63, 81, 85, 86, 90, 107, 114, 115, 157, 274
  and Maori taboo, 124
  critique of Malinowski, 143
  definition of ritual, 88
  Primitive Polynesian Economy, 151
  proportional ritualisation, example of, 124
  response to substantivist school, 32
  role of chief, 104
  The Work of the Gods, 116
fishing
  ritual control of, 116

fissioning, 164, 177, 179, 266
  cost of, 168
  of groups, 76
fitness, 42, 180
Flannery, Kent, 125, 130, 142, 163, 204, 264
Flegr, Jaroslav, 54
fluidity of groups, 167, 182
  example, 183
Fogelin, Lars, 87, 274
Fontana, Marissa, 240
force. *See* coercion, social
formalist school, 31, 59
  influences, 32
  role of force, 58
  summary, 33
formalist-substantivist debate, 34, 35
  as irrelevant, 35
free-rider problem, 2, 3, 7, 8, 9, 41, 50, 77, 81, 115, 132, 155, 156, 159, 167, 169, 176, 181, 182, 237, 266
  in specialized labor group, 95
free-riding, 46, 50, 76, 77, 115
  punishment of, 50, 168
functionalist theory, 270
Futuna Island, 117

game theory, 14, 16, 36, 45, 97, 177. *See* evolutionary game theory, anthropological game theory
  and barter exchange, 136
  and group selection, 181
  classical, 37
  defined, 13
  experimental problems, 51
  of the small group, 6
  to explain co-operation, 36
garden magician
  role in production in Trobriand Islands, 90, 115, 261
Garrard, Andrew, 240
Gavrilets, Sergey, 4, 6
gene-level selection, 38
geoglyphs, 204, 208, 209, 231, 233
  aligned with solstice, 232
  as ritual pathways, 233
  at Caral, 208
  El Gigante, 218
  in Casma valley, 208
  in Chincha, 227, 230
  in Chincha Valley, 224
  in Nasca, 209
  in ritualized landscape, 208
  near Caserones, 217
  near Pircas, 219
Gibson, Alex, 256
Gibson, Jon, 244, 248
gifts, 29, 84, 114, 128, 146, 150
  as obligatory exchange, 94
Gintis, Herb, 3, 35, 36, 39, 43, 50, 158, 276
Glencross, Bonnie, 240
Göbekli Tepe archaeological site, 185, 234, 235, 236, 237, 238, 264, 272, 273
  as center of ritualized economy, 237
  lack of evidence for raiding, 240
Goldman, Irving, 91
Goodenough Island, 74
Goody, Jack, 108
Graeber, David, 284
Graffam, Gray, 219
Gramsch, Alexander, 237
Green, Roger, 262
Greenland, 135
group selection, 39, 182
Guatacondo archaeological site, 219
Gunther, Erna, 261

Haag, William, 245
Haas, Jonathan, 198
Hadza peoples, 131
Håkansson, Thomas, 64
Haley, Kevin, 46
Hallan Çemi archaeological site, 238, 239, 240
Hallum, Anne Motley, 183
Hämäläinen, Pekka, 138
Hambledon Hill archaeological site, 254
Hamilton, W.D., 13
Harris Creek archaeological site, 251
  barter exchange at, 251
Harris, Sandra, 156
Hawaii, 140
  early contact barter fairs described, 140
Hawaiian Islands, 192
Hayden, Brian, 69, 71, 74, 75, 164, 176, 239
Healy, Frances, 254
Heckenberger, Michael, 249
hedonic utility, 285
Hegmon, Michelle, 110
Heizer, Robert, 125
henge monuments, 188, 259
  Neolithic, 254
Henrich, Joseph, 51, 52, 53, 56
Henrich, Natalie, 53
Herbich, Ingrid, 146
Hobbes, Thomas, 11, 13, 15
Hobbesian Rousseauian false dichotomy, 79
Hobbesian tradition, 11, 12, 16, 21, 23, 29, 35, 49, 59, 70, 75
  in formalist-substantivist debate, 32
Hodder, Ian, 256

Hodgson, Geoffrey, 28
Hogbin, H. Ian, 152
Hommon, Robert, 141
Homo economicus, 13, 14, 15, 17, 19, 23, 27, 28, 37, 43, 52, 53, 58, 105, 166
  expectations of, 21
Homo reciprocans, 3, 164
Hopi people, 110, 139
Hose, Charles, 107
household, 97, 99, 110, 155, 177, 194, 266
  as basic economic unit, 80
  as primary unit of co-operation, 104, 108
  as unit of co-operation in Hopi society, 110
  as unit of production, 109
  role of in co-operation, 109
  unit of co-operation, 105
households
  and reciprocal debt obligations, 122
  as unit of production, 108
Huaca La Florida archaeological site, 207
Huaca Soto archaeological site, 229, 230
Humphreys, Paul, 172
hunter-gatherers, 1, 16, 34, 68, 71, 94, 96, 111, 130, 131, 168, 169, 175, 185, 188, 234
  complex, 170, 220, 236
  in canonical model, 174
Huron League, 154
Huron peoples, 139
  feasting as status competition, 154
Huron peoples in 17th century, 67

Iceland, 66
Ifugao society
  importance of religion, 91
India, 54
institutionalism in economic theory, 28
intangible economic utility, 93
Iroquois
  longhouse organisation, 110
Iroquois people, 106
irrational pro-social behavior. *See* social rationality
Island of the Sun, Bolivia, 202
Iterated Prisoner's Dilemma, 44, 48

Jacques, David, 258
Janetski, Joel, 140
Jebel Sahaba archaeological site, 130
Jericho archaeological site, 240
Johnson, Allen, 59
Jones, Dan, 52
Jordan, 238
Junker, Laura, 213

Kaberry, Phyllis, 72, 163
  proportional ritualization, example of, 125
Kalberg, Stephen, 84
Kantner, John, 159, 160
Kantu' people
  households in, 106
Karahan Tepe archaeological site, 237
Kato, Tsuyoshi, 119
Kava ceremony
  in Polynesia, 118
kava feast, 116
Kayan people
  taboo and agricultural cycle, 123
Kayapó peoples, 92, 183
Keeley, Lawrence, 174, 240
Kimbu people, 65
kin groups
  as unit of co-operation, 105
kin-selection, 155
Kirch, Patrick, 117, 118, 122, 164, 193
  and Polynesian time rekoning, 262
Knight, Vernon, 253
Kohler, Stefan, 46
Kolb, Michael, 193
Konkomba people, 64
Kornienko, Tatiana, 238, 239
Kotosh archaeological site, 207
  Temple of the Crossed Hands, at, 207
Kotosh Religious Tradition, 207
Kress, Howard, 126
Kuijt, Ian, 186, 240
Kula exchange, 93
  lifetime partnerships in, 144
  role of domestic trade, 144
Kula ring, 29, 30, 94, 102, 132, 134, 144, 182
  as barter market, 143
  domestic trade, 30
  gift exchange, 146
  gimwali, 93
  gimwali or "barter," 29
  ritualized barter exchange, 144
Kuma people, 74
Kurtz, Donald, 133
Kwakiutl, 150
Kwakiutl peoples, 150

La Laguna archaeological site, 203
labor theory of value, 31, 142
  critique, 142
  in barter exchange, 142, 144
Lake Titicaca Basin, 192, 208
late Paleolithic
  organized conflict in, 130
Lauro, James, 244

Leach, Edmund
and Trobriand calendars, 263
leaders, managerial, 2, 40
accorded ad hoc authority, 61, 100, 115, 129
difference between states and stateless societies, 71
example, 92
reward for, 148
scheduling feasts, 118
to provide group benefits, 75
use of special places, 160
who provide group benefits, 73
leadership, 9, 16
emergence of, 80
failure of, 16, 46
in eastern Ambae society, 73
in stateless societies, 2, 77
successful, 111
Lee, Richard, 105
Lehmann, Geoffrey, 244
leopard-chiefs, Nuer, 64
lack of authority, 64
Levine, Abigail, 127
Levine, David, 48
Lienhardt, Godfrey, 177
Lindstrom, Lamont, 63, 72
Lipson, Ephraim
The Economic History of England, 137
Louisiana, 241
Lowie, Robert, 129
Lubbock, John, 18, 90
Luiseño peoples, 147
Luning, Sabine, 92

MacDonald, Kevin, 181, 182
MaeEnga people, 72, 74, 125
annual ritual cycles, 262
exchange in, 280
magic, 86
coordinating labor, 91
definition of, 89
in Bedouin raiding, 128
in economic co-operation, 112
in Maori life, 90
Mahdi, Ninloufer, 157
Maine, H.J. Sumner, 25
Malaysia, 119
Malinowski, Bronislaw, 27, 33, 60, 63, 81, 84, 86, 102, 112, 114, 115, 129, 143, 144, 182, 274
and Trobriand time marking, 261
contra economic man concept, 29, 30
definition of systematic magic, 91
functionalist theory, 270
link between magic and practical activity, 91
link between ritual and practical activity, 91
nonlethal raiding, example of, 129
proportional ritualization, example of, 123
proportional ritualizing, example of, 124
socially organized labor as Type 1 economy of scale, 99
view of ritual and religion, 90
Malthus, Thomas, 22, 23
Malthusian tradition, 33, 49, 175
Man, Edward Horace, 67
Mandan people, 138
Maori, 90, 107, 112, 114, 124, 157, 274
household as unit of production, 107
pyramid structure as example of a special place, 164
Marcus, Joyce, 125, 130, 142, 163, 181, 204, 229, 264
marginal costs, 99
marginal utility, 95
defined, 136
market exchange, 3, 20, 58, 132
marketplace, 132
defined, 133
markets, 19, 134
as a set of rules, 134
modern, 20, 31
term, 133
Marquesas Islands, 164
Marx, Karl, 60
Maschner, Herbert, 283
Mathew, Sarah, 51, 126
Mauritius, 159
McDougall, William, 107
medium of exchange
in market exchange, 132
Meggitt, Mervyn, 72, 262
proportional ritualisation, example of, 125
Melancon, Mark, 248
men's house, 100, 162
as special place, 162, 163
Mende society, 160
Mercer, Roger, 254
Mesoamerica, 86
methodological individualism, 276
Mexico, 109
Mill, John Stuart, 19, 21, 22, 24, 60
Miller, Bruce, 108
Miller, John, 48
Mirsky, Jeannette, 135
Mitchell, Wesley, 28
Mitchell, William, 119
Monaghan, John, 109
monumental architecture, 199
in special places, 165
Moore, Jerry, 196, 210, 211
Morgan, Henry Lewis, 25, 110

Morgenstern, Oskar, 37
Moseley, Michael, 202, 204, 207
multilevel selection, 39, 56
Myers, Milton, 22

Nasca, 220
Nasca lines, 204
Nash equilibrium, 136
  defined, 37
Nash, John, 37
needs and wants, 33
  assumption of natural abundance, 29
  assumption of universal scarcity, 23, 24
needs versus wants, 23
Neitzel, Jill, 147
neoformalism, 31, 33
Neolithic Demographic Transition, 186
  two-stage NDT, 186, 190
Nettle, Daniel, 45
New Guinea highlands
  household organization in, 108
Nigra, Benjamin, 229
Noble Savage, 82
Nolin, David, 97
Norenzayan, Ara, 53
norms of co-operation, 2, 9, 20, 47, 57, 81, 169, 181, 184, 237
  in reducing transaction costs, 97
  internalized, 8
  transformation to ideologies of power, 267
norms of fairness, 46, 50, 95, 237
  enforcement by group, 156
  in economy of scale organization, 103
  in reducing transaction costs, 97
norms of reciprocity, 40, 150
North, C. Douglass, 8, 28, 97
Northern Panya peoples, 139
Novakova, Julie, 54
Nowak, Martin, 45, 47, 52
Nuer
  Pyramid of Ngundeng, 161
Nuer people, 63, 64, 157
Nuñéz, Lautaro, 217

Oberg, Kalvero, 102
  proportional ritualisation, example of, 124
obsidian
  as controllable preciosity, 144
Oceania, 18
Oliver, Douglas, 150
  and indigenous Tahiti annual time cycles, 261
Orkney Islands, 255
ostracism
  as costly punishment, 157
  as punishment, 51, 179
Ostrom, Elinor, 166
Overing, Joanna, 66
Oxenham, Marc, 187

Page, Karen, 45, 47, 52
Palpa Valley, 225, 230
Papua New Guinea, 66, 67, 71, 72, 74, 100, 118, 125, 163
Paracas
  in Chincha Valley, 222
  platform mounds, 223
Paracas archaeological site, 185
Paracas culture, 220
Pareto optimality, 136
Parker Pearson, Mike, 257
payoff, 38, 39, 155
  defined in game theory, 36
  for co-operation, 7, 15, 35, 41, 52, 76, 80, 94
  for performing costly punishment, 155
  in Prisoner's Dilemma, 45
Pearsall, Deborah, 190
Pecos Pueblo, 139
  barter market at, 139
  fair at, 140
Pérez, Kelita, 229
performance, 266
performance and dance, 229
Perrot, Nicolas, 25
Pesendorfer, Wolfgang, 48
Petersen, Glenn, 66
Philippines, 91
Phillips, Ruth, 160
Piaroa people, 66
Piatek, Bruce, 250
Pinker, Steven, 182
Pircas archaeological site, 216, 217, 219
  ritual objects recovered, 217
Plattner, Stuart, 29
Polanyi, Karl, 31
  The Great Transformation, 133
Polynesia, 118
population density, 3, 22
  increase of promotes social complexity, 265
  promotes co-operation, 267
population pressure, 167, 176, 179
population trends
  mid-Holocene North America, 189
  mid-Holocene of western Africa, 187
  Neolithic Britain, 187
  precontact Hawaii, 193
Posey, Darrell, 92
positivist philosophy, 21
potlatch, 9, 40, 146, 148, 155, 182, 227, 230
  among the Huron, 25
  and debt obligations, 149

potlatch (*cont.*)
  and status rivalry, 154
  as a strategy, 149
  as costly signal, 155
potlatch as exchange system, 93
potlatching
  as universal social technology, 149
Poverty Point archaeological site, 16, 185, 241, 245, 248, 253
  as special place, 248
  astronomical alignments, 246
  Bird Mound at, 241
  excavations at, 244
power. *See* coercion, social
prestige, 5, 9, 10, 16, 40, 73, 112, 127, 149, 152, 180, 280
  acquired by constructing special places, 160
  acquisition in potlatching, 149
  as economic utility, 33
  as intangible utility, 155
  as motivation, 40
  as reward for co-operation, 49
  as reward for leaders in Mae-Enga society, 73
  as reward to leaders, 2
  for leaders, 40
  in Trobriand Islands, 30
  motivation for leaders, 152
Price, Neil, 91
price-making markets, 132, 134, 135, 137, 143, 145
  price fluctuations common in, 145
primitive economy concept, 25, 29, 30, 32, 58, 61, 82, 272, 273
  development of, 18, 19, 92, 274
Prisoner's Dilemma, 44, 48, 159
production and exchange, 3, 80, 85, 93, 105, 149, 167, 171
  of copper, 219
progress
  concept of, 21, 24, 25, 180
proportional rationality
  principle of, 53
proportional ritualisation
  definition of, 123, 125
  principle of, 123, 124
prosocial behavior, 35, 38, 39, 47
  as costly punishment, 4
  by leaders, 9
  payoff for, 41
Proulx, Donald, 204
Provinse, John, 92, 106
Pueblo peoples, 139, 140
Puget Sound chiefdoms, 108
punishment, 46, 50, 145, 155, 156, 167
  as adaptive strategy, 50
  of abusive leaders, 156
  to promote co-operation, 7, 9, 49, 50, 51
  to promote co-operative behavior, 81
Purrington, Robert, 246
Pyramid of Ngundeng, 161
  as a special place, 161

Quilter, Jeffrey, 197, 207
Quinn, Rhonda, 251

Radcliffe-Brown, A.R., 64
raiding, 76, 109, 119, 129, 130, 240
  as exchange mechanism, 93, 128
  as form of exchange, 111
  as production, 125
  as Type 2 economy of scale, 126, 129
  booty as surplus, 111
  endemic in complex stateless societies, 130
  enhances prestige, 130
  in ancient Andes, 210
  in Archaic North America, 253
  lethal, 130
  Neolithic Levant, 240
  nonlethal, 130
  organized by leaders, 74
  promotes co-operation, 8, 76, 126, 127
  to obtain resources, 127
Rand, David, 47
Randall, Asa, 250, 251, 252
Rappaport, Roy, 86, 118
Rashid, Salim, 22
rational actor theory, 14, 19, 21, 23, 26, 45, 70, 82
  expectations not met, 46, 50, 52, 53
rational behavior
  defined, 37
Real Alto archaeological site, 197
reciprocal altruism, 39, 155
reciprocity, 56, 58, 131
  as barter, 95
  as exchange mechanism, 26, 94
  as norm of co-operation, 46, 80, 94
  as norm of fairness, 46, 94
  ideology of, 104
  in Kula exchange, 146
  in potlatch, 149, 155
  labour exchange as form of, 107
  strong versus weak, 50
redistribution, 154
  and barter, 22
  as a reward for co-operation, 268
  as deferred reciprocity, 95
  as strategy to obtain prestige, 152
  in Kula exchange, 146

ritualized, 9, 119
ritually controlled, 104, 115, 117, 118, 266
ritually sanctioned, 112
Reiche, María, 209
Reinhard, Johan, 209
religion, 84
as a set of strategies, 85
as distinct from ritual, 86, 270, 271
as universalizing set of beliefs, 89
definition of, 88
in fairs, 138
misuse of in archaeology, 272, 275
rare in stateless societies, 88
ubiquitous in state societies, 89
religion, concept of
inadequate for stateless societies, 88
Renfrew, Colin, 41, 89, 254, 255
definition of religion, 89
reproductive success, 40, 43
as payoff, 39
reputation, 46, 47, 48, 49, 159
as co-operator, 160
promotes co-operation, 48
retribution. *See* punishment
reward, 46, 155
as payoff for co-operation, 93
for co-operation, 49, 97
to promote co-operation, 9, 49, 76
rewards
costly, 49
to promote co-operation, 49, 81, 93, 94, 268
Ricardo, David, 19, 24
Richards, Colin, 255
Riehl, Simone, 187
ritual, 7, 9, 25, 30, 90, 109, 115, 156, 257
as distinct from religion, 81, 86
as socially rational, 27
defined, 86
in theoretical frameworks, 270
problems with term, 271
promotes co-operation, 159
promotes coordination of tasks, 9, 81, 115, 116
secular, example of, 88
to schedule feasts, 81
to schedule labor, 3, 81, 91, 115, 116, 119, 121
ritual calendar, 81, 123, 138, 169, 247, 261, 262, 263, 264
ritual cycle. *See* ritual calendar
ritual economy
as distinct from ritualized economy, 85
ritual objects
at Caral, 204
at Cerro Gentil, 227
at El Paraiso, 207
at Pircas, 217
at Poverty Point, 244
Mesopotamia, 238
ritualized co-operation. *See* co-operation
ritualized economy, 6, 71, 220, 266
at site of Caral, 198
defined, 7
in ancient Andes, 208
ritualized exchange. *See* barter exchange
ritualized landscape, 233
geoglyphs in, 209
in ancient Peru, 230
ritualized production, 112
rituals, 262
costly, 158
rituals, typology of, 87
Rivera, Mario, 219
Rodríguez, Aurelio, 204
Roscoe, Paul, 7, 41, 69, 108, 110
Rosman, Abraham, 155
Roth, Henry Ling, 116, 148, 149
Rousseau, Jean-Jacques, 11, 12, 15
A Dissertation on the Origin . . . , 12
Rousseauian tradition, 11, 12, 13, 16, 21, 25, 35, 75
in formalist-substantivist debate, 32
Noble Savage, 29
Rowan, Yorke, 86
Rubel, Paula, 155
rudimentary economy of scale. *See* economy of scale
example of, 99
Russo, Michael, 241, 250, 253

Sabloff, Paula, 87, 271, 276
Sahlins, Marshall, 59, 96, 105
Samaniego, Lorenzo, 214
San Jacinto archaeological site, 208
Santa Valley, Peru, 191
Sassaman, Kenneth, 241, 249, 252
Saunders, Joe, 248
Saunders, Rebecca, 251, 253
Scaglion, Richard, 72
Schmidt, Klaus, 234, 236, 272
Schwartz, Douglas, 174
Sechín Bajo
alignments, 206
Sechín Bajo archaeological site, 195, 210, 220
secondary state formation, 71
Sefton, Martin, 49
Shady, Ruth, 200, 201, 204
Shalizi, Cosmo, 16
Shields, Christopher, 11
Shorter, Aylward, 65, 178

Siang people, 107
  household as basic unit of production, 106
Sierra Leone, 160
Sigmund, Karl, 45, 47
Silver Glen Run archaeological site, 251
Silverman, Helaine, 204, 225
Slate site, 244
slavery, 18, 76
  in stateless societies, 77
small groups, 4, 7, 16, 18, 35, 39, 47, 171
  and different payoffs, 40
  and prosocial behavior, 38
  barter and price making, 144
  co-operation in, 55
  different than large groups, 4, 49, 55
  fluidity, 184
  rudimentary economy of scale, 98
small-scale societies. *See* small groups
Smith, Adam, 19, 22, 23
  and labour theory of value, 142
  cause of co-operation, 23
  efficiency of task specialisation, 98
  simple economy of scale described, 98
  Wealth of Nations, 22
Smith, Eric, 43, 158
Smith, John Stuart, 22
Sober, Elliot, 286
social contract theory, 23, 59, 60, 277
social Darwinism, 27
social histories, 4, 47, 48, 49
  in experimental games, 45
  to promote co-operation, 81
social rationality, 3, 4, 5, 6, 14, 30, 33, 56
  definition of, 276
social technologies, 81
  defined, 57
Solomon Islanders, 150, 152
solstice
  as marker of time ritualized landscape, 233
  as time marker, 169, 204, 224
  as time marker at Caral, 205
  building alignments on, 228
  marking as a calender, 264
  to organize ritual processions, 233
solstice marking, 232
  at Poverty Point, 245, 247
  at Stonehenge, 258
  in ancient world, 260
  in Kwakiutl peoples, 261
  in Nootka peoples, 261
Sosis, Richard, 126, 159
South America, 191, 195
special places, 1, 131, 161, 163, 168, 194, 253, 269
  act to materialize norms of co-operation, 160
  as costly signalers, 160
  as promoters of co-operation, 164
  development of, 266
  examples of, 164
  from local to regional, 263
  in Early Holocene, 168
  in Southeast US, 251
  men's houses, for example, 160
  on Futuna Island, 117
  Southeast US, 252
specialized labor, 8, 23, 49, 95, 144, 148, 154, 268
  promotes co-operation, 5, 8
Spencer, Charles, 183
Spencer, Herbert, 24, 60
  evolutionary philosophy of, 24, 25
Speth, John, 130
Spielmann, Katherine, 81, 86, 130, 257
St. Johns River valley, Florida, 250
stakes, 56
  effect in experimental games, 54, 55
Staller, John, 191
stateless societies, 7, 9, 11, 16, 28, 31, 40, 49, 71, 76, 80, 86, 92, 107, 153, 175
  aggrandizers in, 70
  economic production in, 93
  exchange mechanisms, 94
  Hobbesian assumptions, 58
  importance of feasts, 147
  marginal utility in exchange in, 136
  misperceptions about, 61
  misperceptions of, 27
  nonmaterial incentives, 149
  preciosities in, 143
  raiding in, 127
  Trobriand Islands, 30
stateless society
  misperception, 58
status
  acquired by generosity, 153
stelae
  at Gobekli Tepe, 235, 237
stone tools
  at Poverty Point, 244
  production of, at Gobekli Tepe, 236
Stonehenge archaeological site, 16, 257, 272
  solstice alignments at, 258
strategic intermarriage, 146
strategies, 93, 94, 104, 268
  in game theory defined, 37
  of co-operation, 80, 167
  that promote co-operation, 47
strategy selection, 7, 42, 56, 57, 82, 179, 180, 267

defined, 180
distinct from group selection, 181
Strother, Zoë, 65
substantivist school, 31, 32, 59, 133, 143
role of force, 58
Rousseauian assumptions, 31
Rousseauian assumptions, 58
Sudan, 161
Sumner, William, 60
sunken patios
as part of architectural complex, 203
as special place in the Andes, 202
supply and demand, 30, 31, 133, 136, 141, 142, 143
surplus, 155
created in economy of scale, 97
Swat Pathan
political organisation, 162
Sweet, Louise, 128
Swezey, Sean, 125
Symanski, Richard, 133
systematic magic, 91. *See* Malinowski "magic"

taboo, 10, 51, 61, 114, 117, 124, 155, 157, 265
as form of punishment, 9
as informal coercion, 77
as punishment of leaders, 158
by Maori chiefs, 157
in Dusun society, 119
on leaders, 158
to coordinate canoe building, 115
to enforce co-operative norms, 81
to promote co-operation, 82, 90
Tagalog peoples, 213
Tait, David, 64
Tambiah, Stanley, 86
Tantaleán, Henry, 227
Tarapacá Valley, 215
early complex stateless societies, 215
The Mound. *See* Pyramid of Ngundeng
Tick Island archaeological site, 253
Tikopia, 36, 90, 274
first yam ceremony, 116
rite of sacred yams, 115, 117
Tit for Tat strategy, 13, 14, 15, 45
Tlingit peoples, 102, 155
Tomasello, Michael, 42
Tooker, Elisabeth, 154
tragedy of the commons, 7
transaction costs, 97, 103
eliminating redundant, 99
reduction of in co-operative labor, 97
transegalitarian societies, 71, 170
definition, 70
Trigger, Bruce, 67
Trobriand Islanders, 27, 30, 274
exchange, 28
Kula ring, 144
role of ritual in gardening, 90
time reckoning, 261
Trobriand Islands, 63, 123, 134
and calendar, 263
canoe-making as Type 2 economy of scale, 102
fishing, 123
gardening in, 123
Tsembaba people, 67, 118
Tsimshian people, 108
Tung, Tiffany, 213
Tungei people, 66
mining as co-operative labor, 100
Turkana people, 51
Twiss, Katheryn, 177, 239
Tykot, Robert, 191
Tylor, Edward B., 25
Type 1 economy of scale. *See* economy of scale
Type 2 economy of scale. *See* economy of scale

Ultimatum Game, 44, 45, 46, 47, 52, 53, 55
effect of stakes, 54
underlying conditions
for co-operation to emerge, 5
Urton, Gary, 209
U-shaped architectural tradition, 207, 208, 228
U-shaped mounds, 253
in Southeast US, 251
utility, economic, 15, 23, 26, 155
culturally determined, 137
defined, 14
in formalist economic anthropology, 33
in stateless societies, 149
maximisation, 23

Vaughn, Kevin, 159, 160
vernal equinox
alignment at Poverty Point, 246
von Neumann, John, 37
von Rueden, Christopher, 41, 43

Wadi Fainan, 238
Wallerstein, Immanuel, 105
war. *See* raiding
as ritualized activity, 91
Watson Brake archaeological site, 185, 241, 248, 249, 264
Webb, Clarence, 244
Weber, Max, 82, 86

Webster, Hutton, 162
WEIRD societies, 43, 51, 52, 53, 56
Wells, E. Christian, 85
Werner, Dennis, 66, 183
Western Comanche people, 139
Western ethnocentrism
  in perceptions of rank, 61
Wetherill, John, 269
WF 16 archaeological site, 238
Whittle, Alasdair, 254
Wiessner, Polly, 70, 73
Wilmsen, Edwin, 175
Wilshusen, Richard, 92, 161
Wilson, David John, 191
Wilson, David Sloan, 286
Winiata, Maharaia, 112
Winter, Edward, 64
Wohlt, Paul, 73
Woodburn, John, 131
Wright, Henry, 268
Wright, Katherine, 240

Xygalatas, Dimitris, 159

Yang, Mayfair, 84
Yanomami, 130, 178
Yate, William, 164
Young, Michael, 74

Zaggl, Michael, 38
Zagros Mountains, 187
Zheng, Hong-Xiang, 187
Ziegler, Rolf
  on the Kula, 144
Zimbabwe, 46
Zori, Colleen, 216, 219

For EU product safety concerns, contact us at Calle de José Abascal, 56–1°, 28003 Madrid, Spain or eugpsr@cambridge.org.

www.ingramcontent.com/pod-product-compliance
Lightning Source LLC
LaVergne TN
LVHW011528200926
848484LV00001B/189